普通高等教育“十四五”规划教材

物理化学简明双语教程

CONCISE BILINGUAL COURSE OF PHYSICAL CHEMISTRY

（第二版）

何　美　李文泽　张潇飒　周华锋　主编

中国石化出版社
·北京·

内 容 提 要

本书是为了适应物理化学双语教学这一全新的教学模式，解决物理化学双语教材短缺的问题，尽量做到内容精练、简明易懂，并坚持对每一部分内容均采用英、汉双语进行编写。内容包括：气体的 pVT 性质；化学热力学；多组分系统热力学；化学平衡；相平衡；电化学；界面现象；化学动力学。采用二维码将课外拓展资料、例题、典型题和重点知识点的讲解以文件、图片音频、视频等形式加入教材，使教材更加生动、形象、易于理解，提高使用者的学习兴趣。

本书既可作为高等院校双语物理化学教学的教材使用，也可以作为非双语物理化学教学的参考教材。

图书在版编目（CIP）数据

物理化学简明双语教程：英汉对照 / 何美等主编. —2 版. —北京：中国石化出版社，2024.6
普通高等教育"十四五"规划教材
ISBN 978-7-5114-7522-0

Ⅰ.①物… Ⅱ.①何… Ⅲ.①物理化学-双语教学-高等学校-教材-英、汉 Ⅳ.①O64

中国国家版本馆 CIP 数据核字（2024）第 096752 号

未经本社书面授权，本书任何部分不得被复制、抄袭，或者以任何形式或任何方式传播。版权所有，侵权必究。

中国石化出版社出版发行
地址：北京市东城区安定门外大街 58 号
邮编：100011　电话：(010)57512500
发行部电话：(010)57512575
http://www.sinopec-press.com
E-mail:press@sinopec.com
北京科信印刷有限公司印刷
全国各地新华书店经销

*

787 毫米×1092 毫米 16 开本 16.75 印张 392 千字
2024 年 6 月第 2 版　2024 年 6 月第 1 次印刷
定价：45.00 元

前　言

教育部在加强大学本科教学的12项措施中，要求各高校在三年内开设5%~10%的双语课。物理化学作为一门化学化工专业学生必修的专业基础课，开展双语教学是非常必要的。面对国内双语教学刚刚起步、双语教材缺乏的现状，教育部鼓励引进原版教材或自己编写双语教材。但原版物理化学教材一方面价格很昂贵，另一方面难度太大也不完全适合国内大学本科学生的学习。2009年，沈阳化工大学的何美和周华锋老师在开展物理化学双语教学的过程中，根据教学的实际需求共同编写了《物理化学简明双语教程》，该教材的出版填补了国内物理化学双语教学教材的空白。该教材对于教学基本要求所规定的内容采用汉语和英语两种语言形式编写，便于教师和学生理解和学习。事实证明该教材在效度、信度、难度各方面都符合国内双语教学要求，不仅可以减少教师的备课难度和工作量，而且对激发学生专业学习的兴趣也大有帮助。所以教材一经面世就受到广大同行的认可和使用。

目前距离第一版教材出版已经过去15年了，虽然物理化学作为一种基础学科，主要内容并没有发生大的变化，但信息技术的发展和应用已经渗透到我们生活的每一个角落，对人们的生活习惯和思维方式都产生了巨大的影响。将新科技的应用引入教材中，来丰富原有纸质版教材的内容，提升其使用体验就具有非常重要的意义。

本次修订是在第一版《物理化学简明双语教程》的基础上由沈阳化工大学的何美、李文泽、张潇飒、周华锋老师共同完成。教材仍然保留第一版的框架，分为十章：气体的 *pVT* 性质；热力学第一定律；热力学第二定律；多组分系统热力学；化学平衡；相平衡；电解质溶液；电化学系统；界面现象；化学动力学。相比第一版教材，本教材内容兼顾近年来国家对教学提出的新要求，结合上一版教材使用过程中的感受，在内容和形式上都进行了大量的更改。具体体现在各章的写作风格更加统一，每章前面增加了本章的学习目标，各章的内容进一步细化，对原有例题进行删减、替换，同时也增加了部分新的例题。通过添加二维码，将课外拓展资料、部分例题以文件、图片的方式，对典型题和重点知识点的讲解以视频方式加入教材，既丰富教材的内容，又不增加篇幅和成本，使教材更加生动、形象、易于理解，从而增强学习体验，提高学习兴趣，提升学习效果。在拓展资料中加入了思政元素，使思政教育融入日常教学过程中。

另外，与该教材配套的《物理化学双语解题指导》已经由中国石化出版社出版发行，两本书配合使用可以让学生更好适应双语教学，同时也给从事双语教学的老师提供一些参考资料，减轻他们的工作量。

由于编者水平有限，书中错误和不当之处在所难免，欢迎读者批评指正。

CONTENTS

Introduction
(绪论)

0. 1 Physical chemistry (物理化学)

Physical chemistry studies the underlying physical principles that govern the properties and behavior of chemical systems. It is the subject which start with the connection of physical phenomenon and chemical phenomenon, then utilize physical theories and experimental methods to look for the rules of chemical change.

A chemical system can be studied from either a microscopic or a macroscopic viewpoint. The microscopic viewpoint makes explicit use of the concept of molecules. The macroscopic viewpoint studies large-scale properties of matter without explicit use of the molecule concept.

0. 1. 1 Contents of physical chemistry (物理化学的研究内容)

We can divide physical chemistry into four main areas: thermodynamics, quantum chemistry, statistical mechanics, and kinetics.

Thermodynamics (热力学) is a macroscopic science that studies the interrelation-ships among the various equilibrium properties of a system.

Molecules and the electrons and nuclei that compose them do not obey classical (Newtonian) mechanics; instead their motions are governed by the laws of quantum mechanics. Application of quantum mechanics to atomic structure, molecular bonding, and spectroscopy gives us **quantum chemistry** (量子化学).

The macroscopic science of thermodynamics is a consequence of what is happening at a molecular (microscopic) level. The molecular and macroscopic levels are related to each other by the branch of science called **statistical mechanics** (统计力学). Statistical mechanics gives insight into why the laws of thermodynamics hold and allows calculation of macroscopic thermodynamic properties from molecular properties.

Kinetics (动力学) is the study of rate processes such as chemical reactions, diffusion, and the flow of charge in an electrochemical cell. The theory of rate processes is not as well developed as the theories of thermodynamics, quantum mechanics, and statistical mechanics. Kinetics uses relevant portions of thermodynamics, quantum chemistry, and statistical mechanics.

0. 1. 2 Issues to be solved by physical chemistry (物理化学解决的问题)

The principles of physical chemistry provide a framework for all branches of chemistry.

Organic chemists use kinetics studies to figure out the mechanisms of reactions, use quantum-chemistry calculations to study the structures and stabilities of reaction intermediates, use symmetry rules deduced from quantum chemistry to predict the course of many reactions, and use nuclear-magnetic-resonance (NMR) and infrared spectroscopy to help determine the structure of newly synthesized compounds.

Inorganic chemists use quantum chemistry and spectroscopy to study samples of unknown composition.

Analytical chemists use spectroscopy to analyze samples of unknown composition.

Biochemists use kinetics to study rates of enzyme-catalyzed reactions, use thermodynamics to study biological energy transformations, osmosis, and membrane equilibrium, and to determine molecular weights of biological molecules, use spectroscopy to study processed at the molecular level (for example, intramolecular motions in proteins are studied using NMR), and use X-ray diffraction to determine the structures of proteins and nucleic acids.

Chemical engineers use thermodynamics to predict the equilibrium composition of reaction mixtures, use kinetics to calculate how fast products will be formed, and use principles of thermodynamics phase equilibria to design separation procedures such as fractional distillation.

Geologists use thermodynamics phase diagrams to understand processes in the earth.

Polymer chemists use thermodynamics, kinetics, and statistical mechanics to study the kinetics of polymerization, the molecular weights of polymers, the flow of polymer solutions, and the distribution of conformations of a polymer molecule.

0.1.3 Advice for studying physical chemistry (物理化学的学习方法)

Keep up with your studying day to day. New material builds on the old. It is important not to fall behind; if you do, you will find it much harder to follow the lectures and discussions on current topics. Trying to "cram" just before an exam is generally a very ineffective way to study physical chemistry.

Focus your study. The amount of information you will receive in your physical chemistry course and sometimes seem overwhelming. Certainly, there are more facts and details than any student can hope to assimilate in a first course. It is important to focus on the key concepts and skills. Listen intently for what your lecture and discussion leader emphasize. Pay attention to the skills stressed in the sample exercises and homework assignments. Notice the important statements in the text, and study the concepts presented in every chapter.

Keep good lecture notes, so that you have a clear and concise record of the required material. You will find it easier to take useful notes if you skim topics in the text before they are covered in lecture. To skim a chapter, first read the introduction. Then quickly read through the chapter, skipping examples. Pay attention to section heads and subheads, which give you a feeling for the scope of topics. Avoid the feeling that you must learn and understand everything right away.

After lecture, **carefully read the topics covered in class.** Remember, though, that you will need to read assigned material more than once to master it. As you read, pay particular attention to the examples. Once you think you understand an example, test your understanding by working the accompanying exercise.

Finally, **attempt all of the assigned end-of-chapter exercises.** Working out these exercises provides necessary practice in recalling and using the essential ideas of the chapter. You cannot learn merely by observing; you must be a participant. In particular, there is little value in merely copying answers from the solution manual or from another student. If however, you really get stuck on a problem, get help from your instructor, teaching assistant, tutor, or a fellow student. Spending more than 20 minutes on a single exercise is rarely effective unless you know that it is particularly challenging and requires extensive thought and effort.

0.2 Indication and operation of physical quantity (物理量的表示及运算)

Physical quantity(物理量) — physical phenomenon that can be described quantitatively. It is also called **quantity**.

0.2.1 Indication of physical quantity (物理量的表示)

Physical quantity = numeral value × unit, that is, $Q = \{Q\} \times [Q]$.

For instance: $U = 20\text{kJ}$, 20 is numeral value of U, kJ is unit of U.

Symbol of physical quantity is usually one Latin or Greek alphabet, sometimes having subscript or other explaining mark. Italics must be used to print symbol of physical quantity. If subscript is physical quantity, we should print it using italic too. For instance, symbol of temperature is T, symbol of pressure is p, etc.

Symbol of unit is small letter. If name of unit comes from name of person, the first letter is capital letter.

Comment:

① Same physical quantities with same units can be comparable.

② Physical quantity in $\ln x$ or e^x must be pure numeral value.

③ Physical quantity in a plot or a table must be pure numeral value.

④ Identical units (for example, SI unit) should be used in two sides of equality.

In this book, for simplicity and convenience, we can denote logarithm of Q as $\ln Q$, not $\ln(Q/[Q])$.

0.2.2 Operation of physical quantity (物理量的运算)

Operation of physical quantity must satisfy quantity equation, $Q = \{Q\} \times [Q]$. For instance, known conditions are $n = 10\text{mol}$, $T = 300\text{K}$, and $V = 10\text{m}^3$, calculate pressure of system p using quantity equation $pV = nRT$.

Numeral values and units of physical quantities were substituted to quantity equation, we can get:

$$p = \frac{10\text{mol} \times 8.315\text{J} \cdot \text{mol}^{-1} \cdot \text{K}^{-1} \times 300\text{K}}{10\text{m}^3} = 2494.5\text{Pa}$$

For simplicity and convenience, we can calculate pressure using quantity equation $pV = nRT$, and write out the last unit, that is,

$$p = \frac{10 \times 8.315 \times 300}{10} = 2494.5\text{Pa}$$

Chapter 1 The *pVT* properties of gases
(气体的 *pVT* 性质)

Learning objectives:

(1) Write the equation of state for a perfect gas, and use it to predict changes in pressure, volume, and temperature.

(2) Define *partial pressure* of a gas in a mixture and relate it to the *mole fraction* of the component.

(3) Write down the *van der Walls equation of state* and explain the *critical parameters* of the gas.

Matter can exist in three physical states: gas (also known as vapor), liquid, and solid. These states differ in some of their simple observable properties. A gas has no fixed volume or shape; rather, it conforms to the volume and shape of its container. A gas can be compressed to occupy a smaller volume, and it will expand to occupy a large one. A liquid has a distinct volume independent of its container but has no specific shape. It assumes the shape of the portion of the container that it occupies. A solid has both a definite shape and a definite volume; it is rigid. Neither liquids nor solids can be compressed to any appreciable extent.

We are surrounded by an atmosphere composed of a mixture of gases that we refer to as air. The behavior of air determines our weather, and the oxygen, O_2, in air supports life. We also encounter gases in countless other situations. For example, chlorine gas, Cl_2, is used to purify drinking water. Acetylene gas, C_2H_2, is used in welding. Although different gases may vary widely in their chemical properties, they share many physical properties. Our goal in this chapter is to deeper understanding of the *pVT* properties of gases.

1.1 State equation of the perfect gas
(理想气体状态方程)

1.1.1 Empirical law of gas(气体的经验定律)

(1) Boyle's law (波义耳定律)

The volume of a given amount of gas at constant temperature is inversely proportional to the pressure.

Boyle's law can be expressed in mathematical terms:

$$p \propto 1/V, \text{ that is, } pV = \text{constant (at const } n \text{ and } T) \qquad (1.1.1)$$

The value of the constant depends on the temperature and the amount of gas in the sample.

(2) Gay-Lussac's law (盖-吕萨克定律)

The volume of a given amount of gas at constant pressure increase in proportional to the temperature.

Gay-Lussac's law can be expressed in mathematical terms:

$$V \propto T, \quad \text{that is}, \quad V/T = \text{constant (at const } n \text{ and } p\text{)} \tag{1.1.2}$$

The value of the constant depends on the pressure and the amount of gas.

(3) Avogadro's law (阿伏加德罗定律)

The volume of a gas at constant temperature and pressure increase in proportional to the number of moles of the gas.

Avogadro's law can be expressed in mathematical terms:

$$V \propto n, \quad \text{that is}, \quad V/n = \text{constant (at const } T \text{ and } p\text{)} \tag{1.1.3}$$

1.1.2 State equation of the perfect gas (理想气体状态方程)

We can combine above three relationships to make a more general gas law:

$$V \propto \frac{nT}{p}$$

If we call the proportionality constant R, we have

$$V = R\left(\frac{nT}{p}\right)$$

Rearranging, we have this relationship in its more familiar form:

$$pV = nRT \tag{1.1.4}$$

Or

$$pV_m = RT \tag{1.1.5}$$

(1.1.4) and (1.1.5) are known as the **state equation of the perfect gas**.

(1) Meaning of physical quantity

R—molar gas constant (摩尔气体常数), unit: $J \cdot K^{-1} \cdot mol^{-1}$;

V_m—molar volume (摩尔体积), $V_m = \frac{V}{n}$, unit: $m^3 \cdot mol^{-1}$.

(2) Value and unit of R

$R = 8.315 J \cdot K^{-1} \cdot mol^{-1}$(SI) ($p$: Pa; V: m^3; n: mol; T: K);

$R = 0.08206 atm \cdot L \cdot K^{-1} \cdot mol^{-1}$($p$: atm; V: L);

$R = 62.36 mmHg \cdot L \cdot K^{-1} \cdot mol^{-1}$($p$: mmHg; V: L).

(3) Other forms of state equation of the perfect gas

For $n = \frac{m}{M}$, according to the state equation of the perfect gas, we have

$$pV = \frac{m}{M}RT$$

That is,

$$M = \frac{mRT}{pV} \tag{1.1.6}$$

For $\rho = \frac{m}{V}$, according to the state equation of the perfect gas, we have

$$p = \frac{m}{MV}RT = \frac{\rho}{M}RT$$

That is,

$$\rho = \frac{pM}{RT} \tag{1.1.7}$$

The three variables p, V and T may all change for a given sample of gas. Under these circumstance we have,

$$\frac{pV}{T} = nR = \text{constant}$$

Thus, as long as the total quantity of gas, n, is constant, pV/T is a constant. If we represent the initial and final conditions of pressure, temperature, and volume by subscripts 1 and 2, respectively, we can write the following expression:

$$\frac{p_1V_1}{T_1} = \frac{p_2V_2}{T_2} \tag{1.1.8}$$

(4) Applicable condition of state equation of the perfect gas

① The perfect gas;

② Low pressure gas.

1.1.3 Model of the perfect gas (理想气体模型)

(1) Definition of the perfect gas

A gas that obeys state equation $pV = nRT$ at arbitrary temperature and pressure is called as the perfect gas.

(2) Features of the perfect gas

① There are no interactions among the molecules.

② None volume is occupied by the molecules themselves.

1.2 Mixture of the perfect gas (理想气体混合物)

1.2.1 Composition of perfect gas mixture (理想气体混合物的组成)

(1) Mole fraction of B (B 的摩尔分数)

B is the arbitrary component of perfect gas mixture.

$$y_B \overset{\text{def}}{=\!=\!=} n_B / \sum_A n_A \tag{1.2.1}$$

Comment:

① Unit of y_B is 1.

② $\sum y_B = 1$.

For instance, perfect gas mixture (A, B): $y_A = \dfrac{n_A}{n_A + n_B}$

$$y_B = \frac{n_B}{n_A + n_B}$$

So, $y_A + y_B = \dfrac{n_A + n_B}{n_A + n_B} = 1$

(2) Mass fraction of B (B 的质量分数)

$$w_B \overset{\text{def}}{=\!=\!=} m_B / \sum_A m_A \tag{1.2.2}$$

Comment:

① Unit of w_B is 1.

② $\sum w_B = 1$.

(3) Volume fraction of B (B 的体积分数)

$$\varphi_B \xlongequal{\text{def}} x_B V_{m,B}^* / \sum_A x_A V_{m,A}^* \tag{1.2.3}$$

$V_{m,A}^*$ and $V_{m,B}^*$—molar volume of pure perfect gas A and B in T, p of mixture.

Comment:

① Unit of φ_B is 1.

② $\sum \varphi_B = 1$

1.2.2 Dalton's law of partial pressures (道尔顿分压定律)

(1) Partial pressure (分压力)

Suppose that the gas with which we are concerned is not made up of a single kind of gas particle but is rather a mixture of two or more different substances. We expect that the total pressure exerted by the gas mixture is the sum of pressures due to the individual components.

Each of the individual components, if present alone under the same temperature and volume conditions as the mixture, would exert a pressure that we term the **partial pressure**. (混合气体中任一组分 B 单独存在于混合气体的温度体积条件下具有的压力。混合气体中任一组分的分压力是它的摩尔分数与总压力之积。)

$$p_B = \frac{n_B RT}{V}$$

$$p_B = p y_B \tag{1.2.4}$$

(2) Dalton's law of partial pressures

① Statement (语言表述)

The total pressure of a mixture of gases equals the sum of the pressures that each gas would exert if it were present alone. (混合气体的总压力等于各组分单独存在于混合气体的温度、体积条件下产生的压力的总和。)

② Mathematical expression (数学表达式)

$$p = \sum_B p_B \tag{1.2.5}$$

Equation (1.2.5) is the mathematical expression of **Dalton's law of partial pressures**.

In each of the gases obeys the state equation of the perfect gas, we can write

$$p_1 = n_1\left(\frac{RT}{V}\right),\ p_2 = n_2\left(\frac{RT}{V}\right),\ p_3 = n_3\left(\frac{RT}{V}\right),\quad \text{and so forth.}$$

All of the gases experience the same temperature and volume. Therefore, by substituting into equation (1.2.5), we obtain

$$p_t = \frac{RT}{V}(n_1 + n_2 + n_3 + \cdots) \tag{1.2.6}$$

That is, the total pressure at constant temperature and volume is determined by the total number of moles of gas present, whether that total represents just one substance or a mixture.

③ Applicable condition of Dalton's law of partial pressures

ⅰ) The perfect gas;

ⅱ) Low pressure gas.

1.2.3 Amagat's law of partial volumes（阿马加分体积定律）

(1) Partial volume（分体积）

Each of the individual components, if present alone under the same temperature and pressure conditions as the mixture, would occupy a volume that we term the **partial volume**.（混合气体中任一组分B单独存在于混合气体的温度压力条件下占有的体积。混合气体中任一组分的分体积是它的摩尔分数与总体积之积。）

$$V_B = \frac{n_B RT}{p}$$

$$V_B^* = V y_B \tag{1.2.7}$$

(2) Amagat's law of partial volumes

① Statement（语言表述）

The total volume of a mixture of gases equals the sum of the volumes that each gas would occupy if it were present alone.（混合气体的总体积等于各组分单独存在于混合气体的温度、压力条件下产生的体积的总和。）

② Mathematical expression（数学表达式）

$$V = \sum_B V_B^* \tag{1.2.8}$$

Equation (1.2.8) is the mathematical expression of **Amagat's law of partial volumes**.

In each of the gases obeys the state equation of the perfect gas, we can write

$$V_1 = n_1\left(\frac{RT}{p}\right),\ V_2 = n_2\left(\frac{RT}{p}\right),\ V_3 = n_3\left(\frac{RT}{p}\right),\quad \text{and so forth.}$$

All of the gases experience the same temperature and pressure. Therefore, by substituting into equation (1.2.8), we obtain

$$V_t = \frac{RT}{p}(n_1 + n_2 + n_3 + \cdots) \tag{1.2.9}$$

That is, the total volume at constant temperature and pressure is determined by the total number of moles of gas present, whether that total represents just one substance or a mixture.

③ Applicable condition of Amagat's law of partial volumes

ⅰ) The perfect gas;

ⅱ) Low pressure gas.

1.2.4 Mean molar mass of mixed gas（混合气体的平均摩尔质量）

$$M_{mix} = \frac{m}{n} = \frac{m_A + m_B}{n} = \frac{n_A M_A + n_B M_B}{n} = y_A M_A + y_B M_B = \sum_B y_B M_B \tag{1.2.10}$$

For example, air, $y(O_2) = 0.21$, $y(N_2) = 0.79$

Then: $M_{mix}(\text{air}) = y(O_2)\times M(O_2) + y(N_2)\times M(N_2)$

$= 0.21\times 32 + 0.79\times 28$

$= 29\,\text{g}\cdot\text{mol}^{-1}$

1.3 Liquefaction of real gas and critical properties (真实气体的液化及临界性质)

1.3.1 Saturated vapor pressure of liquid (液体的饱和蒸气压)

Vapor-liquid phase equilibrium (气-液相平衡)— the chemical potentials of gas and liquid are equal, which corresponds to gas and liquid existing together in equilibrium, see figure 1.1.

Saturated vapor pressure of liquid (液体的饱和蒸气压)— the pressure corresponding to vapor-liquid equilibrium at a specified temperature.

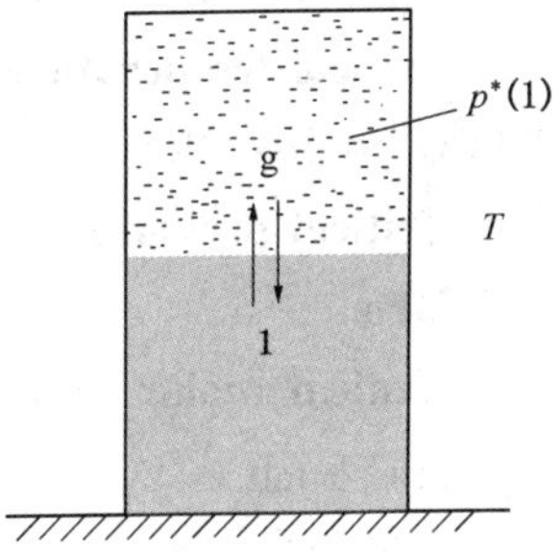

Fig. 1.1 Vapor-liquid phase equilibrium

Boiling point of liquid (液体的沸点)—the temperature at which its equilibrium vapor pressure equals external pressure.

Boiling point of liquid includes normal boiling point and standard boiling point.

Normal boiling point—the temperature at which the liquid's vapor pressure is 101.325kPa.

Standard boiling point—the temperature at which the liquid's vapor pressure is 100kPa.

For instance, H_2O:

the normal boiling point is 100℃, the standard boiling point is 99.67℃.

1.3.2 Liquefaction of real gas and critical properties (真实气体的液化及临界性质)

The perfect gas cannot liquefy. A very significant deviation from the perfect gas behavior is the fact that, provided *T* is below the critical temperature, any real gas condenses to a liquid when the pressure is increased sufficiently.

$$\text{gas} \xrightarrow{\text{increase pressure, decrease temperature}} \text{liquid}$$

Real gas may transform to liquid if the volume of real gas is reduced through increasing pressure or decreasing temperature. But the *p-V-T* relation of the transformation process obeys definite rules, such as liquefaction of CO_2, which is shown in figure 1.2.

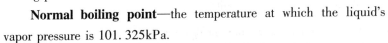

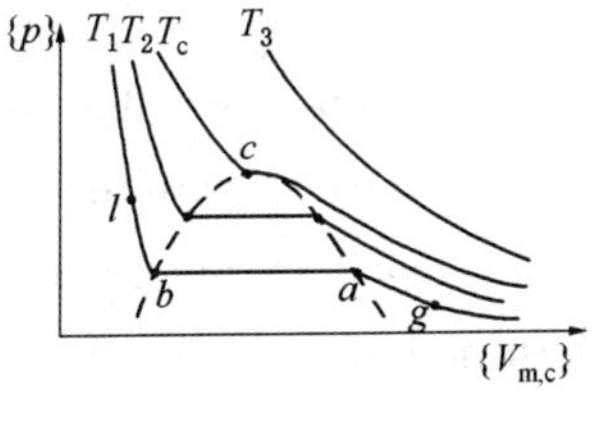

Fig. 1.2 $p-V_{m,c}$ of CO_2

In figure 1.2, each curve is called as *p-V* isotherm, point *c* is called as critical point, the state of point *c* is called as critical state.

From figure 1.2 we can see each $p\text{-}V_m$ isotherm has a level section when the temperature is below T_c.

Note that:

① The level section is called section at constant pressure because of constant pressure.

② The pressure of section at constant pressure will increase with increment of temperature.

③ The length of section at constant pressure will decrease with increment of temperature and become a point at T_c. $[V_m(l)=V_m(g)]$

Let's take T_1 for example. The curve is divided into three sections.

$$g(\text{gas})\xrightarrow[\text{volume is reduced}]{\text{increase pressure}}a(\text{saturated gas})$$

$$a(\text{saturated gass})\xrightarrow[\text{volume is reduced greatly}]{\text{pressure is constant}}b(\text{saturated liquid})$$

$$b(\text{saturated liquid})\xrightarrow[\text{volume is reduced}]{\text{increase pressure}}l(\text{liquid})$$

T_c, p_c, and $V_{m,c}$ are critical parameters.

Critical temperature (T_c)—the highest temperature at which gas can be liquefied. Such as $T_c(CO_2)=304.2K$.

Critical pressure (p_c)—the smallest pressure at the critical temperature. Such as $p_c(CO_2)=7.38MPa$.

Critical molar volume ($V_{m,c}$)—the molar volume of gas at the critical temperature and pressure. Such as $V_{m,c}(CO_2)=94\times10^{-6}m^3\cdot mol^{-1}$.

Critical parameters of some substances are listed in table 1.1.

Tab. 1.1 Critical parameters of some substances

Substance	T_c/K	p_c/MPa	$V_{m,c}/10^{-6}m^3\cdot mol^{-1}$
He	5.26	0.229	58
H_2	33.3	1.30	65
N_2	126.2	3.39	90
O_2	154.4	5.04	74
H_2O	647.4	22.12	56
CH_4	190.7	4.64	99
C_2H_4	283.1	5.12	124
C_6H_6	562.6	4.92	260
C_2H_5OH	516.3	6.38	167

1.4 State equations of real gas
(真实气体状态方程)

1.4.1 The van der Waals equation (范德华方程)

Engineers and scientists who work with gases at high pressures often cannot use the state equation of the perfect gas to predict the pressure-volume properties of gases because departures from the perfect gas behavior are too large. Various equations have been developed to predict more realistically the pressure-volume behavior of real gases. One of the earliest and most useful of these equations was proposed by the Dutch scientist Johannes van der Waals:

$$\left(p+\frac{n^2a}{V^2}\right)(V-nb)=nRT \tag{1.4.1}$$

The van der Waals equation differs from the state equation of the perfect gas by the presence of two correction terms; one corrects the volume, the other modifies the pressure. The term nb in the expression ($V-nb$) is a correction for the finite volume of the gas molecules; the van der Waals constant, b, different for each gas, has units of $L \cdot mol^{-1}$. It is a measure of the actual volume occupied by the gas molecules. Note that b increase with an increase in mass of the molecule or in the complexity of its structure.

The correction to the pressure takes account of the intermolecular attractions between molecules. Notice that it consists of the constant, a, different for each gas, times the quantity $(n/V)^2$. The unit of (n/V) are $mol \cdot L^{-1}$. This quantity is squared because the number of molecular interactions, is proportional to the square of the number of molecules per unit volume. The magnitude of a reflects how strongly gas molecules attract each other. Notice that a, like b, increase with an increase in molar mass and with an increase in complexity of molecular structure.

1.4.2 The virial equation of state (维里方程)

Statistical mechanics shows that the equation of state of a real gas can be expressed as the following power series in $1/V_m$:

$$pV_m = RT\left(1 + \frac{B_2}{V_m} + \frac{B_3}{V_m^2} + \frac{B_4}{V_m^3} + \cdots\right) \tag{1.4.2}$$

This is the **virial equation of state.** The coefficients B_2, B_3, $\cdots$, which are functions of T only, are the second, third, $\cdots$, virial coefficient. They can be determined from experimental p-V-T data of gases. Statistical mechanics gives theoretical expressions for the virial coefficients in terms of the potential energy of intermolecular forces. Unfortunately, not enough is known about these forces to yield accurate theoretical expressions for the virial coefficients of most real gases.

A form of the virial equation equivalent to (1.4.2) uses a power series in p rather than $1/V_m$:

$$pV_m = RT(1 + A_2p + A_3p^2 + A_4p^3 + \cdots) \tag{1.4.3}$$

The coefficients A_2, A_3, $\cdots$, which are functions of T only, are also the second, third, $\cdots$, virial coefficient. At ordinary pressure, terms beyond A_3p^2 in (1.4.3) are usually negligible, and even the A_3p^2 term is small. Thus one cuts off the series after two or three terms. At high pressure, the higher terms become important. At very high pressure, the virial equation fails.

1.5 The law of corresponding states (对应状态原理)

1.5.1 Compression factors (压缩因子)

As a measure of the deviation from the ideality of the behavior of a real gas, we define the **compressibility factor** or **compression factor Z** of a gas as

$$Z = \frac{pV}{nRT} = \frac{pV_m}{RT} \tag{1.5.1}$$

Z is a function of T and p. For a perfect gas, $Z=1$ for all temperatures and pressures. When $Z<1$, the gas exerts a lower pressure than a perfect gas would, at the same time, the real gas is easy to be compressed; When $Z>1$, the real gas is hard to be compressed. We define the critical compression factor Z_c(临界压缩因子) as

$$Z_c = \frac{p_c V_{m,c}}{RT_c} \tag{1.5.2}$$

Of the known Z_c values, 80 percent lie 0.25 and 0.30, significantly less than predicted by the van der Waals equation. The smallest known Z_c is 0.12 for HF; the largest is 0.46 for CH_3NHNH_2.

1.5.2 The law of corresponding states (对应状态原理)

The **reduced pressure p_r**(对比压力), **reduced temperature T_r**(对比温度), and **reduced volume V_r**(对比体积) of a gas in the state (p, V, T) are defined as

$$p_r = p/p_c \tag{1.5.3}$$

$$T_r = T/T_c \tag{1.5.4}$$

$$V_r = V_m/V_{m,c} \tag{1.5.5}$$

Where p_c, T_c, and $V_{m,c}$ are the critical constants of the gas.

van der Waals pointed out that, if one uses reduced variables to express the states of gases, then, to a pretty good approximation, all gases show the same p-V-T behavior; in other words, **if two different gases are each at the same p_r and T_r, they have (nearly) the same V_r values.** This observation is called the **law of corresponding states.** Mathematically,

$$V_r = f(p_r, \ T_r) \tag{1.5.6}$$

Where approximately the same function f applies to any gas.

EXERCISES
(习题)

1.1 A flashbulb of volume 2.6cm^3 contains O_2 gas at a pressure of 233kPa and a temperature of 26℃. How many moles of O_2 does the flashbulb contain?

Answer: 2.4×10^{-4}mol

1.2 The gas pressure in an aerosol can is 1.5atm at 25℃. Assuming that the gas inside obeys the state equation of the perfect gas, what would the pressure be if the can were heated to 450℃?

Answer: 3.6atm

1.3 An inflatable raft is filled with gas at a pressure of 800mmHg at 16℃. When the raft is left in the sun, the gas heats up to 44℃. Assuming no volume change, what is the gas pressure in the raft under these condition?

Answer: 878mmHg

1.4 The mean molar mass of the atmosphere at the surface of Titan, Saturn's largest moon, is 28.6g · mol^{-1}. The surface temperature is 95K, and the pressure is 1.6 earth atm. Assuming ideal behavior, calculate the density of Titan's atmosphere.

Answer: 5.9g · L^{-1}

1.5 What pressure, in atm, is exerted by a mixture of 2.00g of H_2 and 8.00g of N_2 at 273K in a 10. 0L vessel?

Answer: 2. 86atm

1. 6 Cyanogen, a highly toxic gas, is composed of 46. 2 percent C and 53. 8 percent N by mass. At 25℃ and 750mmHg, 1. 05g of cyanogen occupies 0. 50L. What is the molecular formula of cyanogen?

Answer: C_2N_2

1. 7 At 20℃, the gas mixture of $C_2H_6-C_4H_{10}$ is filled into a vacuum vessel of 200cm^3 to make the pressure be 101. 325kPa, the measured mass of mixed gas is 0. 3897g. Calculate the mole fraction and partial pressure of two components of mixed gas.

Answer: $y(C_2H_6) = 0.401$, $p(C_2H_6) = 40.63$kPa; $y(C_4H_{10}) = 0.599$, $p(C_4H_{10}) =$ 60. 69kPa

1. 8 In the gas mixture of C_2H_3Cl, HCl, and C_2H_4, the mole fraction of each component is 0. 89, 0. 09, and 0. 02 respectively. At constant pressure of 101. 325kPa, HCl is absorbed by water, then steam of partial pressure of 2. 67kPa is increased in the gas mixture. Calculate the partial pressure of C_2H_3Cl and C_2H_4 in gas mixture after being washed.

Answer: 96. 487kPa, 2. 168kPa

1. 9 Calculate molar volume of N_2 using the state equation of the perfect gas and the van der Waals equation, the corresponding temperature is 0℃, pressure is 40530Pa.

Given that the experimental value of $V_m(N_2)$ is 70. 3cm^3 · mol^{-1}.

Answer: V_m(pg) = 56. 0cm^3 · mol^{-1}, V_m(van der Waals gas) = 73. 1cm^3 · mol^{-1}

1. 10 A quantity of N_2 gas originally held at 4. 60atm pressure in a 1. 00L container at 26℃ is transferred to a 10. 0L container at 20℃. A quantity of O_2 gas originally at 3. 50atm and 26℃ in a 5. 00L container is transferred to this same container. What is the total pressure in the new container?

Answer: 2. 16atm

Chapter 2 The first law of thermodynamics (热力学第一定律)

Learning objectives:

(1) Define *thermodynamic system*, *surroundings*, *closed system*, and *isolated system*.

(2) Define *energy*, *heat*, and *work*.

(3) State the *First Law of thermodynamics*.

(4) Define *internal energy* and *thermodynamic reversibility*.

(5) Calculate the work during the *isothermal reversible expansion* of a perfect gas.

(6) Define the *heat capacity* of a system.

(7) Define and explain *enthalpy*.

Thermodynamics is an important part of physical chemistry. The basic contents of it are the first law of thermodynamics and the second law of thermodynamics.

Thermodynamics is the study of heat, work, energy, and the changes they produce in the states of systems; in a broader sense, thermodynamics studies the relationships among the macroscopic properties of a system.

2.1 The basic concept of thermodynamics (热力学基本概念)

2.1.1 System and surroundings (系统和环境)

System— the part of the universe under study in thermodynamics. [热力学研究的对象(大量分子、原子、离子等微粒组成的宏观集合体)，即要研究的那部分物质或空间，也称体系或物系。]

Surroundings—the part of the universe that can interact with the system. (是系统以外与之相联系的真实世界，可简称为环境或外界。)

If, for example, we are studying the vapor pressure of water as a function of temperature, we might put a sealed container of water in a constant-temperature bath and connect a manometer to the container to measure the pressure. Here, the system consists of the liquid water and the water vapor in the container, and the surroundings are the constant-temperature bath and the mercury in the manometer.

A system may be separated from its surroundings by various kinds of walls. A wall can be either rigid or nonrigid. A wall may be permeable or impermeable, where by impermeable we mean that it allows no matter to pass through it. Finally, a wall may be adiabatic or nonadiabatic. In plain language, an adiabatic wall is one that does not conduct heat at all, whereas a nonadiabatic wall does conduct heat. However, we have not yet defined heat, and hence to have a logically correct development of thermodynamics, adiabatic or nonadiabatic walls must be defined without reference to heat. This is done as follows.

The connection between system and surroundings consists of transferring of matter and energy.

For convenience, the system is divided into three kinds according to difference of connection between system and surroundings.

A thermodynamic system is either open or closed and is either isolated or nonisolated. It is important to note the kind of system under study, since thermodynamic statements valid for one kind of system may well be invalid for other kinds. Most commonly, we shall deal with **closed system**.

Type of system	Transfer of matter between system and surroundings	Transfer of energy between system and surroundings
open system (敞开系统)	have	have
closed system (封闭系统)	have not	have
isolated system (隔离系统)	have not	have not

2.1.2 Extensive property and intensive property of system (系统的广度性质和强度性质)

Thermodynamic system is a macroscopic aggregation consisted of great amount of molecules, atoms and ion etc. The collective behavior showed by the macroscopic aggregation is called macroscopic property of thermodynamic system, such as p, V, T, U, H, S, A and G. It is also called thermodynamic property or thermodynamic quantity.

Thermodynamic properties are classified as extensive or intensive property.

Extensive property (广度性质)— additive property whose value equal to the sum of its values for the parts of the system. (与系统中所含物质的量有关，有加和性，如 n、V、U、H 等。)

Thus, if we divide system into parts, the mass of the system is the sum of the masses of the parts; mass is an extensive property. So is volume, etc.

Intensive property (强度性质)— other property is independent of the amount of matter in the system. (与系统中所含物质的量无关，无加和性，如 p、T 等。)

Density and pressure are examples of intensive properties. We can take a drop of water or a swimming pool full of water, and both systems will have the same density.

$$\frac{\text{one extensive property}}{\text{another extensive property}}=\text{intensive property, e. g. } V_{\mathrm{m}}=\frac{V}{n},\ \rho=\frac{m}{V},\ \text{etc.}$$

2.1.3 State and state function of system (系统的状态和状态函数)

State of system—the sum of all macroscopic properties which describe the system. (系统所处的样子。用宏观性质描述系统的状态，系统的状态是它所有性质的总体表现，也就是系统性质总和决定了系统状态。)

The state of a system is defined by specifying the values of its thermodynamic properties. However, it is not necessary to specify all the properties to define the state; there is a minimum number of properties, specification of which determines the values of the remaining properties.

State function—a system in a given equilibrium state then has a particular value for each thermodynamic property, these properties are therefore called state function.（系统的宏观性质。当系统的状态发生变化时，它的一系列性质也随之而改变。改变多少，只取决于系统的开始和终了状态，而与变化时所经历的途径无关。无论经历多么复杂的变化，只要系统恢复原状，则这些性质也恢复原状。热力学中，把具有这种特性的物理量叫作状态函数。）

Since the values of state function are functions of the system's state, the terms of thermodynamic variable, thermodynamic property, and state function are synonymous. Note especially that the value of a state function depends only on the present state of a system and not on its past history.

Characteristics of state function:

① For a homogeneous system with a given constant composition, an arbitrary macroscopic property of system is function of two other self-governed macroscopic properties, that is, $z=f(x, y)$. For instance, the perfect gas: $V=nRT/p$, that is, the volume of the prefect gas can be determined by T and p with n is constant.

② If state is constant, state function is constant. State function changes with state, however, the changing value is independent of the way the state was prepared. That is, Δ(state function) = the final value − the initial value.

For instance: $\Delta T=T_2-T_1$, $\Delta U=U_2-U_1$

③ The exact differential can be used in dealing with state function, for instance, dT, dU, dV and dp etc.

If $z=f(x, y)$, the exact differential is: $\mathrm{d}z=\left(\frac{\partial z}{\partial x}\right)_y \mathrm{d}x+\left(\frac{\partial z}{\partial y}\right)_x \mathrm{d}y$

2.1.4 Equilibrium state of thermodynamics（热力学平衡态）

Equilibrium state of thermodynamics—all observed macroscopic properties of system do not change with time after a long chase in a given surroundings condition, then, the system is isolated, all macroscopic properties of system still do not change with time. The state of system in this moment is called **equilibrium state of thermodynamics.**

Conditions should be satisfied for a system in equilibrium state of thermodynamics, that is:

① Thermal equilibrium（热平衡）: temperatures of each several part in a system are equal; temperature of system is equal to temperature of surroundings in an adiabatic condition.

② Mechanical equilibrium（力平衡）: pressures of each several part in a system are equal.

③ Phase equilibrium（相平衡）: every phase of system coexists for a long time and compositions and amounts of each phase do not change with time.

④ Chemical equilibrium（化学平衡）: compositions of system do not change with time, that is, on the view of macroscopic, chemical changes in a system have already ceased.

2.1.5 Process and path（过程与途径）

(1) Process（过程）

Process—system change from initial state to final state in a given surroundings condition.（在一定环境条件下，系统由始态变化到末态的经过。）

Process of system consists of: simple pVT change process（单纯 pVT 变化过程）, phase transformation process（相变化过程）, and chemical change process（化学变化过程）.

(2) Important simple pVT change processes (几种主要的单纯 pVT 变化过程)

initial state 1→final state 2

① Isothermal process (恒温过程)

T is constant throughout the process, that is, $T_1 = T_2 = T_{su} = \text{const}$, $dT=0$, $\Delta T=0$.

② Isobaric process (恒压过程)

p is held constant throughout the process, that is, $p_1 = p_2 = p_{su} = \text{const}$, $dp=0$, $\Delta p=0$.

③ Isochoric process (恒容过程)

V is held constant throughout the process, that is, $V_1 = V_2 = \text{const}$, $dV=0$, $\Delta V=0$.

④ Adiabatic process (绝热过程)

$Q=0$. This can be achieved by surrounding the system with adiabatic walls.

⑤ Cyclic process (循环过程)

The final state of the system is the same as the initial state, so the change in all state functions is zero, that is, Δ(state function) = 0, such as: $\Delta p=0$, $\Delta T=0$, and $\Delta U=0$.

⑥ Expansion against constant external pressure (对抗恒外压过程)

$p_{su} = \text{const}$.

⑦ Free expansion process or expansion into a vacuum process (自由膨胀过程或向真空膨胀过程)

Figure 2.1 is the free expansion process, we have, $W=0$.

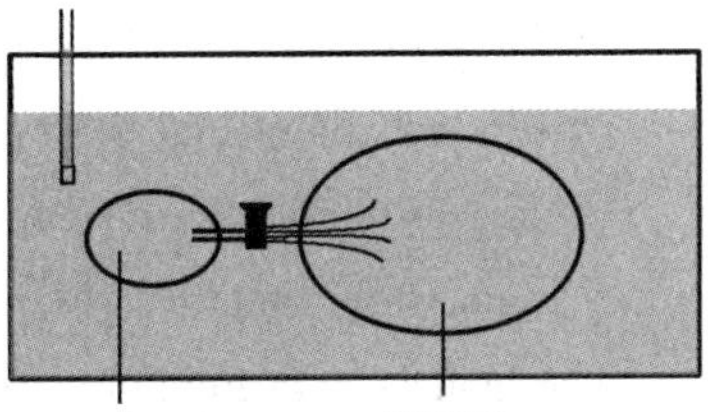

Fig. 2.1 Free expansion process

(3) Path (途径)

Path— the sum of processes going through initial state to final state. (系统从始态变化到末态所经历的过程的总和。)

A path can be made up of one or several processes, that is,

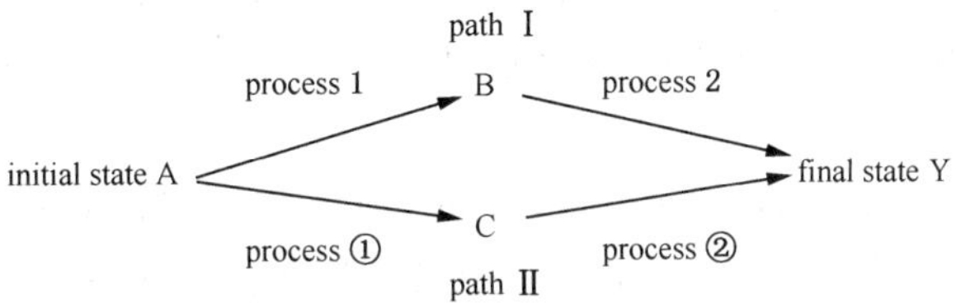

(4) State function method (状态函数法)

若已知过程的始态和末态，需计算过程中某些状态函数的变化，而其进行的条件不明，或计算困难较大，可设始态和末态与实际过程相同的假设途径，假设途径的状态函数的变化，即为实际过程中状态函数的变化。这种利用“状态函数的变化仅取决于始态和末态而与途径无关”的方法，称为状态函数法。状态函数法是热力学中的重要方法。

2.2 The first law of thermodynamics (热力学第一定律)

The essence of the first law of thermodynamics is energy conservation. It shows that thermodynamic energy, heat and work can change each other.

Heat and work are two energy transformation forms between system and surroundings.

The quantities heat and work are not state functions. Given only the initial and final states of the system, we cannot find heat and work. The heat and the work depend on the path used to go from state 1 to state 2, they are path functions.

Path function (途径函数)— function in which the increment depends on the idiographic path and initial and final state of system. (函数增量除与系统始态和末态有关外，还与具体的途径有关。)

Heat and work are path functions.

2.2.1 Heat (热)

Heat—the energy transfer between system and surroundings due to a temperature difference. (系统与环境间由于温度差的存在而引起的能量传递形式。)

Symbol: Q; unit: J.

它是物质热运动的一种表现形式，总是与大量分子无规则运动相联系着。分子无规则运动的强度越大，温度越高。当两个温度不同的物质相接触时，由于无规则运动的混乱程度不同，它们就可能通过分子的碰撞来交换能量。这就是热。

The conventions for the signs of Q is set from the system's viewpoint:

$Q>0$—heat flows into the system from the surroundings during a process (系统从环境吸热);

$Q<0$—an outflow of heat from the system to the surroundings (系统向环境放热).

Q is not state function, its infinitesimal change cannot be noted by $\mathrm{d}Q$, but δQ.

The heat is path function. We can not calculate the heat of process if we know the initial and final state but not the idiographic path.

由于热是系统与环境交换能量的一种形式，所以热必须以实际交换的数值来衡量，如 H_2SO_4 溶于水是放热过程，若过程在绝热容器中进行，环境不会得到或失去热，故此过程的热为零。系统进行的不同过程所伴随的热常冠以不同的名称，如汽化热、标准反应热等。

2.2.2 Work (功)

Work—the energy transfer between system and surroundings due to a macroscopic force acting through a distance. (系统与环境间由于压力差或其他机电“力”的存在引起的能量传递形式。)

Symbol: W; unit: J.

The conventions for the signs of W is also set from the system's viewpoint:

$W>0$—work is done on the system by the surroundings (环境对系统做功);

$W<0$ —work is done on the surroundings by the system (系统对环境做功).

W is not state function, its infinitesimal change cannot be noted by $\mathrm{d}W$, but δW.

The work is also path function. We can not calculate the work of process if we know the initial and final state but not the idiographic path.

The work consists of volume work and non-volume work in thermodynamics.

Volume work—the work transfer between system and surroundings due to volume change, symbol: W;

Non-volume work— the work transfer between system and surroundings due to non-volume change, symbol: W', such as surface work and electrical work.

2.2.3 Calculation of volume work (体积功的计算)

In thermodynamics, the most common way the system and surroundings do work on each other is by a change in the volume of the system.

Volume work is mechanical work in fact. It can be denoted by product of force and displacement.

Consider the system of figure 2.2, where the piston has cross-sectional area A. Let the system be at pressure p_{su}, and let the external pressure on the piston also be p_{su}. Let the volume change of system be dV, and let the distance the piston moves be dl.

For expansion, the system exerted a force on the piston, and the piston moved. The system therefore did an infinitesimal amount of work on its surroundings.

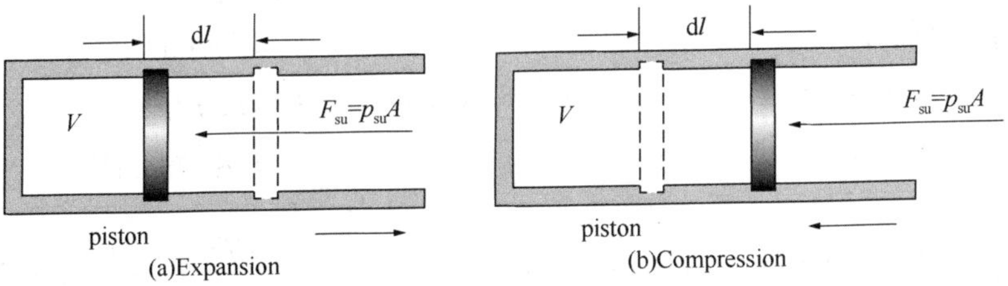

Fig. 2.2 Volume work of expansion and compression process

The work done by the system on its surrounding is

$$\delta W = F_{su}dl = p_{su}Adl = -p_{su}dV$$

Note that δW is negative for the expansion. For expansion, $V>0$, $p_{su}>0$, therefore

$$\delta W = -p_{su}dV \qquad (2.2.1)$$

$$W = -\int_{V_1}^{V_2} p_{su}dV \qquad (2.2.2)$$

(1) Isochoric process: $dV=0$, $W=0$

(2) Free expansion process: $p_{su}=0$ or for the total system $dV=0$, $W=0$

(3) Expansion against constant external pressure:

The work of process of constant external pressure is shown in figure 2.3, that is,

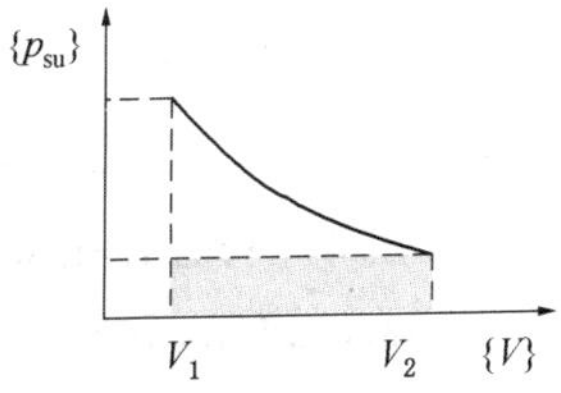

Fig. 2.3 Work of process of $p_{su}=$const

$$W = -\int_{V_1}^{V_2} p_{su}dV = -p_{su}(V_2 - V_1)$$

Comment: *Q* and *W*

① Heat and work are defined only in terms of processes; before and after the process of energy transfer between system and surroundings, heat and work do not exist. So we do not speak of a system as containing a particular amount of "Q and W".

② Unit: J or kJ.

③ They have positive or negative value.

④ They are affected by surroundings.

⑤ They are forms of energy transfer rather than forms of energy.

Example 2. 1

2 mol perfect gas expands isothermally to the same final state $p_2 = 50\text{kPa}$ through the following two different paths, the initial temperature is 300K, pressure is 150kPa. Calculate the work W of each path.

(1) Once expands against constant external pressure $p_{su} = 50\text{kPa}$ to the final state;

(2) At first, expands against constant external pressure $p_{su} = 100\text{kPa}$ to the middle equilibrium state, then expands against constant external pressure $p_{su} = 50\text{kPa}$ to the final state.

***Answer*:**

(1)

$n = 2\text{mol}$ $p_1 = 150\text{kPa}$ $V_1 = nRT/p_1$ $= 33.26\text{dm}^3$	$\xrightarrow[\text{path a: once expansion}]{p_{su} = p_2 = 50\text{kPa}}$	$n = 2\text{mol}$ $p_2 = 50\text{kPa}$ $V_2 = nRT/p_2$ $= 99.78\text{dm}^3$

$$W_a = -\int_{V_1}^{V_2} p_{su}\mathrm{d}V = -p_{su}(V_2 - V_1) = -p_2(V_2 - V_1) = -p_2\left(\frac{nRT}{p_2} - \frac{nRT}{p_1}\right) = -3.326\text{kJ}$$

(2)

$n = 2\text{mol}$ $p_1 = 150\text{kPa}$ $V_1 = nRT/p_1$ $= 33.26\text{dm}^3$	$\xrightarrow[\text{path } b_1]{p_{su} = p' = 100\text{kPa}}$	$n = 2\text{mol}$ $p' = 100\text{kPa}$ $V' = nRT/p'$ $= 49.89\text{dm}^3$	$\xrightarrow[\text{path } b_2]{p'_{su} = p_2 = 50\text{kPa}}$	$n = 2\text{mol}$ $p_2 = 50\text{kPa}$ $V_2 = nRT/p_2$ $= 99.78\text{dm}^3$

path b: twice expansion

$$W_b = W_{b1} + W_{b2} = -p'(V' - V_1) - p_2(V_2 - V') = -4.158\text{kJ}$$

$W_a \neq W_b$, therefore, the work of different paths for the same process is different.

2.2.4 Thermodynamic energy (热力学能)

Thermodynamic energy—internal energy, the total energy possessed by a system except integral kinetic energy and potential energy. (系统内部能量的总和。)

Symbol: U; unit: J.

Thermodynamic energy consists of: molecular translational, rotational, vibrational, and electronic energies, the relativistic rest-mass energy of the electrons and the nuclei; and potential energy of interaction between the molecules.

Comment:

① Thermodynamic energy is state function.

② Thermodynamic energy is extensive property.

③ Unit of thermodynamic energy is J.

④ Absolute value of thermodynamic energy cannot be known, but $\Delta U = U_2 - U_1$ can be calculated.

2.2.5 The first law of thermodynamics (热力学第一定律)

(1) Statement (语言表述)

① The energy of an closed system is converted from one form to another or flows from one part of nature to another, but is neither created nor destroyed in any process. (封闭系统中，能量可以相互转化，但不能凭空产生，也不能自行消失。)

② First kind of perpetual motion machine cannot be made. (第一类永动机是不可能造成的。)

First kind of perpetual motion machine—machine that violates the first law, providing a net output of work on each cycle with no input of energy.

(2) Essence (实质)

Energy conservation. (能量守恒)

Usually, the total energy consists of:

① The macroscopic kinetic energy;

② The macroscopic potential energy;

③ The thermodynamic energy.

In most of the applications of thermodynamics that we shall consider, the system will be at rest and external fields will not be present, the macroscopic kinetic energy and potential energy will be zero. Therefore, the total energy and the thermodynamic energy will be equal.

(3) Mathematical expression (数学表达式)

$$\Delta U = Q + W \text{ (closed system, } W' = 0) \tag{2.2.3}$$

Or

$$dU = \delta Q + \delta W \tag{2.2.4}$$

Comment:

Always think clearly about signs, for instance:

100J of heat has been injected into the system, meanwhile, the system has done 50J of work on surroundings, then:

$$\Delta U = Q + W = 100 + (-50) = 50\text{J}$$

(4) Specific formulas of the first law of thermodynamics (热力学第一定律的特殊式)

① Process of isolated system:

For $Q=0$, $W=0$

So $\Delta U=Q+W=0$

Conclusion:

The thermodynamic energy of an isolated system is constant.

② Cyclic process:

For $\Delta U=0$

So $Q+W=0$, that is, $Q=-W$

③ Adiabatic process:

For $Q=0$

So $W=\Delta U$

2.3 The heat at constant volume, the heat at constant pressure and enthalpy
(恒容热、恒压热与焓)

2.3.1 The heat at constant volume (恒容热)

The heat at constant volume —heat exchanges between system and surroundings at constant volume and $W'=0$. (系统在恒容，且非体积功为零的过程中与环境交换的热。)

Symbol: Q_V

For $\mathrm{d}V=0$, $W=0$

According to the first law of thermodynamics, we can get:

$$Q_V=\Delta U-W=\Delta U(\text{closed system},\ W'=0,\ \mathrm{d}V=0) \tag{2.3.1}$$

Or
$$\delta Q_V=\mathrm{d}U \tag{2.3.2}$$

(2.3.1) and (2.3.2) show that in a constant-volume process with $W'=0$, value of heat absorbed from surroundings by a closed system is equal to increment of thermodynamic energy.

2.3.2 The heat at constant pressure (恒压热)

The heat at constant pressure—heat exchanges between system and surroundings at constant pressure and $W'=0$. (系统在恒压，且非体积功为零的过程中与环境交换的热。)

Symbol: Q_p

For p is constant, that is, $p_1=p_2=p_{su}$

$$W=-p_{su}\Delta V$$

When $W'=0$,

$$\Delta U=Q_p-p_{su}\Delta V$$

That is, $U_2-U_1=Q_p-p_{su}(V_2-V_1)=Q_p-(p_2V_2-p_1V_1)$

So, $Q_p=(U_2+p_2V_2)-(U_1+p_1V_1)=\Delta(U+pV)$

$$H\overset{\text{def}}{=\!=\!=}U+pV$$

H—enthalpy (焓)

Therefore,
$$Q_p=\Delta H(\text{closed system},\ W'=0,\ \mathrm{d}p=0) \tag{2.3.3}$$

Or
$$\delta Q_p=\mathrm{d}H \tag{2.3.4}$$

(2.3.3) and (2.3.4) show that in a constant-pressure process with $W'=0$, value of heat absorbed from surroundings by a closed system is equal to increment of enthalpy.

Comparison of (2.3.3) and (2.3.1) shows that in a constant-pressure process H plays a role analogous to that played by U in a constant-volume process.

2.3.3 Enthalpy (焓)

The enthalpy of a system is defined as

$$H\overset{\text{def}}{=\!=\!=}U+pV \tag{2.3.5}$$

Comment:

① The enthalpy has no definite physical significance.

② The enthalpy is state function.

③ The enthalpy is extensive property.

④ Unit of the enthalpy is J.

⑤ Absolute value of the enthalpy cannot be known, but $\Delta H = H_2 - H_1$ can be calculated.

If the system changes from the initial state 1 to the final state 2, according to the definition of enthalpy, we have

$$\Delta H = \Delta U + \Delta(pV) \tag{2.3.6}$$

In the formula (2.3.6), $\Delta(pV) = p_2V_2 - p_1V_1$

① Perfect gas : $T_1 \to T_2$

$$\Delta(pV) = nRT_2 - nRT_1 = nR\Delta T$$

So,
$$\Delta H = \Delta U + nR\Delta T$$

② Condensed phase (liquid or solid) → gas:

$$\Delta(pV) = p_g V_g - p_{cd} V_{cd} = nRT$$

So,
$$\Delta H = \Delta U + nRT$$

2.3.4 Hess's Law (盖斯定律)

For a volume-constant or a pressure-constant process, according to $Q_V = \Delta U$, $Q_p = \Delta H$, the heats involved depend only on the initial state and the finial state and have nothing to do with the paths. (即反应的热效应只与起始状态和终了状态有关，而与变化的途径无关，这就是盖斯定律。)

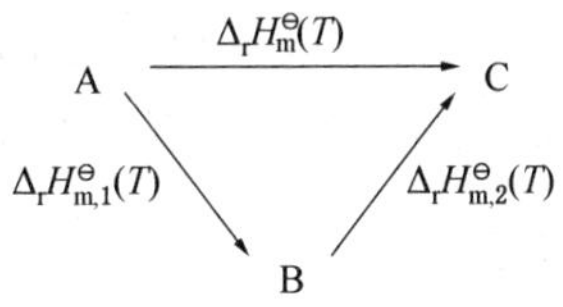

The standard enthalpy change in any reaction can be expressed as the sum of the standard enthalpy changes, at the same temperature of a series of reactions into which the overall reaction may formally be divided.

Therefore, $\Delta_r H_m^\ominus(T) = \Delta_r H_{m,1}^\ominus(T) + \Delta_r H_{m,2}^\ominus(T)$

That is, $Q = Q_1 + Q_2$

Application of Hess's Law:

From known $\Delta_r H_m^\ominus(T)$ to calculate unknown $\Delta_r H_m^\ominus(T)$.

For instance:

① C(graphite) + O_2(g) → CO_2(g), $\Delta_r H_{m,1}^\ominus(T)$

② CO(g) + 1/2 O_2(g) → CO_2(g), $\Delta_r H_{m,2}^\ominus(T)$

③ C(graphite) + 1/2 O_2(g) → CO(g), $\Delta_r H_{m,3}^\ominus(T)$

If $\Delta_r H_{m,1}^\ominus(T)$ and $\Delta_r H_{m,2}^\ominus(T)$ are known, $\Delta_r H_{m,3}^\ominus(T)$ is unknown,

Since, ③ = ① − ②

So, $\Delta_r H_{m,3}^\ominus(T) = \Delta_r H_{m,1}^\ominus(T) - \Delta_r H_{m,2}^\ominus(T)$

2.4 Heat capacity
(热容)

2.4.1 Heat capacity (热容)

Heat capacity—heat absorbed by the system while rising unit temperature and having no phase transition, no chemical change and $W'=0$. [系统在给定条件(定容或定压)下，不发生相变化、化学变化和非体积功为零时，升高单位热力学温度所吸收的热。]

Symbol: $C(T)$; unit: $\mathrm{J \cdot K^{-1}}$.

$$C(T) \xlongequal{\text{def}} \frac{\delta Q}{\mathrm{d}T} \tag{2.4.1}$$

Heat capacity at constant volume, $C_V = \left(\frac{\delta Q}{\mathrm{d}T}\right)_V = \left(\frac{\partial U}{\partial T}\right)_V$ (2.4.2)

Heat capacity at constant pressure, $C_p = \left(\frac{\delta Q}{\mathrm{d}T}\right)_p = \left(\frac{\partial H}{\partial T}\right)_p$ (2.4.3)

Heat capacity is extensive property.

Molar heat capacity (摩尔热容) $C_{\mathrm{m}} \xlongequal{\text{def}} \frac{C}{n} = \frac{1}{n}\frac{\delta Q}{\mathrm{d}T}$ (2.4.4)

Molar heat capacity at constant volume (摩尔定容热容)

$$C_{V,\mathrm{m}} \xlongequal{\text{def}} \frac{C_V}{n} = \frac{1}{n}\frac{\delta Q_V}{\mathrm{d}T} = \left(\frac{\partial U_{\mathrm{m}}}{\partial T}\right)_V \tag{2.4.5}$$

Molar heat capacity at constant pressure (摩尔定压热容)

$$C_{p,\mathrm{m}} \xlongequal{\text{def}} \frac{C_p}{n} = \frac{1}{n}\frac{\delta Q_p}{\mathrm{d}T} = \left(\frac{\partial H_{\mathrm{m}}}{\partial T}\right)_p \tag{2.4.6}$$

将式(2.4.5) 和式(2.4.6)分离变量积分，于是有：

$$\Delta U = \int_{T_1}^{T_2} nC_{V,\mathrm{m}}(T)\,\mathrm{d}T \tag{2.4.7}$$

$$\Delta H = \int_{T_1}^{T_2} nC_{p,\mathrm{m}}(T)\,\mathrm{d}T \tag{2.4.8}$$

Mass heat capacity (质量热容) $c \xlongequal{\text{def}} \frac{C}{m} = \frac{1}{m}\frac{\delta Q}{\mathrm{d}T}$ (2.4.9)

Mass heat capacity at constant volume (质量定容热容)

$$c_V = C_V/m = (\partial u/\partial T)_V \tag{2.4.10}$$

Mass heat capacity at constant pressure (质量定压热容)

$$c_p = C_p/m = (\partial h/\partial T)_p \tag{2.4.11}$$

Molar heat capacity and mass heat capacity are all intensive properties.

The relationship between molar heat capacity at constant pressure and mass heat capacity at constant pressure is: $C_{p,\mathrm{m}} = c_p M$

Where M is the molar mass of substance.

2.4.2 Relation between $C_{p,m}$ and p(摩尔定压热容与压力的关系式)

We can prove later the relation between $C_{p,m}$ and p at a given temperature is:

$$\left(\frac{\partial C_{p,\mathrm{m}}}{\partial p}\right)_T = -T\left(\frac{\partial^2 V_\mathrm{m}}{\partial T^2}\right)_p \tag{2.4.12}$$

Standard molar heat capacity at constant pressure—molar heat capacity at constant pressure of substance in which the pressure is standard pressure $p^\ominus = 100\text{kPa}$.

Symbol: $C^\ominus_{p,\mathrm{m}}$.

气体的标准摩尔定压热容是该气体在标准压力下具有理想气体性质时的摩尔定压热容，即为真实气体在零压下的值。

理想气体的摩尔定压热容与压力无关。

压力变化对于凝聚态物质定压热容的影响非常小，在压力与标准压力相差不大时，完全可以不考虑。

所以可以近似认为：$C_{p,\mathrm{m}} \approx C^\ominus_{p,\mathrm{m}}$

2.4.3 Empirical formula between $C_{p,\mathrm{m}}$ and T(摩尔定压热容与温度的经验关系式)

$$C_{p,\mathrm{m}} = a + bT + cT^2 + dT^3 \tag{2.4.13}$$

or

$$C_{p,\mathrm{m}} = a + bT + c'T^{-2} \tag{2.4.14}$$

a, b, c, c', d—empirical constants (see appendix Ⅷ).

Mean molar heat capacity at constant pressure (平均摩尔定压热容)

$$\bar{C}_{p,\mathrm{m}} = \frac{\int_{T_1}^{T_2} C_{p,\mathrm{m}}\mathrm{d}T}{T_2 - T_1} \tag{2.4.15}$$

Comment:

In different scope of temperature, mean molar heat capacity at constant pressure of a substance is different.

2.4.4 Relation between $C_{p,m}$ and $C_{V,m}$(摩尔定压热容与摩尔定容热容的关系)

What is the relation between $C_{p,m}$ and $C_{V,\mathrm{m}}$?

According to the definition: $C_{p,\mathrm{m}} = \frac{1}{n}\frac{\mathrm{d}H}{\mathrm{d}T} = \left(\frac{\partial H_\mathrm{m}}{\partial T}\right)_p$ and $C_{V,\mathrm{m}} = \frac{1}{n}\frac{\mathrm{d}U}{\mathrm{d}T} = \left(\frac{\partial U_\mathrm{m}}{\partial T}\right)_V$

$$\begin{aligned} C_{p,\mathrm{m}} - C_{V,\mathrm{m}} &= \left(\frac{\partial H_\mathrm{m}}{\partial T}\right)_p - \left(\frac{\partial U_\mathrm{m}}{\partial T}\right)_V \\ &= \left(\frac{\partial (U_\mathrm{m} + pV_\mathrm{m})}{\partial T}\right)_p - \left(\frac{\partial U_\mathrm{m}}{\partial T}\right)_V \\ &= \left(\frac{\partial U_\mathrm{m}}{\partial T}\right)_p + p\left(\frac{\partial V_\mathrm{m}}{\partial T}\right)_p - \left(\frac{\partial U_\mathrm{m}}{\partial T}\right)_V \end{aligned}$$

We expect $\left(\frac{\partial U_\mathrm{m}}{\partial T}\right)_p$ and $\left(\frac{\partial U_\mathrm{m}}{\partial T}\right)_V$ to be related to each other. In $\left(\frac{\partial U_\mathrm{m}}{\partial T}\right)_V$, the thermodynamic energy is taken as a function of T and V: $U_\mathrm{m} = f(T, V_\mathrm{m})$. The total differential of $U_\mathrm{m} = f(T, V_\mathrm{m})$ is

$$\mathrm{d}U_\mathrm{m} = \left(\frac{\partial U_\mathrm{m}}{\partial T}\right)_V \mathrm{d}T + \left(\frac{\partial U_\mathrm{m}}{\partial V_\mathrm{m}}\right)_T \mathrm{d}V_\mathrm{m}$$

p is constant $$\left(\frac{\partial U_{\mathrm{m}}}{\partial T}\right)_p = \left(\frac{\partial U_{\mathrm{m}}}{\partial T}\right)_V + \left(\frac{\partial U_{\mathrm{m}}}{\partial V_{\mathrm{m}}}\right)_T \left(\frac{\partial V_{\mathrm{m}}}{\partial T}\right)_p$$

Then $$C_{p,\mathrm{m}} - C_{V,\mathrm{m}} = \left[\left(\frac{\partial U_{\mathrm{m}}}{\partial V_{\mathrm{m}}}\right)_T + p\right]\left(\frac{\partial V_{\mathrm{m}}}{\partial T}\right)_p \qquad (2.4.16)$$

2.4.5 Joule experiment（焦耳实验）

(1) Experimental process（实验过程）

In 1843, Joule did an experiment that allowed measurement of $\left(\frac{\partial U_{\mathrm{m}}}{\partial V_{\mathrm{m}}}\right)_T$ for a gas; Joule measured the temperature change after free expansion of a gas into a vacuum. A sketch of the setup used by Joule is shown in figure 2.1. Initially, chamber A is filled with air of 2MPa, and chamber B is evacuated. After equilibrium is reached, the temperature of the bath was measured.

James Prescott Joule

(2) Experimental result（实验结果）

The temperature of the bath was not changed, that is, the temperature of air was not changed, $\mathrm{d}T=0$.

(3) Analysis of result（结果分析）

For free expansion process, $\delta W=0$

$\mathrm{d}T=0$, then $\delta Q=0$

Therefore, $\mathrm{d}U=\delta Q+\delta W=0$

If $U=f(T, V)$, the total differential is:

$$\mathrm{d}U = \left(\frac{\partial U}{\partial T}\right)_V \mathrm{d}T + \left(\frac{\partial U}{\partial V}\right)_T \mathrm{d}V$$

According to $\mathrm{d}U=0$, $\mathrm{d}T=0$ and $\mathrm{d}V\neq 0$, we can get

$$\left(\frac{\partial U}{\partial V}\right)_T = 0 \qquad (2.4.17)$$

If $U=f(T, p)$, the total differential is: $\mathrm{d}U = \left(\frac{\partial U}{\partial T}\right)_p \mathrm{d}T + \left(\frac{\partial U}{\partial p}\right)_T \mathrm{d}p$

According to $\mathrm{d}U=0$, $\mathrm{d}T=0$ and $\mathrm{d}p\neq 0$, we can get:

$$\left(\frac{\partial U}{\partial p}\right)_T = 0 \qquad (2.4.18)$$

According to (2.4.17) and (2.4.18), we have

$$U=f(T)\ \text{(perfect gas)} \qquad (2.4.19)$$

The thermodynamic energy of the perfect gas is only function of temperature when the amount of substance is constant.

According to the definition $H=U+pV$, for perfect gas, $pV=nRT=f(T)$, $U=f(T)$, therefore:

$$H=f(T)\ \text{(perfect gas)} \qquad (2.4.20)$$

The enthalpy of the perfect gas is only function of temperature when the amount of substance is constant.

Therefore (2.4.7) and (2.4.8) are suitable for the simple pVT process of the perfect gas because the thermodynamic energy and the enthalpy are only functions of temperature.

At the same time, according to the conclusion of Joule experiment we can get $(C_{p,\mathrm{m}}-C_{V,\mathrm{m}})$ is constant for a perfect gas.

According to (2.4.16), $C_{p,\mathrm{m}} - C_{V,\mathrm{m}} = \left[\left(\frac{\partial U_\mathrm{m}}{\partial V_\mathrm{m}}\right)_T + p\right]\left(\frac{\partial V_\mathrm{m}}{\partial T}\right)_p$

Since $U=f(T)$, therefore: $\left(\frac{\partial U_\mathrm{m}}{\partial V_\mathrm{m}}\right)_T = 0$

For pg, $pV_\mathrm{m}=RT$, therefore: $\left(\frac{\partial V_\mathrm{m}}{\partial T}\right)_p = \frac{R}{p}$

So,
$$C_{p,\mathrm{m}} - C_{V,\mathrm{m}} = p \times \frac{R}{p} = R \text{ or } C_p - C_V = nR \tag{2.4.21}$$

For monatomic molecules (such as He, Ar etc.): $C_{V,\mathrm{m}} = \frac{3}{2}R$, $C_{p,\mathrm{m}} = \frac{5}{2}R$

diatomic molecules (such as N_2, O_2 etc.): $C_{V,\mathrm{m}} = \frac{5}{2}R$, $C_{p,\mathrm{m}} = \frac{7}{2}R$

The molar heat capacity can be calculated by (2.4.22) and (2.4.23) for a perfect gas mixture formed by B, C and D etc.

$$C_{p,\mathrm{m(mix)}} = \sum_\mathrm{B} y(\mathrm{B}) C_{p,\mathrm{m}}(\mathrm{B}) \tag{2.4.22}$$

$$C_{V,\mathrm{m(mix)}} = \sum_\mathrm{B} y(\mathrm{B}) C_{V,\mathrm{m}}(\mathrm{B}) \tag{2.4.23}$$

Example 2.2

4mol Ar(g) and 2mol Cu(s) were sealed in an 0.1m^3 isochoric container, the initial temperature was 0℃. Now the system was heated to 100℃, calculated Q, W, ΔU and ΔH of the process.

Given: $C_{p,\mathrm{m}}(\mathrm{Ar, g, 25℃}) = 20.786\mathrm{J \cdot mol^{-1} \cdot K^{-1}}$,

$C_{p,\mathrm{m}}(\mathrm{Cu, s, 25℃}) = 24.435\mathrm{J \cdot mol^{-1} \cdot K^{-1}}$.

Suppose both $C_{p,\mathrm{m}}$ did not change with temperature.

***Answer*:**

For isochoric process, $W=0$

$$\Delta U = \Delta U(\mathrm{Ar, g}) + \Delta U(\mathrm{Cu, s})$$

$$\Delta U(\mathrm{Ar, g}) = n(\mathrm{Ar, g}) \times C_{V,\mathrm{m}}(\mathrm{Ar, g}) \times (T_2 - T_1)$$

$$\Delta U(\mathrm{Cu, s}) \approx \Delta H(\mathrm{Cu, s})$$
$$= n(\mathrm{Cu, s}) \times C_{p,\mathrm{m}}(\mathrm{Cu, s}) \times (T_2 - T_1)$$

$$\Delta U = \Delta U(\mathrm{Ar, g}) + \Delta U(\mathrm{Cu, s}) = 9.875\mathrm{kJ}$$

$$Q_V = \Delta U = 9.875\mathrm{kJ}$$

$$\Delta H = \Delta U + \Delta(pV) = \Delta U + n(\mathrm{Ar, g})R(T_2 - T_1)$$
$$= 13.201\mathrm{kJ}$$

Example 2.3

There is an adiabatic plank in an adiabatic isochoric container, on the two sides of the plank, there are 3mol monatomic perfect gas A at 0℃, 50kPa and 7mol diatomic perfect gas B at 100℃, 150kPa. Now the plank is drawn away and the two gases are mixed until equilibrium state. Please calculate the T, p of the final state and ΔH, ΔU of the process.

Answer:

A	B		A+B
$n_A = 3\text{mol}$	$n_B = 7\text{mol}$	$Q=0,\ dV=0$ →	$n_A = 3\text{mol},\ n_B = 7\text{mol}$
$T_A = 273.15\text{K}$	$T_B = 373.15\text{K}$	$T=?;\ p=?;\ \Delta H=?$	$T=?\quad p=?$
$p_A = 50\text{kPa}$	$p_B = 150\text{kPa}$		$V = V_A + V_B$

Since the process is adiabatic and isochoric

Therefore, $W=0,\ Q_V = \Delta U = 0$

$$\Delta U = \Delta U(A) + \Delta U(B) = n(A)C_{V,m}(A)(T - T_A) + n(B)C_{V,m}(B)(T - T_B) = 0$$

$$T = \frac{n(A)C_{V,m}(A)T_A + n(B)C_{V,m}(B)T_B}{n(A)C_{V,m}(A) + n(B)C_{V,m}(B)} = 352.70\text{K}$$

$$p = \frac{[n(A) + n(B)]RT}{V} = \frac{[n(A) + n(B)]RT}{\dfrac{n(A)RT_A}{p_A} + \dfrac{n(B)RT_B}{p_B}} = 104.34\text{kPa}$$

$$\begin{aligned}\Delta H &= \Delta H(A) + \Delta H(B)\\ &= n(A)C_{p,m}(A)(T - T_A) + n(B)C_{p,m}(B)(T - T_B)\\ &= 795\text{J}\end{aligned}$$

Or according to the definition of enthalpy,

$$\begin{aligned}H &= U + pV\\ \Delta H &= \Delta U + \Delta(pV) = \Delta(pV)\\ &= pV - (p_A V_A + p_B V_B)\\ &= 795\text{J}\end{aligned}$$

Example 2.4

There is 4mol diatomic perfect gas in an adiabatic container with a piston, the initial temperature is 25℃, pressure is 100kPa. Now the pressure outside the position was decreased from 100kPa to 50kPa, calculate the final temperature T_2 and W, ΔH, ΔU of the process.

Answer:

$n = 4\text{mol}$		$n = 4\text{mol}$
$T_1 = 298.15\text{K}$	$p_{su} = 50\text{kPa}$ →	$p_2 = 50\text{kPa}$
$p_1 = 100\text{kPa}$	$T_2=?;\ W=?;\ \Delta U=?;\ \Delta H=?$	$T_2 = ?$
$V_1 = nRT_1/p_1$		$V_2 = nRT_2/p_2$

Since the process is adiabatic

Therefore, $Q=0$,

$$W = \Delta U = nC_{V,m}(T_2 - T_1)$$

Since the external pressure is const, $p_{su}=p_2$

Therefore, $W=-p_{su}(V_2-V_1)=-p_2(V_2-V_1)=nC_{V,m}(T_2-T_1)$

We can get: $T_2=255.56\text{K}$

So, $W=\Delta U=nC_{V,m}(T_2-T_1)=-3.541\text{kJ}$

$\Delta H=nC_{p,m}(T_2-T_1)=-4.958\text{kJ}$

2.5 Reversible process and equation of adiabatic reversible process of the perfect gas (理想气体的可逆过程和绝热可逆过程方程)

2.5.1 Reversible process (可逆过程)

Reversible process—system and surroundings change respectively from initial state to final state after process L. If we can find process L′, which make both the system and surroundings revert in the reversal process without leaving any perpetual changes. Process L is called reversible process. (过程的进行需要有推动力，若过程的推动力无限小，系统内部及系统和环境间无限接近于平衡态，过程进行得无限缓慢。当系统沿原途径逆向回到初始态时，环境也恢复到原态。这种系统内部及系统与环境间在一系列无限接近平衡条件下进行的过程称为可逆过程。)

A reversible process is one where the system is always infinitesimally close to equilibrium, and an infinitesimal change in conditions can reverse the process to restore both system and surroundings to their initial states. A reversible process is obviously an idealization.

Features of reversible process (可逆过程的特点):

① In a reversible process, the system is always infinitesimally close to equilibrium, that is, $T_{su}=T$, $p_{su}=p$. (在整个过程中，系统内部无限接近于平衡，系统与环境的相互作用无限接近于平衡，过程的进展无限缓慢，$T_{su}=T$，$p_{su}=p$。)

② In a reversible process, an infinitesimal change in conditions can restore both system and surroundings to their initial states. (系统和环境能够由终态，沿着原来的途径从相反方向步步回复，直到都恢复原来的状态。)

③ The maximum work available from a system operating between specified initial and final state, and passing along a specified path, is obtained when it is operating reversibly, that is, $W_r < W_{ir}$. [膨胀时，系统对环境做最大功(绝对值最大)，压缩时，环境对体系做最小功。]

说明：可逆过程是一种理想过程，实际发生的过程应当在有限时间内发生有限的状态变化，如气体自由膨胀过程等，而过程可以无限慢。但热力学是不考虑时间的，所以可逆过程还是一种热力学过程。我们可以把一些无限接近于可逆过程的过程当作可逆过程：如无限接近于相平衡条件下的相变以及无限接近于化学平衡的化学变化。

2.5.2 Isothermal reversible process of the perfect gas (理想气体的等温可逆过程)

Since T is constant, $\Delta U=0$, $\Delta H=0$, $Q=-W$

For reversible process, $p_{su}=p$, so

$$\delta W=-p_{su}\mathrm{d}V=-p\mathrm{d}V$$

Integration of above formula gives

$$W = -\int_{V_1}^{V_2} p\mathrm{d}V$$

For expansion of pg, $pV = nRT$, so

$$W = -\int_{V_1}^{V_2} \frac{nRT}{V}\mathrm{d}V = -nR\int_{V_1}^{V_2} \frac{T}{V}\mathrm{d}V$$

Isothermal expansion of pg, T is constant, so

$$W = -nRT\int_{V_1}^{V_2} \frac{\mathrm{d}V}{V} = -nRT\ln\frac{V_2}{V_1} = nRT\ln\frac{p_2}{p_1} \quad (\text{perfect gas},\ \mathrm{d}T = 0,\ \text{reversible}) \tag{2.5.1}$$

2.5.3 Adiabatic reversible process of the perfect gas（理想气体的绝热可逆过程）

(1) Basic formula of adiabatic process of the perfect gas（理想气体绝热过程的基本公式）

For an adiabatic process, $\delta Q = 0$

According to the first law of thermodynamics, $\mathrm{d}U = \delta Q + \delta W = \delta W$

That is
$$W = \Delta U = \int_{T_1}^{T_2} nC_{V,\mathrm{m}}(T)\mathrm{d}T \tag{2.5.2}$$

If $C_{V,\mathrm{m}}$ is a constant,
$$W = \Delta U = nC_{V,\mathrm{m}}(T_2 - T_1) \tag{2.5.3}$$

(2) Equation of adiabatic reversible process of the perfect gas（理想气体绝热可逆过程方程）

According to the basic formula of adiabatic process of perfect gas, $\delta W = \mathrm{d}U$

For reversible process, $p_{\mathrm{su}} = p$, so

$$\delta W = -p_{\mathrm{su}}\mathrm{d}V = -p\mathrm{d}V = -\frac{nRT}{V}\mathrm{d}V$$

$$\mathrm{d}U = nC_{V,\mathrm{m}}\mathrm{d}T$$

Therefore
$$-nRT\frac{\mathrm{d}V}{V} = nC_{V,\mathrm{m}}\mathrm{d}T$$

$$\frac{\mathrm{d}T}{T} + \frac{R}{C_{V,\mathrm{m}}}\frac{\mathrm{d}V}{V} = 0$$

$$\frac{\mathrm{d}T}{T} + \frac{C_{p,\mathrm{m}} - C_{V,\mathrm{m}}}{C_{V,\mathrm{m}}}\frac{\mathrm{d}V}{V} = 0$$

If
$$C_{p,\mathrm{m}}/C_{V,\mathrm{m}} \overset{\mathrm{def}}{=\!=\!=} \gamma$$

γ—ratio of the molar heat capacities or adiabatic exponent.

Then
$$\frac{\mathrm{d}T}{T} + (\gamma - 1)\frac{\mathrm{d}V}{V} = 0$$

$$\ln\{T\} + (\gamma - 1)\ln\{V\} = C$$

We can get
$$TV^{\gamma-1} = C \tag{2.5.4}$$

According to the state equation of the perfect gas, $pV = nRT$

We can also get
$$pV^{\gamma} = C \tag{2.5.5}$$

$$Tp^{(1-\gamma)/\gamma} = C \tag{2.5.6}$$

Equation (2.5.4), (2.5.5) and (2.5.6) are called **equation of adiabatic process of the**

perfect gas.

The **applicable conditions** of equation of adiabatic process of the perfect gas: closed system, $W'=0$, perfect gas, adiabatic and reversible.

(3) Work of adiabatic reversible process of the perfect gas (理想气体绝热可逆过程的体积功)

For reversible process, $p_{su}=p$, so

$$W = -\int_{V_1}^{V_2} p\mathrm{d}V$$

According to the equation of adiabatic reversible process of the perfect gas,

$$pV^{\gamma} = C$$

We can get

$$W = \frac{p_1 V_1}{\gamma - 1}\left[\left(\frac{V_1}{V_2}\right)^{\gamma-1} - 1\right] \text{ or } W = \frac{p_1 V_1}{\gamma - 1}\left[\left(\frac{p_2}{p_1}\right)^{\frac{\gamma-1}{\gamma}} - 1\right] \tag{2.5.7}$$

But, when we calculate the work of adiabatic process of the perfect gas, we can use the basic formula of adiabatic process of the perfect gas, that is,

$$W = \Delta U = nC_{V,\mathrm{m}}(T_2 - T_1)$$

We might compare an isothermal reversible expansion of a perfect gas with an adiabatic reversible expansion of the gas. Let the gas start from the same initial p_1 and V_1 and go to the same V_2. For the thermal process, $T_2=T_1$. For the adiabatic expansion, we showed that $T_2<T_1$. Hence the final pressure p_2 for the adiabatic expansion must be less than p_2 for the isothermal expansion.

Example 2.5

3 mol perfect gas changes isochorically to $p_2 = 0.10\text{MPa}$, the initial temperature is 409K, pressure is 0.15MPa, please calculate Q, W, ΔU and ΔH of this process.

(the perfect gas: $C_{p,\mathrm{m}}=29.4\text{J}\cdot\text{mol}^{-1}\cdot\text{K}^{-1}$)

***Answer*:**

For isochoric process, $W=0$

$$T_2 = p_2 T_1/p_1 = (0.10 \times 409/0.15) = 273\text{K}$$

$$\Delta U = n C_{V,\mathrm{m}}(T_2 - T_1)$$

$$= 3 \times (29.4 - 8.315)(273 - 409) = -8.635\text{kJ}$$

$$\Delta H = n C_{p,\mathrm{m}}(T_2 - T_1)$$

$$= 3 \times 29.4(273 - 409) = -12.040\text{kJ}$$

$$Q_V = \Delta U = -8.635\text{kJ}$$

Example 2.5

Example 2.6

1mol N_2 expands (1) isothermally and reversibly to twice its volume; (2) adiabatically and reversibly to twice its volume; the initial temperature is 273.15K, pressure is 101.3kPa, please calculate Q, W, ΔU and ΔH of these two processes. (N_2: $C_{V,\mathrm{m}}=5/2R$)

***Answer*:**

(1) For isothermal process, $\mathrm{d}T=0$

Therefore, $\Delta U = \Delta H = 0$

Example 2.6

$$W=-\int_{V_1}^{V_2} p\mathrm{d}V=-nRT\ln\frac{V_2}{V_1}=-1\times 8.315\times 273.15\ln 2=-1.574\mathrm{kJ}$$

$$Q=-W=1.574\mathrm{kJ}$$

(2) For adiabatic and reversible process, $Q=0$

$$T_1V_1^{\gamma-1}=T_2V_2^{\gamma-1},\ \gamma=\frac{7}{5}$$

$$T_2=\left(\frac{V_1}{V_2}\right)^{\gamma-1}\cdot T_1=0.5^{\frac{2}{5}}\times 273.15=207.0\mathrm{K}$$

$$\Delta U=\int_{T_1}^{T_2} nC_{V,\mathrm{m}}\mathrm{d}T=nC_{V,\mathrm{m}}(T_2-T_1)=1\times\frac{5}{2}\times 8.315\times(207.0-273.15)$$

$$=-1.375\mathrm{kJ}$$

$$\Delta H=\int_{T_1}^{T_2} nC_{p,\mathrm{m}}\mathrm{d}T=nC_{p,\mathrm{m}}(T_2-T_1)=1\times\frac{7}{2}\times 8.315\times(207.0-273.15)$$

$$=-1.925\mathrm{kJ}$$

$$W=\Delta U=-1.375\mathrm{kJ}$$

Example 2.7

4mol diatomic molecule perfect gas is condensed adiabatically and reversibly from $p_1=50\mathrm{kPa}$, $V_1=160\mathrm{dm}^3$ to $p_2=200\mathrm{kPa}$, calculate T_2 and Q, W, ΔU, ΔH of this process.

Answer:

Initial state	adiabatic and reversible → ($T_2=?$; $W=?$; $\Delta H=?$; $\Delta U=?$)	Final state
$n=4\mathrm{mol}$ $p_1=50\mathrm{kPa}$ $V_1=160\mathrm{dm}^3$ $T_1=p_1V_1/nR$		$n=4\mathrm{mol}$ $p_2=200\mathrm{kPa}$ $V_2=?$ $T_2=?$

For adiabatic and reversible process, $Q=0$

$$T_1=\frac{p_1V_1}{nR}=\frac{5\times 10^4\times 160\times 10^{-3}}{4\times 8.315}=240.53\mathrm{K}$$

$$\gamma=\frac{C_{p,\mathrm{m}}}{C_{V,\mathrm{m}}}=\frac{7/2}{5/2}=7/5$$

According to the equation of adiabatic reversible process of the perfect gas,

$$T_2=T_1\left(\frac{p_2}{p_1}\right)^{\frac{1-\gamma}{\gamma}}=357.43\mathrm{K}$$

$$\Delta U=nC_{V,\mathrm{m}}(T_2-T_1)=4\times\frac{5}{2}R\times(357.43-240.53)=9.720\mathrm{kJ}$$

$$\Delta H=nC_{p,\mathrm{m}}(T_2-T_1)=4\times\frac{7}{2}R\times(357.43-240.53)=13.608\mathrm{kJ}$$

$$W=\Delta U=9.702\mathrm{kJ}$$

2.6 Phase transformation process
（相变化过程）

2.6.1 Phase and enthalpy of phase transition（相和相变焓）

（1）Phase（相）

Phase—a state of matter that is 'uniform' throughout, not only in chemical composition but also in physical state.（系统内性质完全相同的均匀部分。）

Different uniform parts belong to different phases. There are interfaces among phases. Different phases can be separated using mechanical method in principle.

For instance, there are three phases of gas, liquid and solid when a solid salt coexists with its saturated solution and steam.

（2）Phase transformation process（相变过程）

Phase transformation process—change of substance state of aggregation in given conditions, including evaporation, freezing, melting, fusion, sublimation, condensation and crystal transformation etc.（系统中的同一种物质在不同相之间的转变。）

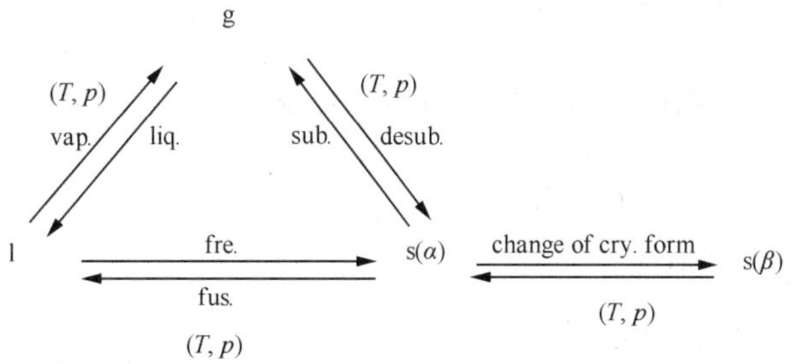

（3）Enthalpy of phase transition（相变焓）

Enthalpy of phase transition—enthalpy change of phase transformation process at equilibrium temperature and pressure.（特指在平衡温度、压力下进行相变的焓变。）

2.6.2 Calculation of ΔH of phase transformation process（相变化过程焓变的计算）

（1）Calculation of ΔH of reversible phase transformation process（可逆相变化过程焓变的计算）

Reversible phase transformation（可逆相变）is phase transformation at equilibrium temperature and pressure in two phase equilibrium.

liquid→gas: vaporization, $\Delta_{vap}H_m$ is known condition, $\Delta H = n\Delta_{vap}H_m$

solid → liquid: fusion, $\Delta_{fus}H_m$ is known condition, $\Delta H = n\Delta_{fus}H_m$

solid → gas: sublimation, $\Delta_{sub}H_m$ is known condition, $\Delta H = n\Delta_{sub}H_m$

（2）Calculation of ΔH of irreversible phase transformation process（不可逆相变化过程焓变的计算）

Irreversible phase transformation（不可逆相变）is phase transformation at non-equilibrium temperature and pressure.

We must design process to calculate ΔH. How to design the process? Let's see examples.

Example 2.8

The normal boiling point of H_2O is 100℃, $C_{p,m}(l) = 75.20\text{J} \cdot \text{mol}^{-1} \cdot \text{K}^{-1}$, $\Delta_{vap}H_m =$

$40.67 \text{kJ} \cdot \text{mol}^{-1}$, $C_{p,m}(g) = 33.57 \text{J} \cdot \text{mol}^{-1} \cdot \text{K}^{-1}$, $C_{p,m}$ and $\Delta_{vap}H_m$ are constants. Please calculate Q, W, ΔU and ΔH of this process:

$1 \text{mol } H_2O(l, 60℃, 101.325 \text{kPa}) \rightarrow H_2O(g, 60℃, 101.325 \text{kPa})$.

Answer:

This process is irreversible process at constant T and p. We must design reversible process to calculate ΔH.

Example 2.8

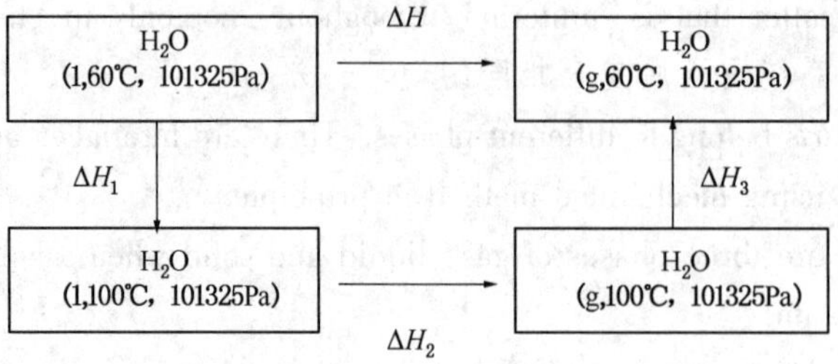

```
         ΔH
H₂O  ───────────►  H₂O
(l,60℃, 101325Pa)   (g,60℃, 101325Pa)
  │                      ▲
  │ ΔH₁                  │ ΔH₃
  ▼         ΔH₂          │
H₂O  ───────────►  H₂O
(l,100℃, 101325Pa)  (g,100℃, 101325Pa)
```

$$\Delta H_1 = \int_{333.15\text{K}}^{373.15\text{K}} nC_{p,m}(H_2O, l)dT = 1 \times 75.2 \times (373.15 - 333.15) = 3008 \text{J}$$

$$\Delta H_2 = n \cdot \Delta_{vap}H_m(H_2O) = 1 \times 40.67 \times 10^3 = 40670 \text{J}$$

$$\Delta H_3 = \int_{373.15\text{K}}^{333.15\text{K}} nC_{p,m}(H_2O, g)dT = 1 \times 33.57 \times (333.15 - 373.15) = -1343 \text{J}$$

$$\Delta H = \Delta H_1 + \Delta H_2 + \Delta H_3 = 42.34 \text{kJ}$$

$$Q_p = \Delta H = 42.34 \text{kJ}$$

$$\alpha(\text{liquid or solid}) \xrightarrow{T, p} \beta(\text{gas})$$

$$W = -p(V_\beta - V_\alpha)$$

Since $V_\beta \gg V_\alpha$,

$$W = -pV_\beta = -nRT = -1 \times 8.315 \times 333.15 = -2.77 \text{kJ}$$

$$\Delta U = Q_p + W = 42.34 - 2.77 = 39.57 \text{kJ}$$

Example 2.9

A 50dm^3 vacuum container is laid in an isothermal flume of $100℃$, there is a glass bottle with 50g water in the bottom of the container. Now break the bottle, water is vaporized under vacuum to equilibrium state. Calculated Q, W, ΔU, ΔH of this process.

Given that:

$$p^*(H_2O, 100℃) = 101.325 \text{kPa}$$

$$\Delta_{vap}H_m(H_2O, 100℃) = 40.668 \text{kJ} \cdot \text{mol}^{-1}$$

Answer:

$m = 50\text{g } H_2O(l)$	liquid expands into a vacuum	$n(g) = ?$
$p_1 = 101.325 \text{kPa}$	$Q=?; W=?; \Delta H=?; \Delta U=?$	$p_2 = 101.325 \text{kPa}$
$V_1 = 50 \text{dm}^3$	──────────────────►	$V_2 = 50 \text{dm}^3$
$T_1 = 373.15 \text{K}$		$T_2 = 373.15 \text{K}$

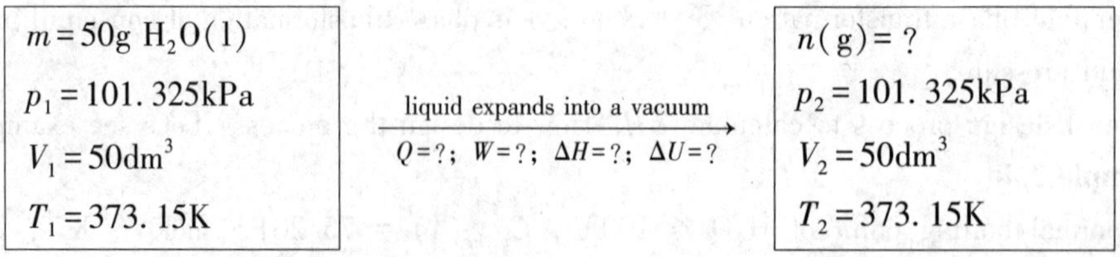

If n mol H_2O (l) is vaporized to H_2O (g), then:

$$n = \frac{p_2V_2}{RT_2} = 1.633\text{mol}$$

$$\Delta H = n\Delta_{\text{Vap}}H_{\text{m}} = 1.633 \times 40.668 = 66.41\text{kJ}$$

$$W = -\int_{V_1}^{V_2} p_{\text{su}}\text{d}V = 0$$

$$Q = \Delta U = \Delta H - \Delta(pV) = \Delta H - p_2V_2 = \Delta H - nRT = 61.35\text{kJ}$$

2.7 Standard molar enthalpy of reaction (标准摩尔反应焓)

2.7.1 Chemical stoichiometric number (化学计量数)

For arbitrary reaction $a\text{A}+b\text{B} = y\text{Y}+z\text{Z}$

We adopt the convention of transposing the reactants to the right side of the equation to get

$$0 = -a\text{A}-b\text{B}+y\text{Y}+\text{Z}z$$

That is $$0 = \sum \nu_{\text{B}}\text{B}$$

Where B denotes molecule, atom, or ion of reaction, ν_{B} is the chemical stoichiometric number of B. It is pure number with no unit. It is negative for reactants and positive for products.

Thus for above reaction, we have $\nu_{\text{A}} = -a$, $\nu_{\text{B}} = -b$, $\nu_{\text{Y}} = y$, $\nu_{\text{Z}} = z$.

2.7.2 Extent of a reaction (反应进度)

Extent of a reaction—the extent that reaction proceeds (化学反应进行的程度).

Symbol: ξ; unit: mol.

Definition formula of extent of a reaction is: $\text{d}\xi = \text{d}n_{\text{B}}/\nu_{\text{B}}$ (2.7.1)

Where n_{B} is amount of substance of arbitrary substance B in reaction equation, ν_{B} is the chemical stoichiometric number of substance B.

Extent of a reaction is unrelated to substance B, that is, extent of a reaction denoted by different substances is identical.

Integral of (2.7.1) can be obtained when ξ_0 corresponds to $n_{\text{B}}(\xi_0)$, and ξ corresponds to $n_{\text{B}}(\xi)$.

$$\Delta\xi = \xi - \xi_0 = \int_{\xi_0}^{\xi} \text{d}\xi = \int_{n_{\text{B}}(\xi_0)}^{n_{\text{B}}(\xi)} \frac{\text{d}n_{\text{B}}}{\nu_{\text{B}}} = \frac{n_{\text{B}}(\xi) - n_{\text{B}}(\xi_0)}{\nu_{\text{B}}}$$

That is, $$\Delta\xi = \frac{\Delta n_{\text{B}}}{\nu_{\text{B}}} \qquad (2.7.2)$$

Comment:

① Unit of ξ is mol.

② ξ is related to the reaction stoichiometric equation.

For instance: $N_2+3H_2 = 2NH_3$ (0.5mol N_2 was dissipated)

$$\text{d}\xi = \frac{-0.5}{-1} = 0.5\text{mol}$$

$$\frac{1}{2}N_2 + \frac{3}{2}H_2 = NH_3(0.5\text{mol } N_2 \text{ was dissipated})$$

$$d\xi = \frac{-0.5}{-0.5} = 1\text{mol}$$

It can clearly be seen that extent of a reaction is related to the reaction stoichiometric equation. We must explain clearly the corresponding stoichiometric equation when we use the concept of extent of a reaction.

2.7.3 Molar enthalpy of reaction (摩尔反应焓)

Enthalpy of reaction (反应焓) —enthalpy difference between products and reactants in chemical reaction at constant temperature and pressure. (在一定的温度压力下，化学反应中生成的产物的焓与反应掉的反应物的焓之差。)

Symbol: $\Delta_r H$

Consider reaction : $aA+bB = yY+zZ$

$$\Delta_r H = n_Y H_m^*(Y) + n_Z H_m^*(Z) - n_A H_m^*(A) - n_B H_m^*(B)$$

Molar enthalpy of reaction (摩尔反应焓)

$$\Delta_r H_m = \frac{\Delta_r H}{\Delta\xi} \tag{2.7.3}$$

Unit : $J \cdot mol^{-1}$.

Consider reaction : $aA+bB = yY+zZ$

$$\Delta_r H_m = \sum_B \nu_B H_m(B) \tag{2.7.4}$$

2.7.4 Standard molar enthalpy of reaction (标准摩尔反应焓)

(1) Thermodynamic standard state of substance (物质的热力学标准态)

According to GB 3102.8, standard pressure is pressure in standard state, that is, $p^\ominus = 100\text{kPa}$.

Thermodynamic standard state of gas (气体的热力学标准态):

T, $p^\ominus = 100\text{kPa}$, hypothetical state of pure gas having features of perfect gas. (不管纯气体B还是气体混合物中的组分B，都是温度T、压力$p^\ominus$下并表现出理想气体特性的气体纯物质B的假想状态。)

Thermodynamic standard state of condensed phase (液体或固体的标准态):

T, $p^\ominus = 100\text{kPa}$, hypothetical state of pure condensed phase. (不管纯液体B还是液体混合物中的组分B，都是温度T、压力$p^\ominus$下液体或固体纯物质B的状态。)

Comment:

Temperature in thermodynamic standard state is arbitrary. However, thermal data of lots of substances in thermodynamic standard state are obtained at $T = 298.15\text{K}$.

(2) Standard molar enthalpy of reaction (标准摩尔反应焓)

Consider reaction : $0 = \sum_B \nu_B B$

$$\Delta_r H_m^\ominus(T) \overset{\text{def}}{=\!=\!=} \sum \nu_B H_m^\ominus(B, \beta, T) \tag{2.7.5}$$

So, for reaction: $aA+bB = yY+zZ$

$$\Delta_r H_m^\ominus(T) = yH_m^\ominus(Y, \beta, T) + z H_m^\ominus(Z, \beta, T) - a H_m^\ominus(A, \beta, T) - b H_m^\ominus(B, \beta, T)$$

$H^\ominus_m(B, \beta, T)$ is molar enthalpy of substance B (reactant or product) at T and $p^\ominus$.

(2.7.5) is only definition formula of standard molar enthalpy of reaction because $H^\ominus_m(B, \beta, T)$ can not be obtained. Therefore, (2.7.5) has no actual meaning in calculation.

In fact, standard molar enthalpy of reaction $\Delta_r H^\ominus_m(T)$ can be calculated by $\Delta_f H^\ominus_m(B, \beta, T)$ and $\Delta_c H^\ominus_m(B, \beta, T)$ of substance B.

2.8 Standard molar enthalpy of formation and standard molar enthalpy of combustion (标准摩尔生成焓与标准摩尔燃烧焓)

2.8.1 Standard molar enthalpy of formation (标准摩尔生成焓)

(1) Formation reaction (生成反应)

Formation reaction—reaction in which compound B is formed from stable elementary substances. (由稳态单质生成化合物 B 的反应。)

Comment:

① Reactants: stable elementary substances, that is, the most stable state at T and p.

For instance:

C→graphite, S→rhombic S, Br_2→Br_2(l), Hg→Hg(l), rare gases→monatomic gases, H_2、F_2、O_2、N_2→diatomic gases, etc., but P is P(s, white phosphorus), not P(s, red phosphorus).

② Products: compound B, $\nu_B=+1$

For instance:

$C(\text{graphite}, 298.15K, p) + 2\ H_2(g, 298.15K, p) + O_2(g, 298.15K, p) = CH_3OH(l, 298.15K, p)$ is a formation reaction.

(2) Standard molar enthalpy of formation (标准摩尔生成焓)

Standard molar enthalpy of formation—the enthalpy changes that occur when unit amount of the compound in its standard state is formed from its elements in their standard states, that is, standard molar enthalpy change of formation reaction is the standard molar enthalpy of formation of product B. (在温度为 T 的标准态下，由稳定相态的单质生成 1mol β 相态化合物 B 的焓变即为化合物 B(β)在温度 T 下的标准摩尔生成焓。)

Symbol: $\Delta_f H^\ominus_m(B, \beta, T)$; unit: $J \cdot mol^{-1}$ or $kJ \cdot mol^{-1}$.

Obviously, $\Delta_f H^\ominus_m(T)$ of stable elementary substance is zero at arbitrary temperature, that is, $\Delta_f H^\ominus_m(\text{stable elementary substance}, T) = 0$.

For instance: $\Delta_f H^\ominus_m(C, \text{graphite}, T) = 0$.

We can get $\Delta_f H^\ominus_m(B, \beta, T)$ of substance B from handbook or textbook, such as appendix IX in this textbook.

(3) Calculate $\Delta_r H^\ominus_m(T)$ from $\Delta_f H^\ominus_m(B, \beta, T)$

$$\Delta_r H^\ominus_m(T) = \sum_B \nu_B \Delta_f H^\ominus_m(B, \beta, T) \tag{2.8.1}$$

Or

$$\Delta_r H^\ominus_m(298.15K) = \sum_B \nu_B \Delta_f H^\ominus_m(B, \beta, 298.15K) \tag{2.8.2}$$

For reaction: $aA(g) + bB(s) = yY(g) + zZ(s)$

$$\Delta_r H_m^\ominus(298.15K) = y\,\Delta_f H_m^\ominus(Y,\ g,\ 298.15K) + z\,\Delta_f H_m^\ominus(Z,\ s,\ 298.15K) - a\,\Delta_f H_m^\ominus(A,\ g,\ 298.15K) - b\,\Delta_f H_m^\ominus(B,\ s,\ 298.15K)$$

The formula shows that $\Delta_r H_m^\ominus(298.15K)$ can be calculated if we know $\Delta_f H_m^\ominus(298.15K)$ of every substance in a chemical reaction.

2.8.2 Standard molar enthalpy of combustion (标准摩尔燃烧焓)

(1) Combustion reaction (燃烧反应)

Combustion reaction—reaction accompanying total oxidation of 1 mol material B. (化学反应方程式中 B 的化学计量数 $\nu_B = -1$ 时，B 完全氧化成相同温度下指定产物的反应。)

Comment:

① Reactants : material B, $\nu_B = -1$.

② Products: total oxidation, for instance: $C \rightarrow CO_2$, $H \rightarrow H_2O(l)$, $N \rightarrow N_2$, $S \rightarrow SO_2(g)$, $Cl \rightarrow HCl(aq)$ etc.

For instance:

$C(\text{graphite},\ 298.15K,\ p) + O_2(g,\ 298.15K,\ p) = CO_2(g,\ 298.15K,\ p)$ and $CO(g,\ 298.15K,\ p) + 1/2O_2(g,\ 298.15K,\ p) = CO_2(g,\ 298.15K,\ p)$ are combustion reactions.

(2) Standard molar enthalpy of combustion (标准摩尔燃烧焓)

Standard molar enthalpy of combustion—standard molar enthalpy change of combustion reaction is the standard molar enthalpy of combustion of reactant B. (在温度为 T 的标准态下，1mol 物质 B 与氧气进行完全燃烧反应，生成相同温度下指定产物时的标准摩尔反应焓即为反应物 B(β)在温度 T 下的标准摩尔燃烧焓。)

Symbol: $\Delta_c H_m^\ominus(B,\ \beta,\ T)$; unit: $J \cdot mol^{-1}$ or $kJ \cdot mol^{-1}$.

Obviously, $\Delta_c H_m^\ominus(T)$ of substance of total oxidation is zero at arbitrary temperature, that is, $\Delta_c H_m^\ominus(\text{substance of total oxidation},\ T) = 0$.

For instance: $\Delta_c H_m^\ominus(CO_2,\ g,\ T) = 0$

$\Delta_c H_m^\ominus(H_2O,\ l,\ T) = 0$

We can get $\Delta_c H_m^\ominus(B,\ \beta,\ T)$ of substance B from handbook or textbook, such as appendix X in this textbook.

(3) Calculate $\Delta_r H_m^\ominus(T)$ from $\Delta_c H_m^\ominus(B,\ \beta,\ T)$

$$\Delta_r H_m^\ominus(T) = -\sum_B \nu_B \Delta_c H_m^\ominus(B,\ \beta,\ T) \tag{2.8.3}$$

Or

$$\Delta_r H_m^\ominus(298.15K) = -\sum_B \nu_B \Delta_c H_m^\ominus(B,\ \beta,\ 298.15K) \tag{2.8.4}$$

For reaction: $aA(g) + bB(s) = yY(g) + zZ(s)$

$$\Delta_r H_m^\ominus(298.15K) = a\,\Delta_c H_m^\ominus(A,\ g,\ 298.15K) + b\,\Delta_c H_m^\ominus(B,\ s,\ 298.15K) - y\,\Delta_c H_m^\ominus(Y,\ g,\ 298.15K) - z\Delta_c H_m^\ominus(Z,\ s,\ 298.15K)$$

The formula shows that $\Delta_r H_m^\ominus(298.15K)$ can be calculated if we know $\Delta_c H_m^\ominus(298.15K)$ of every substance in a chemical reaction.

Conclusion:

$$\Delta_f H_m^\ominus(CO_2,\ g,\ T) = \Delta_c H_m^\ominus(C,\ \text{graphite},\ T)$$

$$\Delta_f H_m^\ominus(H_2O,\ l,\ T) = \Delta_c H_m^\ominus(H_2,\ g,\ T)$$

Value of $\Delta_c H_m^\ominus(T)$ is larger so that it can be determined easily and accurately. Therefore, $\Delta_c H_m^\ominus(T)$ can be used as basic thermodynamic data. We can calculate $\Delta_r H_m^\ominus(T)$ and $\Delta_f H_m^\ominus(T)$ using $\Delta_c H_m^\ominus(T)$.

Example 2.10

At 25℃, $\Delta_c H_m^\ominus(C_2H_5OH) = -1366.8\text{kJ}\cdot\text{mol}^{-1}$, please calculate $\Delta_f H_m^\ominus(C_2H_5OH)$ at the same temperature.

Given: $\Delta_f H_m^\ominus(CO_2,\ g) = -393.51\text{kJ}\cdot\text{mol}^{-1}$, $\Delta_f H_m^\ominus(H_2O,\ l) = -285.83\text{kJ}\cdot\text{mol}^{-1}$.

Answer:

The combustion reaction of alcohol is:

$$C_2H_5OH(l) + 3O_2(g) = 2CO_2(g) + 3H_2O(l)$$

According to the definition of $\Delta_c H_m^\ominus(B,\ \beta,\ T)$, we can get:

$$\Delta_r H_m^\ominus(T) = \Delta_c H_m^\ominus(C_2H_5OH,\ l) = -1366.8\text{kJ}\cdot\text{mol}^{-1}$$

$\Delta_r H_m^\ominus(T)$ of the same combustion reaction can also be calculated from $\Delta_f H_m^\ominus(B,\ \beta,\ T)$, that is

$$\Delta_r H_m^\ominus(T) = 2\Delta_f H_m^\ominus(CO_2, g) + 3\Delta_f H_m^\ominus(H_2O, l) - \Delta_f H_m^\ominus(C_2H_5OH, l) = \Delta_c H_m^\ominus(C_2H_5OH, l)$$

Therefore,

$$\Delta_f H_m^\ominus(C_2H_5OH,\ l) = 2\Delta_f H_m^\ominus(CO_2,\ g) + 3\Delta_f H_m^\ominus(H_2O,\ l) - \Delta_c H_m^\ominus(C_2H_5OH,\ l)$$
$$= -277.71\text{kJ}\cdot\text{mol}^{-1}$$

2.8.3 The temperature—dependence of $\Delta_r H_m^\ominus(T)$—Kirchhoff's formula (标准摩尔反应焓与温度的关系)

$\Delta_f H_m^\ominus(T)$ and $\Delta_c H_m^\ominus(T)$ reported in references are usually obtained at 25℃, therefore, $\Delta_r H_m^\ominus(T)$ can be calculated by $\Delta_f H_m^\ominus(T)$ and $\Delta_c H_m^\ominus(T)$ only at 25℃. But we often need $\Delta_r H_m^\ominus(T)$ of other temperatures, so we should know the relation between $\Delta_r H_m^\ominus(T)$ and T.

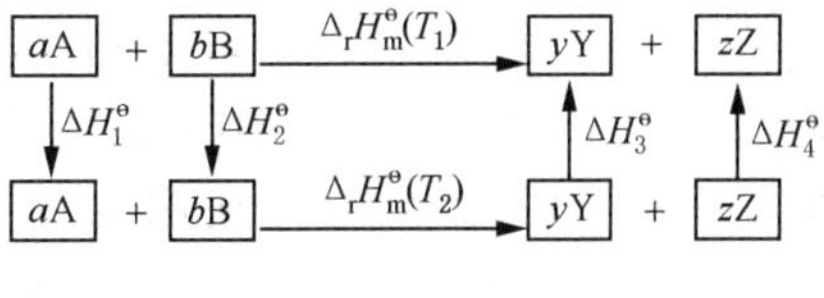

$$\Delta_r H_m^\ominus(T_1) = \Delta H_1^\ominus + \Delta H_2^\ominus + \Delta_r H_m^\ominus(T_2) + \Delta H_3^\ominus + \Delta H_4^\ominus$$

$$\Delta H_1^\ominus = a\int_{T_1}^{T_2} C_{p,\ m}(A)\,dT,\quad \Delta H_2^\ominus = b\int_{T_1}^{T_2} C_{p,\ m}(B)\,dT$$

$$\Delta H_3^\ominus = -y\int_{T_1}^{T_2} C_{p,\ m}(Y)\,dT,\quad \Delta H_4^\ominus = -z\int_{T_1}^{T_2} C_{p,\ m}(Z)\,dT$$

Therefore $\quad \Delta_r H_m^\ominus(T_2) = \Delta_r H_m^\ominus(T_1) + \int_{T_1}^{T_2} \sum \nu_B C_{p,\ m}(B)\,dT$

If $T_2 = T$, $T_1 = 298.15\text{K}$, then:

$$\Delta_r H_m^\ominus(T) = \Delta_r H_m^\ominus(298.15\text{K}) + \int_{298.15\text{K}}^{T} \sum \nu_B C_{p,m}(\text{B})\,dT \qquad (2.8.5)$$

——基希霍夫公式(**Kirchhoff's formula, 298.15K ~ *T*: no phase transition**).

2.8.4 The relation between Q_p and Q_V(恒容反应热与恒压反应热之间的关系)

$$Q_p = \Delta_r H_m(\text{closed system},\ W'=0,\ dp=0),$$

$$Q_V = \Delta_r U_m(\text{closed system},\ W'=0,\ dV=0)$$

$$\Delta_r H_m(T) = \sum \nu_B H_m(\text{B}, \beta, T) = \sum \nu_B U_m(\text{B}, \beta, T) + \sum \nu_B [p V_m(\text{B}, T)]$$

So: ① Condensed phase: $\Delta_r H_m(T, \text{l or s}) \approx \Delta_r U_m(T, \text{l or s})$

② Perfect gas: $\Delta_r H_m(T) = \Delta_r U_m(T) + RT\sum \nu_B(\text{g})$

Therefore $$\Delta_r H_m(T) = \Delta_r U_m(T) + RT\sum \nu_B(\text{g}) \qquad (2.8.6)$$

Or $$\Delta_r H_m^\ominus(T) = \Delta_r U_m^\ominus(T) + RT\sum \nu_B(\text{g}) \qquad (2.8.7)$$

2.8.5 Maximum temperature of explosion and flame reaction(火焰和爆炸反应的最高温度)

原来的恒温恒压放热反应，若反应很快，传热很慢，则可认为是绝热的，产物温度就要升高。例如恒压燃烧反应，它可达到的最高温度，称为最高火焰温度。

同样，原来恒温恒容下的放热反应，若反应瞬间完成，可认为它是绝热的，热量不能散发，产物温度升高，系统内压力也达到最大。例如爆炸反应，它也有一个爆炸反应最高温度与最大压力的计算问题。

很明显，计算恒压燃烧反应最高火焰温度的依据是 $Q_p = \Delta H = 0$，计算恒容爆炸反应的最高温度的依据是 $Q_V = \Delta U = 0$。

Explosion reaction—combustion reaction at constant volume.

Flame reaction—combustion reaction at constant pressure.

Maximum temperature of explosion: adiabatic and isochoric, that is, $Q_V = \Delta U = 0$.

Maximum temperature of flame: adiabatic and isobaric, that is, $Q_p = \Delta H = 0$.

Example 2.11

Please calculate $\Delta_r H_m^\ominus(393\text{K})$ and $\Delta_r U_m^\ominus(393\text{K})$ of reaction:

$$C_2H_5OH(g) + HCl(g) = C_2H_5Cl(g) + H_2O(g)$$

substance	$\Delta_f H_m^\ominus(298\text{K})/\text{kJ}\cdot\text{mol}^{-1}$	$C_{p,m}/\text{J}\cdot\text{K}^{-1}\cdot\text{mol}^{-1}$
C_2H_5Cl (g)	−105.0	$13.07+188.5\times10^{-3}(T/\text{K})$
H_2O (g)	−241.84	$30.00+10.71\times10^{-3}(T/\text{K})$
C_2H_5OH (g)	−235.3	$19.07+212.7\times10^{-3}(T/\text{K})$
HCl(g)	−92.31	$26.53+4.62\times10^{-3}(T/\text{K})$

Answer:

$$\Delta_r H_m^\ominus(298\text{K}) = \sum \nu_B \Delta_f H_m^\ominus(\text{B}, \beta, 298\text{K})$$
$$= \Delta_f H_m^\ominus(C_2H_5Cl, g, 298\text{K}) + \Delta_f H_m^\ominus(H_2O, g, 298\text{K}) - \Delta_f H_m^\ominus(C_2H_5OH, g, 298\text{K}) - \Delta_f H_m^\ominus(HCl, g, 298\text{K})$$
$$= -19.23\text{kJ}\cdot\text{mol}^{-1}$$

$$\Delta_r H_m^\ominus(393\text{ K}) = \Delta_r H_m^\ominus(298\text{K}) + \int_{298\text{K}}^{393\text{K}} \sum \nu_B C_{p,m}(\text{B})\,dT$$

$$\sum \nu_B C_{p,m}(B) = -2.53 - 18.09 \times 10^{-3} T$$

Then,

$$\Delta_r H_m^\ominus(393K) = -19.23 kJ \cdot mol^{-1} + \int_{298K}^{393K} (-2.53 - 18.09 \times 10^{-3} T) dT$$
$$= -20.06 kJ \cdot mol^{-1}$$
$$\Delta_r U_m^\ominus(393K) = \Delta_r H_m^\ominus(393K) - RT \sum \nu_B(g)$$
$$= \Delta_r H_m^\ominus(393K)$$
$$= -20.06 kJ \cdot mol^{-1}$$

2.9 Throttling process and Joule-Thomson effect (节流过程与 Joule-Thomson 效应)

2.9.1 Joule-Thomson experiment (焦耳–汤姆生实验)

In 1853 Joule and William Thomson did an experiment similar to the Joule experiment but allowing far more accurate results to be obtained. The Joule-Thomson experiment involves the slow throttling of a gas through a rigid, porous plug. An idealized sketch of the experiment is shown in figure 2.4. The system is enclosed in adiabatic walls. The left piston is held at a fixed pressure p_1; the right piston is held at a fixed pressure $p_2 < p_1$. The partition B is porous but not greatly so; this allows the gas to be slowly forced from one chamber to the other. Because the throttling process is slow, pressure equilibrium is maintained in each chamber. Essentially all the pressure drop from p_1 to p_2 occurs in the porous plug.

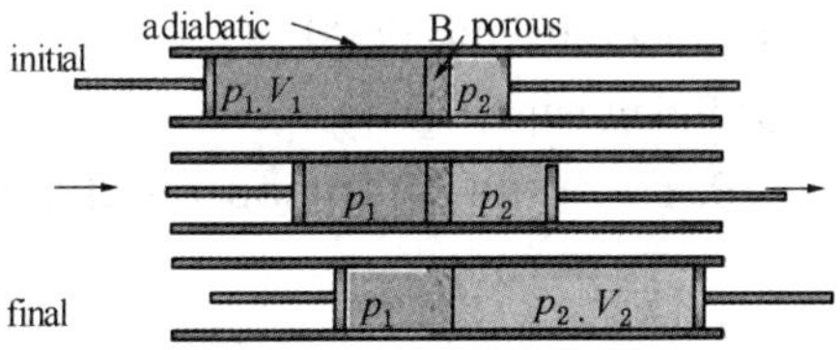

Fig. 2.4 Joule-Thomson experiment ($p_1 > p_2$)

Throttling process (节流过程) —expansion process in which pressures of initial and final state are constant in adiabatic condition. (绝热条件下，气体始末态压力分别保持恒定条件下的膨胀过程。)

2.9.2 Experiment result (实验结果)

The Joule-Thomson experiment shows:

① 在室温常压下，多数气体经节流膨胀后，温度下降，产生制冷效应；

② 而氢、氦等少数气体，经节流膨胀后，温度升高，产生制热效应；

③ 理想气体经节流膨胀后，温度基本不变。

可见：一定量的理想气体，其热力学能 U 与焓 H 仅是温度的函数。只要温度不变，即使气体的体积或压力变化了，U 与 H 的值依然不变；反之，若理想气体的 U、H 不变，即使 p、V 改变，温度 T 也应当不变。

实际气体分子间有相互作用力，因而不服从理想气体状态方程，不再有 $U=f(T)$ 和 $H=f(T)$ 的关系，而是：$U=f(T, V)$ 和 $H=f(T, p)$。

2.9.3 Result analysis (结果分析)

We calculate W firstly, the work done on the gas in throttling it through the plug. The overall process is irreversible since exceeds p_2 by a finite amount, and an infinitesimal change in pressure cannot reverse the process. However, the pressure drop occurs almost completely in the plug; the plug is rigid, and the gas does no work on the plug, or vice versa. The exchange of work between system and surroundings occurs solely at the two pistons. Since pressure equilibrium is maintained at each piston, we can use $W = -\int_{V_1}^{V_2} p_{su}\mathrm{d}V$ to calculate the work at each piston.

According to $W = -\int_{V_1}^{V_2} p_{su}\mathrm{d}V = -p(V_2 - V_1)$

Therefore $\quad W = -p_1(0 - V_1) + [-p_2(V_2 - 0)]$

$= p_1 V_1 - p_2 V_2$

For adiabatic process, $Q = 0$

So $\quad \Delta U = Q + W = W$

$$U_2 - U_1 = p_1 V_1 - p_2 V_2$$

$$U_2 + p_2 V_2 = U_1 + p_1 V_1$$

That is $\quad H_2 = H_1 \qquad (2.9.1)$

2.9.4 Characteristics of throttling process (节流过程的特征)

① Adiabatic $Q = 0$

② Underpressure $\mathrm{d}p < 0$

③ Isoenthalpy $H_2 = H_1$

2.9.5 Joule-Thomson coefficient (焦-汤系数)

$$\mu_{J\text{-}T} \overset{\text{def}}{=\!=} \left(\frac{\partial T}{\partial p}\right)_H \qquad (2.9.2)$$

$\mu_{J\text{-}T}$ is the change rate of temperature with pressure for a throttling process in isoenthalpic condition.

Most gases: $\mu_{J\text{-}T} > 0$, for $\mathrm{d}p < 0$, so $T_2 < T_1$—cooling process

H_2, He, Ne: $\mu_{J\text{-}T} < 0$, for $\mathrm{d}p < 0$, so $T_2 > T_1$—heating process

Perfect gas: $\mu_{J\text{-}T} = 0$, so $T_2 = T_1$—isothermal process

Supplementary Examples of Chapter 2

EXERCISES

(习题)

1. 10mol pg whose initial state is $T_1 = 25℃$, $p_1 = 10^6$Pa expands to the final state $T_2 = 25℃$, $p_2 = 10^5$Pa through the following different processes. Calculate W, Q, ΔU and ΔH of each process.

(1) Free expansion process;

(2) Expands against constant external pressure $p_{su} = 10^5$Pa;

(3) Expands isothermally and reversibly.

Answer: (1) 0, 0, 0, 0; (2) −22.3kJ, 22.3kJ, 0, 0; (3) −57.1kJ, 57.1kJ, 0, 0

2. 2mol monatomic perfect gas expands adiabatically against constant external pressure p_{su} = 100kPa to equilibrium, the initial state is $T_1 = 600$K, $p_1 = 1.00$MPa, calculate W, Q, ΔU and ΔH

of this process.

Answer: −5. 39kJ, 0, −5. 39kJ, −8. 98kJ

3. 1mol monatomic perfect gas whose initial state is $T_1 = 298.15\text{K}$, $p_1 = 6 \times 101.3\text{kPa}$ expands adiabatically to the final state $p_2 = 101.3\text{kPa}$ through the following two different processes. Calculate the final temperature T_2, W, ΔU and ΔH of each process.

(1) Expands reversibly;

(2) Expands against constant external pressure $p_{su} = 101.3\text{kPa}$.

Answer: (1) 145. 6K, −1. 902kJ, −1. 902kJ, −3. 171kJ

(2) 198. 8K, −1. 239kJ, −1. 239kJ, −2. 065kJ

4. 2 mol perfect gas with $C_{p,\text{ m}} = 3.5R$ whose initial state is $p_1 = 100\text{kPa}$, $V_1 = 50\text{dm}^3$ is heated firstly isochoricly to 200kPa, then cooled isobaricly to 25dm^3, calculate W, Q, ΔU and ΔH of the whole process.

Answer: 5. 00kJ, −5. 00 kJ, 0, 0

5. 4 mol perfect gas with $C_{p,\text{ m}} = 2.5R$ whose initial state is $p_1 = 100\text{kPa}$, $V_1 = 100\text{dm}^3$ is heated firstly isobaricly to 150dm^3, then heated isochoricly to 150kPa, calculate W, Q, ΔU and ΔH of the whole process.

Answer: −5. 00kJ, 23. 75 kJ, 18. 75kJ, 31. 25kJ

6. 5 mol diatomic perfect gas whose initial state is $T_1 = 300\text{K}$, $p_1 = 200\text{kPa}$ first expands isothermally and reversibly to 50kPa, then compresses adiabaticly and reversibly to the final pressure of 200kPa. Calculate the final temperature T_2 and Q, W, ΔU, ΔH of the whole process.

Answer: 445. 80K, 17. 29kJ, −2. 14kJ, 15. 15kJ, 21. 21kJ

7. 2 mol monatomic perfect gas A and 3 mol diatomic perfect gas B are mixed to form a perfect gas mixture system, the initial state is $T_1 = 350\text{K}$, $V_1 = 72.75\text{dm}^3$, the system reaches to its equilibrium state through the following different processes. Calculate Q, W, ΔU, and ΔH of each process.

(1) Expands isothermally and reversibly to 120dm^3;

(2) Expands isothermally and against constant external pressure of 121. 25kPa to 120dm^3;

(3) Expands adiabaticly and reversibly to 120dm^3;

(4) Expands adiabaticly and against constant external pressure of 121. 25kPa to equilibrium.

Answer: (1) 7282J, −7282J, 0, 0; (2) 5729J, −5729J, 0, 0; (3) 0, −6479. 6J, −6479. 6J, −9565. 1J; (4) 0, −3881J, −3881J, −5729. 8 J

8. 2mol monatomic perfect gas whose initial state is $T_1 = 273\text{K}$, $p_1 = 202.6\text{kPa}$ is heated reversibly to the final state $p_2 = 405.2\text{kPa}$ through the path of $pV^{-1} = p_1 V_1^{-1}$, calculate Q, W, ΔU, and ΔH of this process.

Answer: 27. 24kJ, −6. 81kJ, 20. 429kJ, 34. 048kJ

9. Calculate Q, W, ΔU, and ΔH of the following process:

1kg H_2O(g, 100℃, 101. 325kPa)→H_2O(l, 100℃, 101. 325kPa).

Given that: $p^*(H_2O,\ 100℃) = 101.325\text{kPa}$, $\Delta_{vap}H_m(H_2O,\ 100℃) = 40.668\text{kJ} \cdot \text{mol}^{-1}$.

Suppose the steam obeys the state equation of perfect gas.

Answer: -2257kJ, 172.2kJ, -2085kJ, -2257kJ

10. (1) 1mol water is vaporized to steam (pg) at 100℃ and 101.325kPa, the heat absorbed in this process is 40.67kJ/mol. Calculate Q, W, ΔU, and ΔH of the above process;

(2) The initial state is same as (1), firstly the water is vaporized isothermally and at constant external pressure of 50kPa, then the steam is compressed isothermally and reversibly from 100℃, 50kPa to 100℃, 101.325kPa. Calculate Q, W, ΔU, and ΔH of the whole process;

(3) 1mol water is vaporized to vacuum to turn into steam, the initial and final temperature is 100℃, pressure is 101.325kPa, calculate Q, W, ΔU, and ΔH of the process.

Answer: (1) 40.67kJ, -3.102kJ, 37.57kJ, 40.67kJ; (2) 38.48kJ, -0.911kJ, 37.57kJ, 40.67kJ; (3) 37.57kJ, 0, 37.57kJ, 40.67kJ

11. 2mol liquid benzene of 60℃ and 100kPa is vaporized completely to vapor of 60℃ and 24kPa at constant external pressure, calculate the Q, W, ΔU, and ΔH of the process.

Given: $p^*(C_6H_6, 40℃) = 24kPa$, $\Delta_{vap}H_m(C_6H_6, 40℃) = 33.43kJ \cdot mol^{-1}$

$C_{p,m}(C_6H_6, l) = 141.5J \cdot K^{-1} \cdot mol^{-1}$, $C_{p,m}(C_6H_6, g) = 94.12J \cdot K^{-1} \cdot mol^{-1}$

both $C_{p,m}(C_6H_6, l)$ and $C_{p,m}(C_6H_6, g)$ don't change with temperature, the volume of liquid can be omitted.

Answer: 64.96kJ, -5.54kJ, 59.42kJ, 64.96kJ

12. Calculate $\Delta_r H_m^\ominus(298.15K)$ of the following reaction:

$C(graphite) + 2H_2O(g) \longrightarrow 2H_2(g) + CO_2(g)$

Given: at 25℃, $\Delta_c H_m^\ominus(C, graphite) = -393.51kJ \cdot mol^{-1}$; $\Delta_c H_m^\ominus(H_2, g) = -285.84\ kJ \cdot mol^{-1}$; $\Delta_{vap}H_m(H_2O, 25℃) = 40.0kJ \cdot mol^{-1}$

Answer: 98.17kJ · mol^{-1}

13. Gas reaction $A(g) + B(g) \rightarrow Y(g)$ proceeds at 500℃, 100kPa. The relative data are shown in the following table:

Substance	$\Delta_f H_m^\ominus(298.15K)/kJ \cdot mol^{-1}$	$C_{p,m}^\ominus/J \cdot mol^{-1} \cdot K^{-1}$(25℃~500℃)
A(g)	-235	19.1
B(g)	52	4.2
Y(g)	-241	30.0

Calculate $\Delta_r H_m^\ominus(298.15K)$、$\Delta_r H_m^\ominus(773.15K)$ and $\Delta_r U_m^\ominus(773.15K)$ of the above reaction.

Answer: -58kJ · mol^{-1}, -54.82kJ · mol^{-1}, -48.39kJ · mol^{-1}

14. At 25℃ and in a tightly closed isochoric container, 10g $C_{10}H_8(s)$ was completely combusted into $CO_2(g)$ and $H_2O(l)$ in excess $O_2(g)$, releasing heat of 401.727kJ in this process. Please calculate:

(1) Extent of reaction $C_{10}H_8(s) + 12O_2(g) = 10CO_2(g) + 4H_2O(l)$;

(2) $\Delta_c U_m^\ominus$ of $C_{10}H_8(s)$;

(3) $\Delta_c H_m^\ominus$ of $C_{10}H_8(s)$.

Answer: 78.019mmol, -5149.1kJ · mol^{-1}, -5154.1kJ · mol^{-1}

Chapter 3 The second law of thermodynamics (热力学第二定律)

Learning objectives:

(1) State the criterion for the direction of *spontaneous change.*

(2) Explain the basis of the *statistical definition* of entropy.

(3) State the *Second Law of thermodynamics.*

(4) State the *Second Law of thermodynamics* in terms of the entropy.

(5) Calculate the change of entropy.

(6) Define the *Helmholtz function* and the *Gibbs function* of a system and use them as criteria for the direction of spontaneous change.

(7) State the *Third Law of thermodynamics* and explain the experimental evidence for it.

A major application of thermodynamics to chemistry is to provide information about equilibrium in chemical systems. The first law assures us that the total energy of system plus surroundings remains constant during the reaction, but the first law cannot say what the final equilibrium concentrations will be. We shall see that the second law provides such information. The second law leads to the existence of the state function S, which possesses the property that for an isolated system the equilibrium position corresponds to maximum entropy. The remainder of this chapter shows how to calculate entropy changes in processes, and leads to two new state functions Helmholtz function and Gibbs function. Combine the first law and the second law of thermodynamics will give the fundamental equation of thermodynamics. We can deduce Clapeyron equation and Clausius-Clapeyron equation from the fundamental equation of thermodynamics.

3.1 Carnot cycle (卡诺循环)

3.1.1 Efficiency of heat engine (热机效率)

Forms of energy defined in thermodynamics are two kinds, that is, heat and work. Transformation process of substance is related to mutual transition of heat and work.

We can convert work into heat totally. But we can not convert heat into work totally. There are fixed restrictions in converting of heat into work. The fixed restrictions make state change of substance have fixed direction and limit.

Direction and limit of process can be researched by the converting restrictions of heat and work using the second law of thermodynamics.

Heat engine—engine that converts some of the random molecular energy of heat flow into macroscopic mechanical energy, such as: internal combustion engine, steam engine etc. (热机是

指通过工质，从高温热源吸热，向低温热源放热，并对环境做功的循环操作的机器。）

The working substance (for example, steam in a steam engine) is heated in a cylinder, and its expansion causes a piston to move, thereby doing mechanical work. If the engine is to operate continuously, the working substance has to be cooled back to its original state and the piston has to return to its original location before we can heat the working substance again and get another work-producing expansion. Hence the working substance goes through a **cyclic process.** The essentials of the cycle are the absorption of heat Q_1 by the working substance from a hot body, the performance of work W by the working substance on the surroundings, and the emission of heat Q_2 by the working substance to a cold body, with the working substance returning to its original state at the end of the cycle. The system is the working substance.

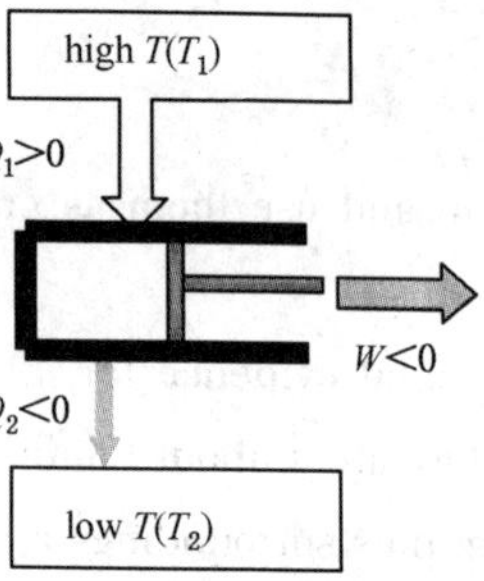

Fig. 3. 1 Energy flow of the heat engine

The energy flow schematic diagram of the heat engine is shown in figure 3. 1.

For a heat engine, Q_1 means heat added to the system, $Q_1>0$, the quantity W is negative because the engine does work on its surroundings, Q_2 is negative because heat flows out of the system to the cold body.

The efficiency of heat engine is the fraction of energy input that appears as useful energy output, that is, that appears as work. The energy input per cycle is the heat input Q_1 to the engine. We have

$$\eta \overset{\text{def}}{=\!=} \frac{-W}{Q_1} \tag{3.1.1}$$

η—efficiency of heat engine.

For cyclic process: $\Delta U = 0$

Hence $Q_1+Q_2+W = 0$

So that $Q_1+Q_2=-W$

Therefore

$$\eta \overset{\text{def}}{=\!=} \frac{-W}{Q_1}=\frac{Q_1+Q_2}{Q_1} \tag{3.1.2}$$

Since Q_2 is negative and Q_1 positive, the efficiency of heat engine is less than 1.

3. 1. 2 Carnot cycle (卡诺循环)

Carnot heat engine—reversible engine that absorbs heat from high-temperature heat source and liberates heat to low-temperature heat source and does work on surroundings.

Sadi Carnot

In 1824, French engineer N. L. S. Carnot designed a Carnot heat engine, in Carnot heat engine, the working substance is the perfect gas.

A Carnot cycle is defined as a reversible cycle that consists of two isothermal steps at different temperatures and two adiabatic steps, which is shown in figure 3. 2.

① Isothermal and reversible expansion process (AB).

② Adiabatic and reversible expansion process (BC).

③ Isothermal and reversible compression process (CD).

④ Adiabatic and reversible compression process (DA).

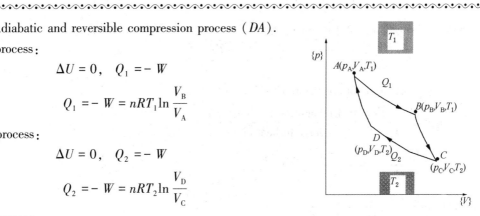

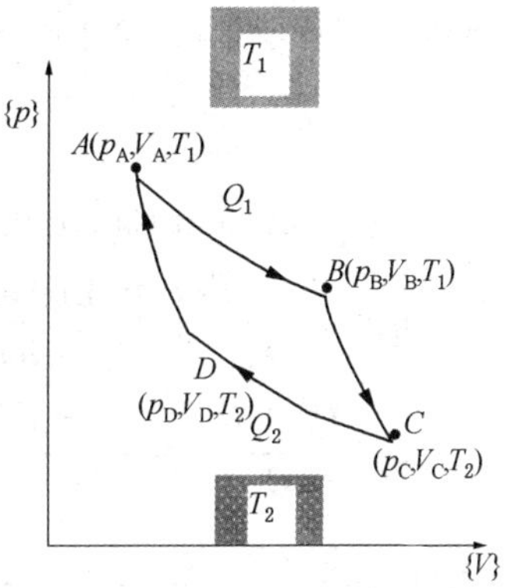

Fig. 3. 2 Carnot cycle

AB process:

$$\Delta U = 0, \quad Q_1 = -W$$

$$Q_1 = -W = nRT_1 \ln \frac{V_B}{V_A}$$

CD process:

$$\Delta U = 0, \quad Q_2 = -W$$

$$Q_2 = -W = nRT_2 \ln \frac{V_D}{V_C}$$

BC process:

$$T_1 V_B^{\gamma-1} = T_2 V_C^{\gamma-1}$$

DA process:

$$T_1 V_A^{\gamma-1} = T_2 V_D^{\gamma-1}$$

According to BC process and DA process, we can get

$$\frac{V_B}{V_A} = \frac{V_C}{V_D}$$

Therefore, $Q_1 + Q_2 = nRT_1 \ln \frac{V_B}{V_A} - nRT_2 \ln \frac{V_B}{V_A} = nR(T_1 - T_2) \ln \frac{V_B}{V_A}$

$$\eta \overset{\text{def}}{=\!=\!=} \frac{-W}{Q_1} = \frac{Q_1 + Q_2}{Q_1} = \frac{nR(T_1 - T_2)\ln \frac{V_B}{V_A}}{nRT_1 \ln \frac{V_B}{V_A}} = \frac{T_1 - T_2}{T_1} \tag{3.1.3}$$

We derived (3. 1. 3) using a perfect gas as the working substance, but since we earlier pointed that η is independent of the working substance, (3. 1. 3) must hold for any working substance undergoing a Carnot cycle.

Conclusion:

① η of Carnot heat engine is independent of the nature of the working substance. It depends only on T_1 and T_2.

$$\eta = \frac{T_1 - T_2}{T_1}$$

② For $\eta = \frac{Q_1 + Q_2}{Q_1} = \frac{T_1 - T_2}{T_1}$, $1 + \frac{Q_2}{Q_1} = 1 - \frac{T_2}{T_1}$

Therefore
$$\frac{Q_1}{T_1} + \frac{Q_2}{T_2} = 0$$

That is
$$\frac{Q_{1,\,r}}{T_1} + \frac{Q_{2,\,r}}{T_2} = 0 \tag{3.1.4}$$

3. 1. 3 Carnot theorem (卡诺定理)

(1) Carnot theorem

Efficiency of Carnot heat engine is the largest in all heat engines working between high-temperature heat source and low-temperature heat source. That is, $\eta_r > \eta_{ir}$.

$$\eta_r = \frac{T_1 - T_2}{T_1}$$

(2) Deduction of Carnot theorem

$$① \eta \leqslant \frac{T_1 - T_2}{T_1} \begin{pmatrix} <, \text{ irreversible heat engine} \\ =, \text{ reversible heat engine} \end{pmatrix} \quad (3.1.5)$$

$$② \eta = \frac{Q_1 + Q_2}{Q_1} \leqslant \frac{T_1 - T_2}{T_1}$$

Therefore, $1 + \frac{Q_2}{Q_1} \leqslant 1 - \frac{T_2}{T_1}$

That is, $\frac{Q_1}{T_1} + \frac{Q_2}{T_2} \leqslant 0 \begin{pmatrix} <, \text{ irreversible heat engine} \\ =, \text{ reversible heat engine} \end{pmatrix}$ (3.1.6)

3.2 The second law of thermodynamics (热力学第二定律)

3.2.1 Spontaneous process (自发过程)

All processes that can proceed:

nonequilibrium state $\xrightarrow{\text{spontaneous}}$ equilibrium state

For instance:

① Transmission of Q:

high temperature (T_1) $\xrightarrow{\text{spontaneous}}$ low temperature (T_2), until $T_1' = T_2'$, the reverse process can not be spontaneous.

② Expansion of gas:

high pressure (p_1) $\xrightarrow{\text{spontaneous}}$ low pressure (p_2), until $p_1' = p_2'$, the reverse process can not be spontaneous.

③ Mixing of water and alcohol:

water+alcohol $\xrightarrow{\text{spontaneous}}$ solution, until uniform, the reverse process can not be spontaneous.

From above examples, we can get the common features of spontaneous processes: irreversible, the reverse process of every spontaneous process is non-spontaneous.

3.2.2 Statements of the second law of thermodynamics (热力学第二定律的语言表述)

(1) Clausius saying (克劳休斯说法)

It is impossible for a system to undergo a cyclic process whose sole effects are the flow of heat into the system from a cold reservoir and the flow of an equal amount of heat out of the system into a hot reservoir. (不可能把热由低温物体转移到高温物体，而不留下其他变化。)

(2) Kelvin saying (开尔文说法)

No process is possible in which the sole result is the absorption of heat from a reservoir and its conversion into work. (不可能从单一热源吸热使之完全变为功，而不留下其他变化。)

For instance：

Expansion of perfect gas, $dT=0$, reversible: $\Delta U=0$, that is, $Q=-W$.

理想气体定温可逆膨胀，系统从单一热源吸的热全转变为对环境做的功，但系统的状态发生了变化(膨胀了)。

(3) Second kind of perpetual motion machine can not be made. (第二类永动机是不可能造成的。)

Second kind of perpetual motion machine—machine that can totally transfer heat absorbing from single heat source into work without other changes.

3.2.3 Essence of the second law of thermodynamics (热力学第二定律的实质)

Processes that are spontaneous in one direction are not spontaneous in the reverse direction.

3.3 Entropy and the principle of the increase of entropy (熵和熵增原理)

3.3.1 Entropy (熵)

Arbitrary large reversible cycle = $\sum$ many small carnot cycle.

Figure 3.3 shows that an arbitrary reversible cycle can be divided into an infinite number of infinitesimal strips, each strip being a Carnot cycle.

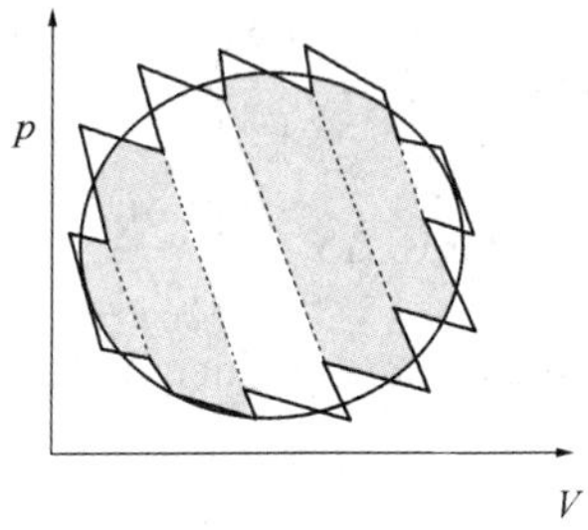

Fig. 3.3 An arbitrary reversible cycle

For any closed system that undergoes a small carnot cycle, equation (3.1.4) shows that the sum of $\frac{\delta Q_r}{T}$ around the cycle is zero, that is

$$\frac{\delta Q_1}{T_1}+\frac{\delta Q_2}{T_2}=0$$

$$\frac{\delta Q'_1}{T_1}+\frac{\delta Q'_2}{T_2}=0$$

$$\cdots \quad \cdots$$

So

$$\sum \frac{\delta Q_r}{T}=0$$

That is

$$\oint \frac{\delta Q_r}{T}=0$$

Integral theorem: Integrand must be the exact differential of a state function if the cycle-integral of a closed curve is zero.

Then

$$dS \xlongequal{\text{def}} \frac{\delta Q_r}{T} \tag{3.3.1}$$

S—entropy(熵).

Comment:

① Entropy is a state function.

② Entropy is an extensive property.

③ Unit of entropy is $J \cdot K^{-1}$.

④ Entropy has definite physical significance.

⑤ Absolute value of entropy can be known.

The entropy change on going from state 1 to state 2 is given by the integral of (3.3.1):

$$\int_{S_1}^{S_2} dS = S_2 - S_1 = \Delta S = \int_1^2 \frac{\delta Q_r}{T} \tag{3.3.2}$$

3.3.2 Clausius inequality (克劳休斯不等式)

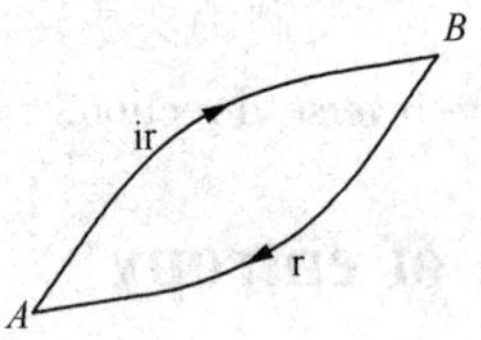

Fig. 3.4 Irreversible cyclic process

An irreversible cycle has two parts as shown in figure 3.4:

For irreversible cyclic process:

$$\int_A^B \frac{\delta Q_{ir}}{T_{su}} + \int_B^A \frac{\delta Q_r}{T} < 0$$

So,

$$\int_A^B \frac{\delta Q_{ir}}{T_{su}} < \int_A^B \frac{\delta Q_r}{T} = \Delta S$$

While, $\Delta S = \int_A^B \frac{\delta Q_r}{T}$

Therefore, $\Delta S \geqslant \int_A^B \frac{\delta Q}{T_{su}}$

We get Clausius inequality (formula of the second law of thermodynamics):

$$\Delta S \geqslant \int_A^B \frac{\delta Q}{T_{su}} \quad \begin{matrix} >, \text{ ir} \\ =, \text{ r} \end{matrix} \tag{3.3.3}$$

Or

$$dS \geqslant \frac{\delta Q}{T_{su}} \quad \begin{matrix} >, \text{ ir} \\ =, \text{ r} \end{matrix} \tag{3.3.4}$$

3.3.3 The principle of the increase of entropy and entropy criterion of equilibrium (熵增原理和平衡熵判据)

(1) The principle of the increase of entropy (熵增原理)

For adiabatic process: $\delta Q = 0$

According to Clausius inequality, we can get:

$$\Delta S_{adia} \geqslant 0 \quad \begin{matrix} >, \text{ ir} \\ =, \text{ r} \end{matrix} \tag{3.3.5}$$

Or

$$dS_{adia} \geqslant 0 \quad \begin{matrix} >, \text{ ir} \\ =, \text{ r} \end{matrix} \tag{3.3.6}$$

(3.3.5) and (3.3.6) are called **the principle of the increase of entropy.**

It shows that the entropy increase for an adiabatic and irreversible process and remains the same for an adiabatic and reversible process; process in which the entropy decrease is impossible. This is the principle of the increase of entropy. Adiabatic and reversible process is process of constant entropy.

(2) Entropy criterion of equilibrium (平衡熵判据)

An isolated system is necessarily closed, and any process in an isolated system must be adiabatic (since no heat can flow between the isolated system and its surroundings).

So for isolated system: $\delta Q=0$.

According to Clausius inequality, we can get:

$$\Delta S_{iso} \geqslant 0 \quad \begin{matrix} >, & \text{ir} \\ =, & \text{r} \end{matrix} \tag{3.3.7}$$

Or

$$dS_{iso} \geqslant 0 \quad \begin{matrix} >, & \text{ir} \\ =, & \text{r} \end{matrix} \tag{3.3.8}$$

(3.3.7) and (3.3.8) are called **entropy criterion of equilibrium.** It shows that S_{iso} increase for an irreversible process and remains the same for a reversible process.

That is, ① When infinitesimal change operates in isolated system, the isolated system is at equilibrium if $dS_{iso}=0$.

② The process of $\Delta S_{iso}>0$ or $dS_{iso}>0$ may be spontaneous.

Thermodynamic equilibrium in an isolated system is reached when the system's entropy is maximized.

$$dS_{iso}=dS+dS_{su}, \quad dS \overset{\text{def}}{=\!=} \frac{\delta Q_r}{T}$$

So if we want to use entropy criterion of equilibrium to judge the spontaneous process, we must calculate dS_{su}.

(3) Calculation of entropy change of surroundings (环境熵变的计算)

Since system and surroundings are isolated from the rest of the world (将系统与环境合在一起，形成一个大隔离系统，该隔离系统自然与外界绝热), we have

$$\delta Q_{su}=-\delta Q$$

Therefore

$$dS_{su}=\frac{\delta Q_{su}}{T_{su}}=\frac{-\delta Q}{T_{su}}$$

Surroundings can be seen as origin of heat for a closed system. if we suppose every origin of heat is large enough, and the volume and temperature are fixed, that is, change of origin of heat is always reversible, we can consider temperature of surroundings T_{su} is constant.

Then, we have:

$$\Delta S_{su}=-\int\frac{\delta Q}{T_{su}}=-\frac{Q}{T_{su}}$$

(4) Physical significance of entropy (熵的物理意义)

① Physical significance of entropy

It is often said that entropy is a measure of the molecular disorder of a state; increasing entropy means increasing disorder. That is, the more disordered a system, the larger its entropy. (熵是量度系统无序程度的函数，系统内部物质分子运动越激烈，运动自由度越大，无序程度越大，熵越大。)

② Effecting factors(T, p, β)

i) β (same T, p) : $S_g > S_l > S_s$

ii) T(same p, β) : $S_{\text{high } T} > S_{\text{low } T}$

iii) p (same T, β) : $S_{\text{high } p} < S_{\text{low } p}$

3.4 Calculation of entropy change of system
(系统熵变的计算)

3.4.1 Calculation of ΔS of system in simple pVT process (单纯 pVT 变化过程系统熵变的计算)

The entropy change on going from state 1 to state 2 is given by (3.3.2) as

$$\Delta S = S_2 - S_1 = \int \frac{\delta Q_r}{T} = \int \frac{dU - \delta W}{T} = \int \frac{dU + pdV}{T}$$

Let us calculate ΔS for some common processes. (Note that, as before, all state functions refer to the system and ΔS means ΔS_{syst}. Equation (3.3.2) gives ΔS_{syst} and does not include any entropy changes that may occur in the surroundings.)

(1) Isothermal process

① The perfect gas(abbreviated as pg): $dT=0$, so $dU=0$, and $p=nRT/V$

$$\Delta_T S = \int \frac{pdV}{T} = \int \frac{(nRT/V)\,dV}{T} = nR\int \frac{dV}{V} = nR\ln \frac{V_2}{V_1} = nR\ln \frac{p_1}{p_2} \qquad (3.4.1)$$

(pg, $dT = 0$, reversible, simple pVT process)

② Condensed phase: $dT=0$, so $dU=0$, and $dV=0$

$$\Delta_T S = 0 \qquad (3.4.2)$$

For instance, $1\text{mol pg}(p_1, 298.15\text{K}) \xrightarrow[\text{ir}]{\text{r}} 1\text{mol pg}(p_2, 298.15\text{K})$ $\quad \Delta S' = ?$

According to (3.4.1), $\Delta S' = \Delta_T S = nR\ln \dfrac{V_2}{V_1} = nR\ln \dfrac{p_1}{p_2}$

So, simple pVT process can personally be seen as reversible process.

(2) Isochoric process

$$\Delta_V S = \int \frac{\delta Q_V}{T} = \int \frac{dU}{T} = \int_{T_1}^{T_2} \frac{nC_{V,m}dT}{T} \qquad (3.4.3)$$

$$\Delta_V S = nC_{V,m}\ln \frac{T_2}{T_1}\ (C_{V,m} \text{ is const}) \qquad (3.4.4)$$

(3) Isobaric process

$$\Delta_p S = \int \frac{\delta Q_p}{T} = \int \frac{dH}{T} = \int_{T_1}^{T_2} \frac{nC_{p,m}dT}{T} \qquad (3.4.5)$$

$$\Delta_p S = nC_{p,m}\ln \frac{T_2}{T_1}\ (C_{p,m} \text{ is const}) \qquad (3.4.6)$$

(4) pVT change process

$$\Delta S = \Delta_V S + \Delta_T S = n(C_{V,\mathrm{m}}\ln\frac{T_2}{T_1} + R\ln\frac{V_2}{V_1}) \tag{3.4.7}$$

Or

$$\Delta S = \Delta_p S + \Delta_T S = n\left(C_{p,\mathrm{m}}\ln\frac{T_2}{T_1} + R\ln\frac{p_1}{p_2}\right) \tag{3.4.8}$$

Or

$$\Delta S = \Delta_V S + \Delta_p S = n\left(C_{V,\mathrm{m}}\ln\frac{p_2}{p_1} + C_{p,\mathrm{m}}\ln\frac{V_2}{V_1}\right) \tag{3.4.9}$$

(5) Mixing process of pg

① Isothermal mixing process of different gas

2mol O_2 300K	3mol N_2 300K

dummy plate

For adiabatic vessel $Q=0$

$$dV = 0, \quad W = 0$$

According to the first law of thermodynamics, $\Delta U = Q+W=0$

So that, for the perfect gases, there are no intermolecular interactions either before or after the partition is removed; hence the total thermodynamic energy is unchanged on mixing, and T is unchanged on mixing.

That is, the mixing process of different gas is isothermal process, $dT=0$.

Therefore, $\Delta_{\mathrm{mix}}S = \Delta_T S_{O_2} + \Delta_T S_{N_2}$

$$\Delta_T S_{O_2} = nR\ln\frac{V_2}{V_1} = 2R\ln\frac{2V}{V} = 2R\ln 2$$

$$\Delta_T S_{N_2} = nR\ln\frac{V_2}{V_1} = 3R\ln\frac{2V}{V} = 3R\ln 2$$

$$\Delta_{\mathrm{mix}}S = \Delta_T S_{O_2} + \Delta_T S_{N_2} = 5R\ln 2$$

Since the system can be seen as isolated system, $\Delta S_{\mathrm{iso}} = \Delta_{\mathrm{mix}}S > 0$.

We can deduce the isothermal mixing process of different gas is spontaneous process.

② Isothermal mixing process of same gas

2mol O_2 300K	3mol O_2 300K

dummy plate

For adiabatic vessel, $Q=0$

$$dV = 0, \quad W = 0$$

According to the first law of thermodynamics, $\Delta U=Q+W=0$

So the mixing process of different gas is isothermal process, $dT=0$.

Therefore, $\Delta_{\mathrm{mix}}S = \Delta_T S_{O_2} = 5R\ln\frac{2V}{2V} = 0$

Example 3. 1

2mol diatomic perfect gas whose initial state is $T_1=300\text{K}$, $V_1=50\text{dm}^3$ first heats isochorically to 400 K, then heats isobaricly to 100dm^3. Calculate Q, W, ΔU, ΔH and ΔS of the whole process.

Answer:

$n=2\text{mol}$, $C_{V,\text{m}}=2.5R$, $C_{p,\text{m}}=3.5R$

$$\begin{array}{ccc} 2\text{mol} & & 2\text{mol} & & 2\text{mol} \\ T_1=300\text{K} & \xrightarrow{dV=0} & T_0=400\text{K} & \xrightarrow{dp=0} & T_2=? \\ 50\text{dm}^3,\ p_1 & & 50\text{dm}^3,\ p_0 & & 100\text{dm}^3,\ p_0 \end{array}$$

$p_1=2RT_1/V_1=2\times8.315\times300/(50\times10^{-3})=99780\text{Pa}$

$p_0=p_1T_0/T_1=99780\times400/300=133040\text{Pa}$

$T_2=p_0V_2/(nR)_1=133040\times100\times10^{-3}/(2\times8.315)=800.0\text{K}$

$$\Delta U=nC_{V,\text{m}}(T_2-T_1)=2\times\frac{5}{2}R\times(800.0-300)=20788\text{J}=20.79\text{kJ}$$

$$\Delta H=nC_{p,\text{m}}(T_2-T_1)=2\times\frac{7}{2}R\times(800.0-300)=29103\text{J}=29.10\text{kJ}$$

$W_1=0$; $W_2=-p_{\text{su}}(V_2-V_0)=-133040\times(100-50)\times10^{-3}=-6652\text{J}$

So $W=W_2=-6.652\text{kJ}$

$Q=\Delta U-W=20.79+6.65=27.44\text{kJ}$

$$\Delta S=\Delta_V S+\Delta_p S=nC_{V,\text{m}}\ln\frac{T_0}{T_1}+nC_{p,\text{m}}\ln\frac{T_2}{T_0}$$

$$=2\times\frac{5}{2}R\ln\frac{400}{300}+2\times\frac{7}{2}R\ln\frac{800.0}{400}$$

$$=52.29\text{J}\cdot\text{K}^{-1}$$

Example 3. 2

At 0℃, pg of monatomic molecule expands adiabatically to $p_2=0.1\text{MPa}$, the initial volume is 10dm^3, pressure is 1MPa. Please calculate Q, W, ΔU, ΔH, ΔS and state the reversibility of these processes: (1) $p_{\text{su}}=0.1\text{MPa}$, (2) $p_{\text{su}}=0$.

Answer:

The final sates of these processes are not different.

(1) pg of monatomic molecule: $C_{V,\text{m}}=3/2R$, $C_{p,\text{m}}=5/2R$

For adiabatic process, $Q=0$

$$\Delta U=W$$

$$nC_{V,\text{m}}(T_2-T_1)=-p_{\text{su}}(V_2-V_1)$$

$$n(3R/2)(T_2-T_1)=-p_{\text{su}}nR(T_2/p_2-T_1/p_1)$$

$$3(T_2-T_1)/2=-(T_2-T_1/10),\ T_1=273.2\text{K}$$

$$T_2=174.8\text{K}$$

$$n=\frac{p_1V_1}{RT_1}=4.403\text{mol}$$

Therefore, $W = \Delta U = nC_{V,\ m}(T_2 - T_1) = -5403\text{J}$

$$\Delta H = nC_{p,\ m}(T_2 - T_1) = -9006\text{J}$$

$$\Delta S = nC_{p,\ m}\ln\frac{T_2}{T_1} + nR\ln\frac{p_1}{p_2} = 43.43\text{J}\cdot\text{K}^{-1} > 0$$

So this process is irreversible.

(2) For adiabatic process, $Q=0$

$$p_{su} = 0,\quad W = 0$$

According to the first law of thermodynamics, $\Delta U = Q + W = 0$

So the process is isothermal process, $dT = 0$, $T_2 = T_1 = 273.2\text{K}$.

Therefore, $\Delta H = 0$

$$\Delta S = nR\ln(p_1/p_2) = 84.29\text{J}\cdot\text{K}^{-1} > 0$$

So this process is irreversible.

Example 3.3

There are Ar(g) at 0 ℃, 100 kPa and 500g Cu(s) at 100 ℃ in a 200 dm^3 adiabatic container, calculate T, p of the system when it reaches equilibrium and ΔH, ΔS of the process.

Given: $C_{p,\ m}(\text{Ar, g}) = 20.786\text{J}\cdot\text{mol}^{-1}\cdot\text{K}^{-1}$, $C_{p,\ m}(\text{Cu, s}) = 24.435\text{J}\cdot\text{mol}^{-1}\cdot\text{K}^{-1}$, suppose that both $C_{p,\ m}(\text{Ar, g})$ and $C_{p,\ m}(\text{Cu, s})$ do not change with temperature, Ar(g) obeys the state equation of pg.

Answer:

Initial state	adiabatic and isochoric →	Final state
$n(\text{Cu}) = 7.868\text{mol}$ $n(\text{Ar}) = p_1V_1/RT(\text{Ar})$ $= 8.806\text{mol}$ $p_1 = 100\text{kPa}$ $V_1 = 200\text{dm}^3$ $T_1(\text{Cu}) = 373.15\text{K}$ $T_1(\text{Ar}) = 273.15\text{K}$	adiabatic and isochoric $p_2=?$; $T_2=?$ $\Delta H=?$; $\Delta S=?$	$n(\text{Cu}) = 7.868\text{mol}$ $n(\text{Ar}) = 8.806\text{mol}$ $V_2 = 200\text{dm}^3$ $T_2 = ?$ $p_2 = ?$

For adiabatic and isochoric process, $Q_V = \Delta U = 0$

$$\Delta U(\text{Ar}) = \int_{T_1}^{T_2} n(\text{Ar})C_{V,\ m}(\text{Ar})dT = n(\text{Ar})C_{V,\ m}(\text{Ar})[T_2 - T_1(\text{Ar})]$$

$$\Delta U(Cu) = \Delta H = \int_{T_1}^{T_2} n(\text{Cu})C_{p,\ m}(\text{Cu})dT = n(\text{Cu})C_{p,\ m}(\text{Cu})[T_2 - T_1(\text{Cu})]$$

$$\Delta U(\text{Ar}) + \Delta U(\text{Cu}) = 0$$

$$C_{V,\ m}(\text{Ar}) = C_{p,\ m}(\text{Ar}) - R = 12.471\text{J}\cdot\text{mol}^{-1}\cdot\text{K}^{-1}$$

We can calculate T_2, $T_2 = 336.79\text{K}$

Since the process is isochoric, we can regard the volume of Ar(g) as constant, then:

$$p_2 = \frac{T_2}{T_1(\text{Ar})}p_1 = \frac{336.79}{273.15} \times 100 = 123.30\text{kPa}$$

$$\Delta H = \Delta U + \Delta(pV) = \Delta(pV) = V\Delta p = V[p_2 - p_1(\text{Ar})] = 4.66\text{kJ}$$

$$\Delta S(\text{Ar}) = \int_{T_1(\text{Ar})}^{T_2} n(\text{Ar})C_{V,\text{m}}(\text{Ar})\frac{dT}{T} = n(\text{Ar})C_{V,\text{m}}(\text{Ar})\ln\frac{T_2}{T_1(\text{Ar})} = 23.00\text{J}\cdot\text{K}^{-1}$$

$$\Delta S(\text{Cu}) = \int_{T_1(\text{Cu})}^{T_2} n(\text{Cu})C_{p,\text{m}}(\text{Cu})\frac{dT}{T} = n(\text{Cu})C_{p,\text{m}}(\text{Cu})\ln\frac{T_2}{T_1(\text{Cu})} = -19.71\text{J}\cdot\text{K}^{-1}$$

$$\Delta S = \Delta S(\text{Cu}) + \Delta S(\text{Ar}) = 3.29\text{J}\cdot\text{K}^{-1}$$

Example 3.4

2 mol monatomic molecule pg B at 0 ℃, 100 kPa and 5 mol diatomic molecule pg C at 150℃, 100kPa are mixed adiabatically and isobaricly until equilibrium state, $p=100$kPa. Please calculate W, Q, ΔU and ΔS of the process.

Answer:

Initial state:
- $n(\text{B}) = 2$mol
- $p_\text{B} = 100$kPa
- $T_1(\text{B}) = 273.15$K
- $C_{V,\text{m}}(\text{B}) = 3/2R$
- $n(\text{C}) = 5$mol
- $p_\text{C} = 100$kPa
- $T_1(\text{C}) = 423.15$K
- $C_{V,\text{m}}(\text{C}) = 5/2R$

adiabatic and isobaric, $W=?$; $T_2=?$ $\Delta U=?$; $\Delta S=?$

Final state:
- $n(\text{B}) = 2$mol
- $n(\text{C}) = 5$mol
- $T_2 = ?$
- $p_2 = 100$kPa

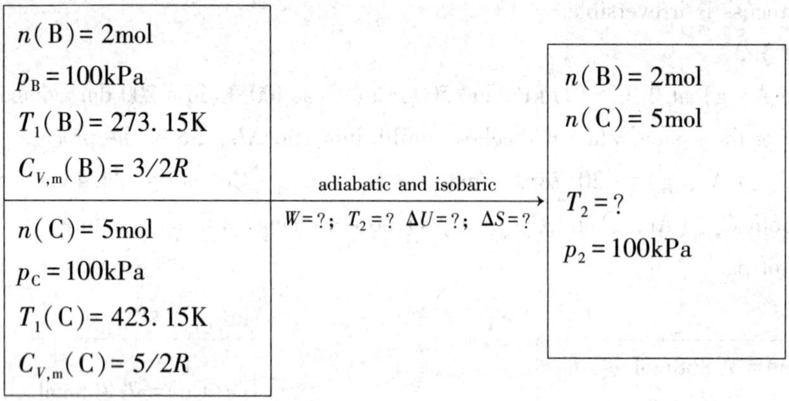

For adiabatic and isobaric process, $Q_p = \Delta H = 0$

$$\Delta H(\text{B}) + \Delta H(\text{C}) = 0$$

$$\Delta H(\text{B}) = n_\text{B}C_{p,\text{m}}(\text{B})[T_2 - T_1(\text{B})]$$

$$\Delta H(\text{C}) = n_\text{C}C_{p,\text{m}}(\text{C})[T_2 - T_1(\text{C})]$$

$$T_2 = \frac{n_\text{B}C_{p,\text{m}}(\text{B})T_1(\text{B}) + n_\text{C}C_{p,\text{m}}(\text{C})T_1(\text{C})}{n_\text{B}C_{p,\text{m}}(\text{B}) + n_\text{C}C_{p,\text{m}}(\text{C})} = 389.82\text{K}$$

$$W = \Delta U = \Delta U(\text{B}) + \Delta U(\text{C})$$
$$= \int_{T_1(\text{B})}^{T_2} n(\text{B})C_{V,\text{m}}(\text{B})dT + \int_{T_1(\text{C})}^{T_2} n(\text{C})C_{V,\text{m}}(\text{C})dT = -554\text{J}$$

According to the temperature and partial pressure of initial and final state, we can calculate ΔS.

$$p_2(\text{B}) = \frac{n_\text{B}}{n_\text{C} + n_\text{B}}p_2 = 28.57\text{kPa}$$

$$p_2(\text{C}) = \frac{n_\text{C}}{n_\text{C} + n_\text{B}}p_2 = 71.43\text{kPa}$$

$$\Delta S = \Delta S(\text{B}) + \Delta S(\text{C})$$

$$\Delta S(\text{B}) = n_\text{B}C_{p,\text{m}}(\text{B})\ln\left(\frac{T_2}{T_1(\text{B})}\right) - n_\text{B}R\ln\left(\frac{p_2(\text{B})}{p_1(\text{B})}\right) = 35.62\text{J}\cdot\text{K}^{-1}$$

$$\Delta S(\mathrm{C}) = n_{\mathrm{C}} C_{p,\mathrm{m}}(\mathrm{C}) \ln\left(\frac{T_2}{T_1(\mathrm{C})}\right) - n_{\mathrm{C}} R \ln\left(\frac{p_2(\mathrm{C})}{p_1(\mathrm{C})}\right) = 2.05\mathrm{J} \cdot \mathrm{K}^{-1}$$

$$\Delta S = 37.67\mathrm{J} \cdot \mathrm{K}^{-1}$$

3.4.2 Calculation of ΔS in phase transformation process (相变化过程熵变的计算)

If we want to calculate ΔS in phase transformation process, we must determine whether the process is reversible phase transformation process.

The relation of functions between temperature and pressure is determinated at phase equilibrium of pure materials. If pressure is fixed, temperature is fixed, or vice versa.

(1) Calculation of ΔS in reversible phase transformation process

$$\Delta S = \int \frac{\delta Q_{\mathrm{r}}}{T} = \int \frac{\delta Q_p}{T} = \int \frac{\mathrm{d}H}{T} = \frac{\Delta H}{T} = \frac{n\Delta_{相变} H_{\mathrm{m}}}{T_{相变}} \tag{3.4.10}$$

(2) Calculation of ΔS in irreversible phase transformation process

If we want to calculate ΔS in irreversible phase transformation process, we must design a process, use the reversible path for this phase transformation.

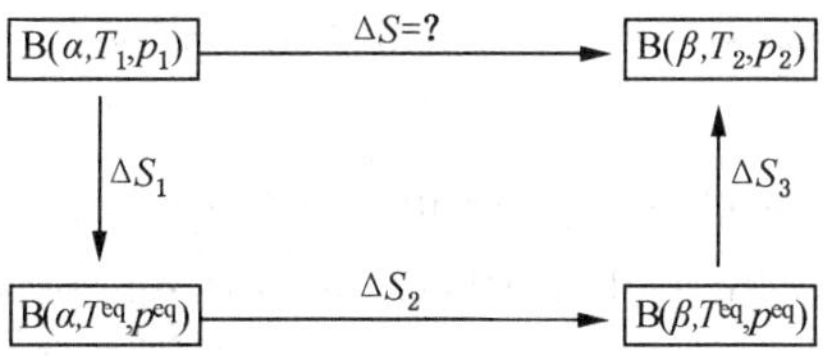

Then: $\Delta S = \Delta S_1 + \Delta S_2 + \Delta S_3$

ΔS for the irreversible phase transformation equals the sum of the entropy changes for the three reversible steps, since the irreversible process and the reversible process each start from the same state and each end at the same state.

Example 3.5

The normal boiling point of H_2O is 100℃, $\Delta_{\mathrm{vap}} H_{\mathrm{m}} = 40.67\mathrm{kJ} \cdot \mathrm{mol}^{-1}$, $C_{p,\mathrm{m}}(\mathrm{l}) = 75.20\mathrm{J} \cdot \mathrm{mol}^{-1} \cdot \mathrm{K}^{-1}$, $C_{p,\mathrm{m}}(\mathrm{g}) = 33.57\mathrm{J} \cdot \mathrm{mol}^{-1} \cdot \mathrm{K}^{-1}$, $C_{p,\mathrm{m}}$ and $\Delta_{\mathrm{vap}} H_{\mathrm{m}}$ are constants. Please determine whether the following process can actually take place at constant T and p: 1mol H_2O(l, 90℃, 101.325kPa) → 1mol H_2O (g, 90℃, 101.325kPa).

***Answer*:**

This process is irreversible process *at constant T and p*. We must design reversible process to calculate ΔS.

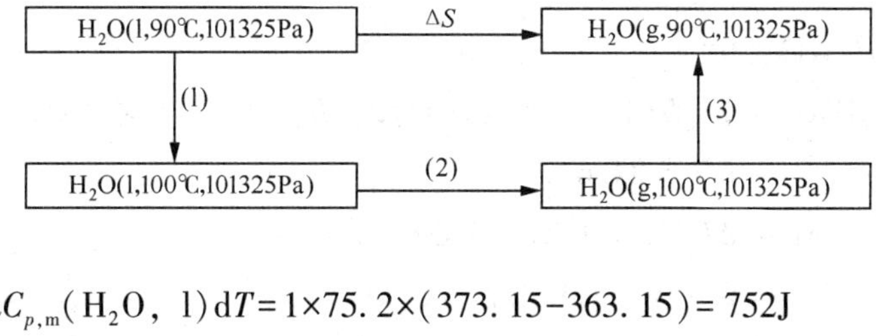

$$\Delta H_1 = \int_{333.15\mathrm{K}}^{373.15\mathrm{K}} nC_{p,\mathrm{m}}(\mathrm{H_2O},\ \mathrm{l})\,\mathrm{d}T = 1 \times 75.2 \times (373.15 - 363.15) = 752\mathrm{J}$$

$$\Delta H_2 = n \cdot \Delta_{\mathrm{vap}} H_{\mathrm{m}}(\mathrm{H_2O}) = 1 \times 40.67 \times 10^3 = 40670\mathrm{J}$$

$$\Delta H_3 = \int_{373.15\mathrm{K}}^{333.15\mathrm{K}} nC_{p,\mathrm{m}}(\mathrm{H_2O},\ \mathrm{g})\mathrm{d}T = 1\times33.57\times(363.15-373.15) = -335.7\mathrm{J}$$

Therefore, $\Delta H = \Delta H_1 + \Delta H_2 + \Delta H_3 = 41.09\mathrm{kJ}$

$Q_p = \Delta H = 41.09\mathrm{kJ}$

$$\Delta S_1 = nC_{p,\mathrm{m}}(\mathrm{H_2O},\ \mathrm{l})\ln\frac{T_2}{T_1} = 1\times75.2\times\ln\frac{373.15}{363.15} = 2.04\mathrm{J}\cdot\mathrm{K}^{-1}$$

$$\Delta S_2 = \frac{\Delta H_2}{T_2} = \frac{40670}{373.15} = 109.0\mathrm{J}\cdot\mathrm{K}^{-1}$$

$$\Delta S_3 = nC_{p,\mathrm{m}}(\mathrm{H_2O},\ \mathrm{g})\ln\frac{T_2}{T_1} = 1\times33.57\times\ln\frac{363.15}{373.15} = -0.91\mathrm{J}\cdot\mathrm{K}^{-1}$$

Therefore, $\Delta S = \Delta S_1 + \Delta S_2 + \Delta S_3 = 110.1\mathrm{J}\cdot\mathrm{K}^{-1}$

$$\Delta S_{\mathrm{su}} = \frac{-Q_p}{T} = \frac{-41090}{363.15} = -113.1\mathrm{J}\cdot\mathrm{K}^{-1}$$

Therefore, $\Delta S_{\mathrm{iso}} = \Delta S + \Delta S_{\mathrm{su}} = -3.0\mathrm{J}\cdot\mathrm{K}^{-1}$

For $\Delta S_{\mathrm{iso}} < 0$, the above process can not actually take place.

Example 3.6

There are 1mol $N_2(g)$ and a glass bottle with 4mol $H_2O(l)$ in a cylinder with a piston. The cylinder was placed in a thermostatic bath of 100℃, the external pressure is remained 150kPa. Now the bottle is broken and the water vaporized to equilibrium, calculate Q, W, ΔU, ΔH, ΔS of the process and ΔS_{iso}.

Given: at 100℃, the saturated pressure of water is 101.325 kPa, and $\Delta_{\mathrm{vap}}H_{\mathrm{m}} = 40.668\mathrm{kJ}\cdot\mathrm{mol}^{-1}$ in the condition.

***Answer*:**

The external pressure of system is 150kPa, which is larger than the saturated pressure of water at 100℃. Therefore, water can vaporize when the partial pressure of steam is smaller than the saturated pressure of water at 100℃ because of existence of $N_2(g)$. Since $n(\mathrm{H_2O},\ \mathrm{l})$ is much more, part of water will vaporize until the steam is saturated.

The vaporized water is:

$$n(\mathrm{H_2O},\ \mathrm{g}) = \frac{p(\mathrm{H_2O},\ \mathrm{g})}{p(\mathrm{N_2})}n(\mathrm{N_2}) = \frac{101.325}{150-101.325}\times1 = 2.082\mathrm{mol}$$

$$Q_p = \Delta H = \Delta H(\mathrm{N_2}) + \Delta H(\mathrm{H_2O})$$

$$\Delta H(\mathrm{N_2}) = 0$$

$$\Delta H(\mathrm{H_2O}) = \Delta_{\mathrm{vap}}H(\mathrm{H_2O}) = n(\mathrm{H_2O})\Delta_{\mathrm{vap}}H_{\mathrm{m}}(\mathrm{H_2O}) = 84.671\mathrm{kJ}$$

$$Q_p = \Delta H = 84.671\mathrm{kJ}$$

$$\begin{aligned}\Delta U &= \Delta U(\mathrm{N_2}) + \Delta U(\mathrm{H_2O})\\ &= 0 + \Delta_{\mathrm{vap}}H(\mathrm{H_2O}) - \Delta(pV)\\ &= n(\mathrm{H_2O})\Delta_{\mathrm{vap}}H_{\mathrm{m}}(\mathrm{H_2O}) - n(\mathrm{H_2O},\ \mathrm{g})RT\\ &= 78.211\mathrm{kJ}\end{aligned}$$

$$\Delta S = \Delta S(N_2) + \Delta S(H_2O)$$

$$\Delta S(N_2) = n(N_2)R\ln\left(\frac{p_1(N_2)}{p_2(N_2)}\right) = 9.36\text{J}\cdot\text{K}^{-1}$$

$$\Delta S(H_2O) = \frac{\Delta H(H_2O)}{T} = \frac{84671}{373.15} = 226.91\text{J}\cdot\text{K}^{-1}$$

$$\Delta S = \Delta S(N_2) + \Delta S(H_2O) = 236.27\text{J}\cdot\text{K}^{-1}$$

ΔS_{su} can be calculated by: $\Delta S_{su} = \frac{Q_{su}}{T_{su}} = -\frac{Q_p}{T_{su}} = -226.91\text{J}\cdot\text{K}^{-1}$

ΔS of isolated system is:

$$\Delta S_{iso} = \Delta S + \Delta S_{su} = 236.27 - 226.91 = 9.36\text{J}\cdot\text{K}^{-1}$$

$\Delta S_{iso} > 0$, so the process is irreversible.

3.5 The Third law of thermodynamics and calculation of the entropy change of reaction (热力学第三定律和化学变化过程熵变的计算)

3.5.1 The Third law of thermodynamics (热力学第三定律)

(1) Statement of the third law of thermodynamics(热力学第三定律的语言表述)

① Nernst saying (Nernst heat theorem) (能斯特热定理)

The entropy change of isothermal process in condensed system approaches zero as the temperature approaches zero.

② Planck saying (普朗克说法)

The entropy become zero for pure materials in condensed phase at T=0K.

③ Corrected Planck saying (修正的普朗克说法)

The entropy become zero for all perfect crystalline materials at T=0K.

Perfect crystal(完美晶体)—a regular array of individual particles.

例如，NO 分子晶体中分子的规则排列应为 NO NO NO NO……，但若有的分子反向排列，成为 NO NO ON……，会使熵增大。前一种排列的晶体为完美晶体，后一种不规则排列不是完美晶体。

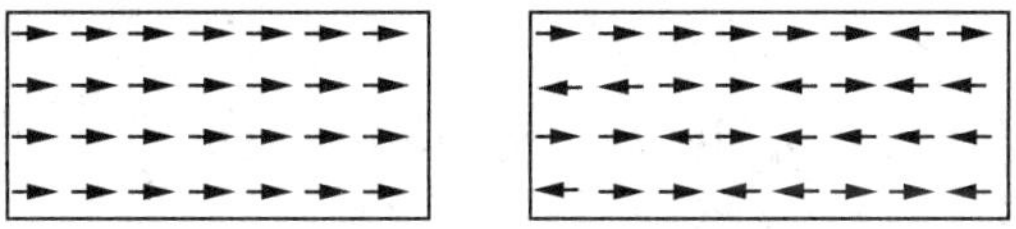

完美晶体 非完美晶体

(2) Mathematical expression of the third law of thermodynamics(热力学第三定律的数学表达式)

① Nernst saying (Nernst heat theorem) (能斯特热定理)

$$\lim_{T\to 0\text{K}} \Delta_T S(\text{凝聚相}) = 0$$

② Planck saying (普朗克说法)

$$S^*(\text{凝聚相}, 0\text{K}) = 0$$

③ Corrected Planck saying (修正的普朗克说法)

$$S^*(\text{完美晶体}, 0\text{K}) = 0$$

3.5.2 Conventional molar entropy and standard molar entropy(规定摩尔熵和标准摩尔熵)

(1) Conventional molar entropy(规定摩尔熵)

According to the third law of thermodynamics, $S^*(\text{完美晶体}, 0\text{K}) = 0$, we can calculate $S_m(B, T)$ of one state. $S_m(B, T)$ is the **conventional molar entropy.**

$$\Delta S = S_m(B, T) - S_m(B, 0\text{K}) = \int_{0\text{K}}^{T} \frac{\delta Q_{r,m}}{T}$$

$$S_m(B, T) = \int_{0\text{K}}^{T} \frac{\delta Q_{r,m}}{T} \qquad (3.5.1)$$

(2) Standard molar entropy(标准摩尔熵)

Standard molar entropy(标准摩尔熵)—conventional molar entropy in standard state ($p^\ominus$ = 100kPa), denoted by $S_m^\ominus(B, \beta, T)$.

$$S_m^\ominus(B, \beta, T) = \int_{0\text{K}}^{T} \frac{\delta Q_{p,m}}{T} = \int_{0\text{K}}^{T} \frac{dH_m^\ominus}{T} = \int_{0\text{K}}^{T} \frac{C_{p,m}(B)\,dT}{T} \qquad (3.5.2)$$

In order to calculate $S_m^\ominus(B, \beta, T)$, we must use Debye-T^3 formula. Debye's statistical-mechanical theory of solids and experimental data show that specific heats of nonmetallic solids at very low temperature (0~15K) have the form

$$C_{p,m} \approx C_{V,m} = aT^3 \qquad (3.5.3)$$

Where a is a constant characteristic of the substance.

For metals, a statistical-mechanical treatment and experimental data show that at very low temperatures

$$C_{p,m} \approx C_{V,m} = aT^3 + bT \qquad (3.5.4)$$

Where a and b are constants. One uses measured values of $C_{p,m}$ at very low temperatures to determine the constants in (3.5.3) and (3.5.4). Then one uses (3.5.3) and (3.5.4) to extrapolate $C_{p,m}$ to T=0K.

If we choose the initial temperature to be 298.15K, the calculation of $S_m^\ominus(B, \beta, T)$ will be easy.

$$S_m^\ominus(B, \beta, T) - S_m^\ominus(B, \beta, 298.15\text{K}) = \int_{298.15\text{K}}^{T} \frac{C_{p,m}(B)\,dT}{T}$$

That is $$S_m^\ominus(B, \beta, T) = S_m^\ominus(B, \beta, 298.15\text{K}) + \int_{298.15\text{K}}^{T} \frac{C_{p,m}(B)\,dT}{T} \qquad (3.5.5)$$

(simple pVT change)

We can get $S_m^\ominus(B, \beta, 298.15\text{K})$ and $C_{p,m}(B)$ from appendix IX.

3.5.3 Calculation of $\Delta_r S_m^\ominus(T)$(标准摩尔反应熵的计算)

Consider reaction: $0 = \sum_B \nu_B B$

$$\Delta_r S_m^\ominus(T) = \sum_B \nu_B S_m^\ominus(B, \beta, T) \qquad (3.5.6)$$

When $T=298.15\text{K}$, $\Delta_r S_m^\ominus(298.15\text{K})=\sum_B \nu_B S_m^\ominus(\text{B},\ \beta,\ 298.15\text{K})$

So that one can calculate $\Delta_r S_m^\ominus(298.15\text{K})$ for a chemical reaction from the tabulated standard entropy values.

For reaction: $a\text{A}+b\text{B} = y\text{Y}+z\text{Z}$

$$\Delta_r S_m^\ominus(T)=yS_m^\ominus(\text{Y},\ \beta,\ T)+zS_m^\ominus(\text{Z},\ \beta,\ T)-aS_m^\ominus(\text{A},\ \beta,\ T)-bS_m^\ominus(\text{B},\ \beta,\ T)$$

Since $$S_m^\ominus(\text{B},\ \beta,\ T)=S_m^\ominus(\text{B},\ \beta,\ 298.15\text{K})+\int_{298.15\text{K}}^{T}\frac{C_{p,m}(\text{B})\text{d}T}{T}$$

$$\Delta_r S_m^\ominus(T)=\Delta_r S_m^\ominus(298.15\text{K})+\int_{298.15\text{K}}^{T}\frac{\sum \nu_B C_{p,m}(\text{B})\text{d}T}{T} \quad (3.5.7)$$

3.6 Helmholtz function and Gibbs function (亥姆霍兹函数和吉布斯函数)

The entropy criterion of equilibrium can be used to judge the reversibility of process. We must calculate the entropy change both of system and of surroundings if we want to use the entropy criterion of equilibrium.

Most chemical reactions proceed in the condition of $W'=0$ and $\text{d}T=0$, $\text{d}V=0$ or $\text{d}T=0$, $\text{d}p=0$. We can get two new criterions from the entropy criterion of equilibrium in these two conditions to avoid calculating the entropy change of surroundings. And we have two new state functions, that is, Helmholtz function and Gibbs function.

3.6.1 Helmholtz function (亥姆霍兹函数)

According to Clausius inequality, $\Delta S \geqslant \int_A^B \frac{\delta Q}{T_{su}}$

For isothermal process $$\Delta S \geqslant \frac{Q}{T}$$

$$T\Delta S \geqslant Q,\quad Q=\Delta U - W$$

$$T(S_2-S_1) \geqslant \Delta U - W$$

So $$TS_2 - TS_1 = \Delta(TS) \geqslant \Delta U - W$$

That is $$\Delta(U-TS) \leqslant W \quad (3.6.1)$$

Definite $A \overset{\text{def}}{=\!=} U - TS$, A is called Helmholtz function or Helmholtz free energy.

Comment:

① Helmholtz function is a state function.

② Helmholtz function is an extensive property.

③ Unit of Helmholtz function is J.

④ Absolute value of Helmholtz function can not be known, but ΔA can be calculated.

3.6.2 Helmholtz function criterion (亥姆霍兹函数判据)

We can illustrate Helmholtz function criterion by combining the definition of Helmholtz function and formula (3.6.1)

$$\Delta A_T \leqslant W \quad \begin{matrix} <, \text{ ir} \\ =, \text{ r} \end{matrix}$$

$dV=0$, $W=0$, then $\qquad \Delta A_{T,V} \leqslant W'$

If $W'=0$, $$\Delta A_{T,V} \leqslant 0 (dT=0,\ dV=0,\ W'=0) \quad \begin{matrix} <, \text{ ir} \\ =, \text{ r} \end{matrix} \tag{3.6.2}$$

Or $$dA_{T,V} \leqslant 0 (dT=0,\ dV=0,\ W'=0) \quad \begin{matrix} <, \text{ ir} \\ =, \text{ r} \end{matrix} \tag{3.6.3}$$

(3.6.2) and (3.6.3) are called **Helmholtz function criterion**, that is, during the isothermal, isochoric and no non-volume work process, the spontaneous change always proceeds toward the direction that Helmholtz function A decreases, Helmholtz function A is unchanged and reaches a minimum when the process reaches equilibrium, the process in which Helmholtz function A increases can not happen. (在恒温恒容且非体积功为零的条件下，系统亥姆霍兹函数减小的过程能够自动进行，亥姆霍兹函数不变时处于平衡态，不可能发生亥姆霍兹函数增大的过程。)

3.6.3 Gibbs function (吉布斯函数)

According to Clausius inequality, $\Delta S \geqslant \int_A^B \frac{\delta Q}{T_{su}}$

For isothermal process $\Delta S \geqslant \frac{Q}{T}$

$$T\Delta S \geqslant Q, \quad Q = \Delta U - W$$

$$T(S_2 - S_1) \geqslant \Delta U - W$$

So $$TS_2 - TS_1 = \Delta(TS) \geqslant \Delta U - W$$

For isobaric process, $W = -\int_{V_1}^{V_2} p_{su} dV = -p(V_2 - V_1) = -(p_2V_2 - p_1V_1) = -\Delta(pV)$

$$\Delta(TS) \geqslant \Delta U + \Delta(pV) - W'$$

$$\Delta U + \Delta(pV) - \Delta(TS) = \Delta(U + pV - TS) \leqslant W'$$

Therefore $$\Delta(H - TS) \leqslant W' \tag{3.6.4}$$

Definite $G \overset{\text{def}}{=\!=\!=} H - TS$, G is Gibbs function or Gibbs free energy.

Comment:

① Gibbs function is a state function.

② Gibbs function is an extensive property.

③ Unit of Gibbs function is J.

④ Absolute value of Gibbs function can not be known, but ΔG can be calculated.

3.6.4 Gibbs function criterion (吉布斯函数判据)

We can illustrate Gibbs function criterion by combining the definition of Gibbs function and formula (3.6.4),

$$\Delta G_{T,p} \leqslant W' \quad \begin{matrix} <, \text{ ir} \\ =, \text{ r} \end{matrix}$$

If $W' = 0$ $$\Delta G_{T,p} \leqslant 0 (dT = 0,\ dp = 0,\ W' = 0) \quad \begin{matrix} <, \text{ ir} \\ =, \text{ r} \end{matrix} \tag{3.6.5}$$

Or $$\mathrm{d}G_{T,p} \leqslant 0(\mathrm{d}T=0,\ \mathrm{d}p=0,\ W'=0)\ \begin{matrix} <,\ \text{ir} \\ =,\ \text{r} \end{matrix} \tag{3.6.6}$$

(3.6.5) and (3.6.6) are called **Gibbs function criterion**, that is, during the isothermal, isobaric and no non-volume work process, the spontaneous change always proceeds toward the direction that Gibbs function G decreases, Gibbs function G is unchanged and reaches a minimum when the process reaches equilibrium, the process in which Gibbs function G increases can not happen. (在恒温恒压且非体积功为零的条件下，系统吉布斯函数减小的过程能够自动进行，吉布斯函数不变时处于平衡态，不可能发生吉布斯函数增大的过程。)

Summarizing, we have shown that:

In a closed system capable of doing only p-V work, the constant-T-and-V equilibrium condition is the minimization of the Helmholtz function A, and the constant-T-and-p equilibrium condition is the minimization of the Gibbs function G.

3.7 Calculation of ΔA and ΔG (亥姆霍兹函数变和吉布斯函数变的计算)

If calculate Q, W, ΔU, ΔH, ΔS, ΔA and ΔG, we can use the definition formula to calculate A and G. That is,

According to the definition $A=U-TS$, we can get $\Delta A=\Delta U-\Delta(TS)$

For isothermal process, $\Delta A=\Delta U-T\Delta S$

Similarly, according to the definition $G=H-TS$, we can get $\Delta G=\Delta H-\Delta(TS)$

For isothermal process, $\Delta G=\Delta H-T\Delta S$

3.7.1 Simple pVT change process (单纯 pVT 变化过程)

According to Clausius inequality, $\Delta S \geqslant \int_A^B \frac{\delta Q}{T_{\text{su}}}$

For isothermal process, $\Delta S \geqslant \frac{Q}{T}$

$$T\Delta S \geqslant Q = \Delta U - W$$

$$\Delta(U-TS) \leqslant W$$

That is $$\Delta A_T \leqslant W\ \begin{matrix} <\ \text{ir} \\ =,\ \text{r} \end{matrix}$$

If the process is reversible, $\Delta A_T = W_{\text{r}} = -\int_{V_1}^{V_2} p\mathrm{d}V$

For pg, $$\Delta A_T = -\int_{V_1}^{V_2} \frac{nRT}{V}\mathrm{d}V = -nRT\ln\frac{V_2}{V_1} = nRT\ln\frac{p_2}{p_1} \tag{3.7.1}$$

(closed system, $W'=0$; simple pVT change; pg, $\mathrm{d}T=0$, reversible)

According to the definition $G=H-TS$, we can get, $G=U+pV-TS=A+pV$

$$\mathrm{d}G = \mathrm{d}A + p\mathrm{d}V + V\mathrm{d}p = -p\mathrm{d}V + p\mathrm{d}V + V\mathrm{d}p = V\mathrm{d}p$$

So $$\Delta G = \int \mathrm{d}G = \int_{p_1}^{p_2} V\mathrm{d}p$$

For pg $$\Delta G_T = \int_{p_1}^{p_2} \frac{nRT}{p} \mathrm{d}p = nRT\ln \frac{p_2}{p_1} = - nRT\ln \frac{V_2}{V_1} \quad (3.7.2)$$

(closed system, $W'=0$; simple pVT change; pg, $\mathrm{d}T=0$, reversible)

Conclusion:

$$\Delta G_T = \Delta A_T = W_r = - nRT\ln \frac{V_2}{V_1} = nRT\ln \frac{p_2}{p_1} \quad (3.7.3)$$

(closed system, $W'=0$; simple pVT change; pg, $\mathrm{d}T=0$, reversible)

The change in Gibbs free energy and Helmholtz free energy for a process at constant temperature equals the maximum useful work that can be done by the system on its surroundings.

Example 3.7

$S_m^{\ominus}(298\text{K})$ of 2 mol diatomic perfect gas is 205.1J·mol^{-1}·K^{-1}, the initial state is $T_1=298\text{K}$, $p_1=100\text{kPa}$. It is compressed reversibly to the final state of 200kPa along the path $pT=C$ (constant). Calculate Q, W, ΔU, ΔH, ΔS and ΔG of this process.

Answer:

$p_2T_2=p_1T_1$

$$T_2=\frac{p_1}{p_2}T_1=\frac{100}{200}\times298=149\text{K}$$

$$\Delta U=nC_{V,m}(T_2-T_1)=2\times\frac{5}{2}\times8.315\times(149-298)=-6.195\text{kJ}$$

$$\Delta H=nC_{p,m}(T_2-T_1)=2\times\frac{7}{2}\times8.315\times(149-298)=-8.672\text{kJ}$$

$$V=\frac{nRT}{p}=\frac{nRT}{C/T}=\frac{nRT^2}{C}$$

$$\mathrm{d}V=\frac{nR}{C}2T\mathrm{d}T$$

$$W=-\int p\mathrm{d}V=-\int \frac{C}{T}\mathrm{d}V=-2nR\int \mathrm{d}T=-2\times2\times8.315\times(149-298)=4.956\text{kJ}$$

$$Q=\Delta U-W=-6.194-4.956=-11.15\text{kJ}$$

$$\Delta S=\Delta_T S+\Delta_p S=nR\ln\frac{p_1}{p_2}+nC_{p,m}\ln\frac{T_2}{T_1}$$

$$=2\times8.315\times\ln\frac{100}{200}+2\times\frac{7}{2}\times8.315\times\ln\frac{149}{298}=-51.87\text{J}\cdot\text{K}^{-1}$$

$$S_2=\Delta S+S_1=\Delta S+nS_{m,1}^{\ominus}=-51.87+2\times205.1=358.33\text{J}\cdot\text{K}^{-1}$$

$$\Delta G=\Delta H-\Delta(TS)=\Delta H-(T_2S_2-T_1S_1)=60.2\text{kJ}$$

3.7.2 Phase transformation process (相变化过程)

(1) Reversible phase transformation process ($\mathrm{d}T=0$, $\mathrm{d}p=0$)

According to the definition $G=H-TS$, we can get, $\Delta G=\Delta H-\Delta(TS)$

For isothermal process, $\Delta G=\Delta H-T\Delta S=T\Delta S-T\Delta S=0 \quad (3.7.4)$

If the reversible phase transformation process is:

$$\alpha(\text{condensed phase}) \xrightarrow{T,\ p} \beta(\text{gas})$$

According to $G = A+pV$

$$\Delta A = \Delta G - \Delta(pV) = -\Delta(pV) = -nRT \qquad (3.7.5)$$

(2) Irreversible phase transformation process ($dT=0$, $dp=0$)

We must design process.

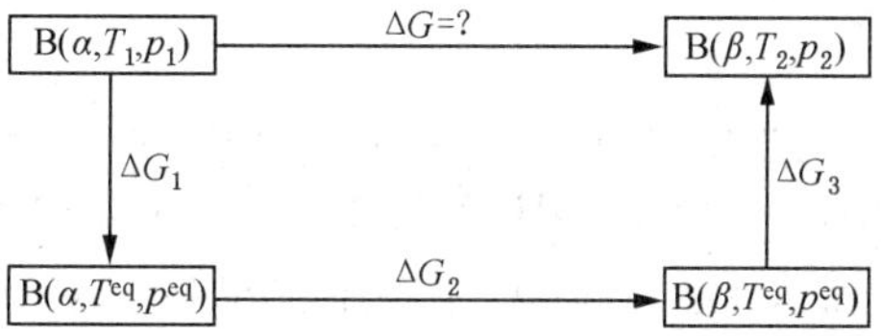

Then: $\Delta G = \Delta G_1 + \Delta G_2 + \Delta G_3$

ΔG for the irreversible phase transformation equals the sum of the Gibbs function changes for the three reversible steps, since the irreversible process and the reversible process each start from the same state and each end at the same state.

3.7.3 Chemical change process (化学变化过程)

$\Delta_r G_m^\ominus(T)$ is the standard molar Gibbs function change for the reaction whose reactants and products are in standard states.

(1) Calculate $\Delta_r G_m^\ominus(T)$ from $\Delta_f G_m^\ominus(B, \beta, T)$

① Definition of standard molar Gibbs function change of formation $\Delta_f G_m^\ominus(B, \beta, T)$ — the Gibbs function changes that occur when unit amount of the compound in its standard state is formed from its elements in their standard states, that is, standard molar Gibbs function change of formation reaction is the standard molar enthalpy of formation of product B.

Comment:

$$\Delta_f G_m^\ominus(\text{steady elementary substance}, T) = 0 \qquad (3.7.6)$$

We can get $\Delta_f G_m^\ominus(B, \beta, 298.15K)$ from appendix Ⅸ.

② Calculate $\Delta_r G_m^\ominus(T)$ from $\Delta_f G_m^\ominus(B, \beta, T)$

$$\Delta_r G_m^\ominus(T) = \sum_B \nu_B \Delta_f G_m^\ominus(B, \beta, T) \qquad (3.7.7)$$

(2) Calculate $\Delta_r G_m^\ominus(T)$ from $\Delta_f H_m^\ominus(B, \beta, T)$, $\Delta_c H_m^\ominus(B, \beta, T)$, $S_m^\ominus(B, \beta, T)$ and $C_{p,m}(B, \beta, T)$

① Calculate $\Delta_r H_m^\ominus(T)$

$$\Delta_r H_m^\ominus(298.15K) = \sum_B \nu_B \Delta_f H_m^\ominus(B, \beta, 298.15K)$$

Or

$$\Delta_r H_m^\ominus(298.15K) = -\sum_B \nu_B \Delta_c H_m^\ominus(B, \beta, 298.15K)$$

According to **Kirchhoff formula**, we can calculate $\Delta_r H_m^\ominus(T)$

$$\Delta_r H_m^\ominus(T) = \Delta_r H_m^\ominus(298.15K) + \int_{298.15K}^{T} \sum \nu_B C_{p,m}(B)\,dT$$

② Calculate $\Delta_r S_m^\ominus(T)$

$$\Delta_r S_m^\ominus(298.15\text{K}) = \sum_B \nu_B S_m^\ominus(\text{B}, \beta, 298.15\text{K})$$

$$\Delta_r S_m^\ominus(T) = \Delta_r S_m^\ominus(298.15\text{K}) + \int_{298.15\text{K}}^{T} \frac{\sum \nu_B C_{p,m}(\text{B})\,dT}{T}$$

③ Calculate $\Delta_r G_m^\ominus(T)$

$$\Delta_r G_m^\ominus(T) = \Delta_r H_m^\ominus(T) - T\Delta_r S_m^\ominus(T) \tag{3.7.8}$$

Example 3.8

Calculate the ΔG of the following process and judge if the process is spontaneous.

1mol H_2O(g, 110℃, 101.325kPa)→H_2O(l, 110℃, 101.325kPa).

Given: the boiling point of water is 100℃ at 101.325kPa, $\Delta_{vap}H_m = 40.67$ kJ · mol^{-1}, $C_{p,m}$(H_2O, l) = 75.2J · K^{-1} · mol^{-1}, $C_{p,m}$(H_2O, g) = 33.57J · K^{-1} · mol^{-1}.

***Answer*:**

The phase transformation process: 1mol H_2O(g, 110℃, 101.325kPa)→H_2O(l, 110℃, 101.325kPa) is irreversible phase transformation process, while 1mol H_2O(g, 100℃, 101.325kPa)→H_2O(l, 100℃, 101.325kPa) is reversible phase transformation process. We can design the process with $m(H_2O)$ = 1mol.

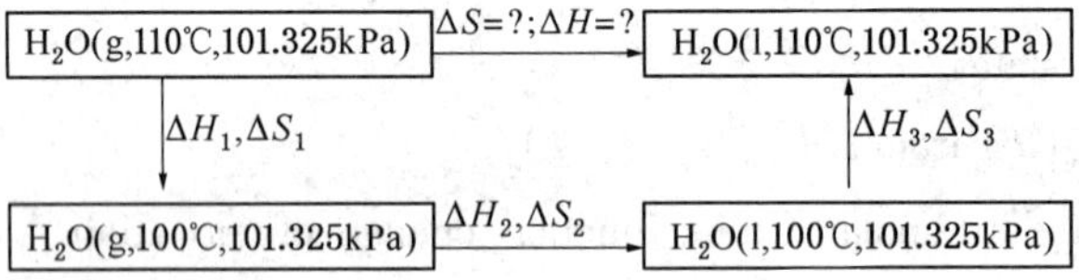

(1) Isobaric and reversible process of steam:

$$\Delta H_1 = nC_{p,m}(H_2O, g)\Delta T = -343.7\text{J}$$

$$\Delta S_1 = nC_{p,m}(H_2O, g)\ln\frac{T_2}{T_1} = 1\times 33.57\times\ln\frac{373.15}{383.15} = -0.910\text{J}\cdot\text{K}^{-1}$$

(2) Reversible phase transformation process:

$$\Delta H_2 = -40670\text{J}$$

$$\Delta S_2 = \frac{n\Delta_{vap}H_m}{T} = \frac{-1\times 40.67\times 10^3}{373.15} = -109.0\text{J}\cdot\text{K}^{-1}$$

(3) Isobaric and reversible process of water:

$$\Delta H_3 = n\,C_{p,m}(H_2O, l)\Delta T = 753\text{J}$$

$$\Delta S_3 = nC_{p,m}(H_2O, l)\ln\frac{T_2}{T_1} = 1\times 75.2\times\ln\frac{383.15}{373.15} = 1.99\text{J}\cdot\text{K}^{-1}$$

$$\Delta S = \Delta S_1 + \Delta S_2 + \Delta S_3 = -107.87\text{J}\cdot\text{K}^{-1}$$

$$\Delta H = \Delta H_1 + \Delta H_2 + \Delta H_3 = -40.23\text{kJ}$$

$$\Delta G = \Delta H - T\Delta S = -40.23 + (110 + 273.15)\times 107.87\times 10^{-3} = 1.08\text{kJ}$$

Since $\Delta G > 0$, the process is not spontaneous, that is, the reverse process:

1mol H_2O(l, 110℃, 101.325kPa)→H_2O(g, 110℃, 101.325kPa) is spontaneous.

3.8 The fundamental equation of thermodynamics (热力学基本方程)

3.8.1 The fundamental equation of thermodynamics (热力学基本方程)

According to the first law of thermodynamics, $dU = \delta Q + \delta W$

For reversible process, the relation $dS = \dfrac{\delta Q_r}{T}$ gives $\delta Q_r = TdS$

If only $p-V$ work is possible, and if the work is done reversibly, then

$$\delta W_r = -pdV$$

Hence, under these conditions, $dU = TdS - pdV$ (3.8.1)

This is the first fundamental equation; it combines the first and second laws.

According to the definition $H = U + pV$

$$dH = dU + pdV + Vdp = TdS - pdV + pdV + Vdp$$

So that $$dH = TdS + Vdp \tag{3.8.2}$$

According to the definition $A = U - TS$

$$dA = dU - TdS - SdT = TdS - pdV - TdS - SdT$$

So that $$dA = -SdT - pdV \tag{3.8.3}$$

According to the definition $G = H - TS$

$$dG = dH - TdS - SdT = TdS + Vdp - TdS - SdT$$

So that $$dG = -SdT + Vdp \tag{3.8.4}$$

(3.8.1), (3.8.2), (3.8.3) and (3.8.4) are **the fundamental equation of thermodynamics.**

Comment about applicable conditions of the fundamental equation of thermodynamics:

① Phase transformation process and chemical change process

closed system, $W' = 0$, reversible

② Simple pVT change process

closed system, $W' = 0$

3.8.2 The relation of characteristic function (特性函数关系式)

Characteristic function—state function that describes property of system.

According to the fundamental equation of thermodynamics (3.8.1), $dU = TdS - pdV$

Put $dV = 0$ and $dS = 0$ respectively, we can get

$$T = \left(\frac{\partial U}{\partial S}\right)_V, \quad -p = \left(\frac{\partial U}{\partial V}\right)_S \tag{3.8.5}$$

According to the fundamental equation of thermodynamics (3.8.2), $dH = TdS + Vdp$

Put $dp = 0$ and $dS = 0$ respectively, we can get

$$T = \left(\frac{\partial H}{\partial S}\right)_p, \quad V = \left(\frac{\partial H}{\partial p}\right)_S \tag{3.8.6}$$

According to the fundamental equation of thermodynamics (3.8.3), $dA = -SdT - pdV$

Put $dV = 0$ and $dT = 0$ respectively, we can get

$$-S=\left(\frac{\partial A}{\partial T}\right)_V,\quad -p=\left(\frac{\partial A}{\partial V}\right)_T \qquad (3.8.7)$$

According to the fundamental equation of thermodynamics (3.8.4), $dG=-SdT+Vdp$

Put $dp=0$ and $dT=0$ respectively, we can get

$$-S=\left(\frac{\partial G}{\partial T}\right)_p,\quad V=\left(\frac{\partial G}{\partial p}\right)_T \qquad (3.8.8)$$

(3.8.5), (3.8.6), (3.8.7) and (3.8.8) are **the relation of characteristic function.**

3.8.3 Gibbs-Helmholtz equation (吉布斯-亥姆霍兹方程)

According to relations of characteristic function (3.8.8), $-S=\left(\frac{\partial G}{\partial T}\right)_p$

We have

$$\left[\frac{\partial(G/T)}{\partial T}\right]_p=\frac{1}{T}\left(\frac{\partial G}{\partial T}\right)_p-\frac{G}{T^2}=-\frac{S}{T}-\frac{G}{T^2}=-\frac{(TS+G)}{T^2}=-\frac{H}{T^2} \qquad (3.8.9)$$

According to relations of characteristic function (3.8.7), $-S=\left(\frac{\partial A}{\partial T}\right)_V$

We have

$$\left[\frac{\partial(A/T)}{\partial T}\right]_V=\frac{1}{T}\left(\frac{\partial A}{\partial T}\right)_V-\frac{A}{T^2}=-\frac{S}{T}-\frac{A}{T^2}=-\frac{(TS+A)}{T^2}=-\frac{U}{T^2} \qquad (3.8.10)$$

(3.8.9) and (3.8.10) are **Gibbs-Helmholtz equation.**

3.8.4 Maxwell's relations (麦克斯韦关系式)

If $z=f(x,y)$, $dz=\left(\frac{\partial z}{\partial x}\right)_y dx+\left(\frac{\partial z}{\partial y}\right)_x dy$

And z has continuous second order differential, then:

$$\left[\frac{\partial}{\partial y}\left(\frac{\partial z}{\partial x}\right)_y\right]_x=\left[\frac{\partial}{\partial x}\left(\frac{\partial z}{\partial y}\right)_x\right]_y$$

That is

$$\frac{\partial^2 z}{\partial x\partial y}=\frac{\partial^2 z}{\partial y\partial x}$$

Apply the result to the fundamental equation of thermodynamics, we have:

$$dU=TdS-pdV$$

$dS=0$ ↙ ↘ $dV=0$

$$\left(\frac{\partial U}{\partial V}\right)_S=-p \qquad \left(\frac{\partial U}{\partial S}\right)_V=T$$

$dV=0$ ↓ 对 S 微分 $\qquad$ $dS=0$ ↓ 对 V 微分

$$\left(\frac{\partial^2 U}{\partial V\partial S}\right)=-\left(\frac{\partial p}{\partial S}\right)_V=\left(\frac{\partial^2 U}{\partial S\partial V}\right)=\left(\frac{\partial T}{\partial V}\right)_S$$

↓

$$-\left(\frac{\partial p}{\partial S}\right)_V=\left(\frac{\partial T}{\partial V}\right)_S$$

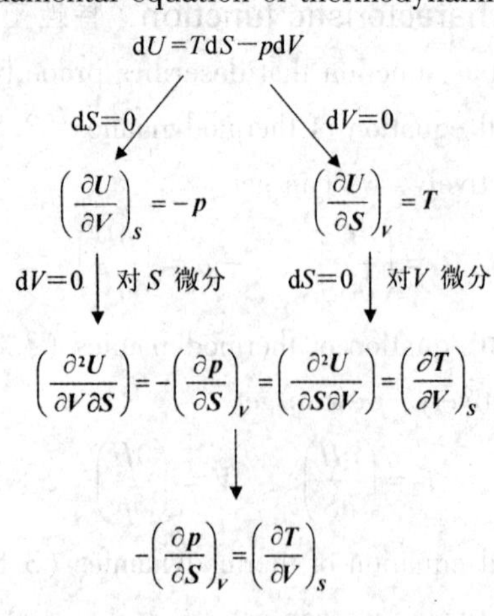

That is, according to the fundamental equation of thermodynamics (3.8.1), $dU = TdS - pdV$, we can get:

$$\left(\frac{\partial T}{\partial V}\right)_S = -\left(\frac{\partial p}{\partial S}\right)_V \quad (3.8.11)$$

According to the fundamental equation of thermodynamics (3.8.2), $dH = TdS + Vdp$

We can get:

$$\left(\frac{\partial T}{\partial p}\right)_S = \left(\frac{\partial V}{\partial S}\right)_p \quad (3.8.12)$$

According to the fundamental equation of thermodynamics (3.8.3), $dA = -SdT - pdV$

We can get:

$$\left(\frac{\partial S}{\partial V}\right)_T = \left(\frac{\partial p}{\partial T}\right)_V \quad (3.8.13)$$

According to the fundamental equation of thermodynamics (3.8.4), $dG = -SdT + Vdp$

We can get:

$$-\left(\frac{\partial S}{\partial p}\right)_T = \left(\frac{\partial V}{\partial T}\right)_p \quad (3.8.14)$$

(3.8.11), (3.8.12), (3.8.13) and (3.8.14) are **Maxwell's relations.**

3.8.5 Thermodynamic equation of state (热力学状态方程)

According to the fundamental equation of thermodynamics, $dU = TdS - pdV$, we can get:

$$\frac{dU}{dV_T} = T\frac{dS}{dV_T} - p$$

That is
$$\left(\frac{\partial U}{\partial V}\right)_T = T\left(\frac{\partial S}{\partial V}\right)_T - p$$

According to Maxwell's relations (3.8.13), $\left(\frac{\partial S}{\partial V}\right)_T = \left(\frac{\partial p}{\partial T}\right)_V$, we have:

$$\left(\frac{\partial U}{\partial V}\right)_T = T\left(\frac{\partial p}{\partial T}\right)_V - p \quad (3.8.15)$$

According to the fundamental equation of thermodynamics, $dH = TdS + Vdp$, we can get:

$$\frac{dH}{dp_T} = T\frac{dS}{dp_T} + V$$

That is
$$\left(\frac{\partial H}{\partial p}\right)_T = T\left(\frac{\partial S}{\partial p}\right)_T + V$$

According to Maxwell's relations (3.8.14), $-\left(\frac{\partial S}{\partial p}\right)_T = \left(\frac{\partial V}{\partial T}\right)_p$, we have:

$$\left(\frac{\partial H}{\partial p}\right)_T = -T\left(\frac{\partial V}{\partial T}\right)_p + V \quad (3.8.16)$$

(3.8.15) and (3.8.16) are **thermodynamic equation of state.**

Example 3.9

Please prove $\left(\frac{\partial H}{\partial V}\right)_T = T\left(\frac{\partial p}{\partial T}\right)_V + V\left(\frac{\partial p}{\partial V}\right)_T$

Answer:

According to the fundamental equation of thermodynamics, $dH=TdS+Vdp$

If $dT=0$, $\left(\frac{\partial H}{\partial V}\right)_T=T\left(\frac{\partial S}{\partial V}\right)_T+V\left(\frac{\partial p}{\partial V}\right)_T$

According to Maxwell's relations, $\left(\frac{\partial S}{\partial V}\right)_T=\left(\frac{\partial p}{\partial T}\right)_V$

We can get: $\left(\frac{\partial H}{\partial V}\right)_T=T\left(\frac{\partial p}{\partial T}\right)_V+V\left(\frac{\partial p}{\partial V}\right)_T$

Example 3.10

Please prove: for pg $\left(\frac{\partial H}{\partial p}\right)_T=0$

Answer:

According to the fundamental equation of thermodynamics, $dH=TdS+Vdp$

If $dT=0$, $\left(\frac{\partial H}{\partial p}\right)_T=T\left(\frac{\partial S}{\partial p}\right)_T+V$

According to Maxwell's relations, $-\left(\frac{\partial S}{\partial p}\right)_T=\left(\frac{\partial V}{\partial T}\right)_p$

We can get: $\left(\frac{\partial H}{\partial p}\right)_T=-T\left(\frac{\partial V}{\partial T}\right)_p+V$

According to the state equation of pg, $V=\frac{nRT}{p}$

$$\left(\frac{\partial V}{\partial T}\right)_p=\frac{nR}{p}=\frac{V}{T}$$

So for pg, $\left(\frac{\partial H}{\partial p}\right)_T=-T\times\frac{V}{T}+V=0$

Example 3.11

Please prove:

(1) $dU=C_VdT+\left\{T\left(\frac{\partial p}{\partial T}\right)_V-p\right\}dV$

(2) perfect gas $\left(\frac{\partial U}{\partial V}\right)_T=0$

(3) van der waals gas $\left(\frac{\partial U}{\partial V}\right)_T=\frac{a}{V^2}$

Answer:

(1) If $U=f(T, V)$, $dU=\left(\frac{\partial U}{\partial T}\right)_V dT+\left(\frac{\partial U}{\partial V}\right)_T dV$

According to the fundamental equation of thermodynamics, $dU=TdS-pdV$

$$\left(\frac{\partial U}{\partial V}\right)_T=T\left(\frac{\partial S}{\partial V}\right)_T-p$$

According to Maxwell's relations, $\left(\frac{\partial S}{\partial V}\right)_T = \left(\frac{\partial p}{\partial T}\right)_V$

We can get: $\left(\frac{\partial U}{\partial V}\right)_T = T\left(\frac{\partial p}{\partial T}\right)_V - p$

Since $\left(\frac{\partial U}{\partial T}\right)_V = C_V$

So, $\mathrm{d}U = C_V\mathrm{d}T + \left\{T\left(\frac{\partial p}{\partial T}\right)_V - p\right\}\mathrm{d}V$

(2) From $\left(\frac{\partial U}{\partial V}\right)_T = T\left(\frac{\partial p}{\partial T}\right)_V - p$

According to the state equation of pg, $p = \frac{nRT}{V}$

$$\left(\frac{\partial p}{\partial T}\right)_V = \frac{nR}{V} = \frac{p}{T}$$

So for pg, $\left(\frac{\partial U}{\partial V}\right)_T = T \times \frac{p}{T} - p = 0$

(3) From $\left(\frac{\partial U}{\partial V}\right)_T = T\left(\frac{\partial p}{\partial T}\right)_V - p$

According to van der waals equation, $p = \frac{nRT}{V - nb} - \frac{a}{V^2}$

$$\left(\frac{\partial p}{\partial T}\right)_V = \frac{nR}{V - nb}$$

So for van der waals gas, $\left(\frac{\partial U}{\partial V}\right)_T = T(\frac{nR}{V - nb}) - p = \frac{a}{V^2} + p - p = \frac{a}{V^2}$

3.9 Clapeyron equation
(克拉佩龙方程)

3.9.1 Condition for phase equilibrium of one-component system (单组分系统的相平衡条件)

For phase equilibrium of one-component system at T, p:

$$\mathrm{B}^*(\alpha) \xrightleftharpoons{T,\ p} \mathrm{B}^*(\beta)$$

If $\mathrm{d}n_\mathrm{B}$ of B is transformed from α phase to β phase,

$$\mathrm{d}G_{T,\ p} = \mathrm{d}n_\mathrm{B}[G_\mathrm{m}^*(\mathrm{B},\ \beta,\ T,\ p) - G_\mathrm{m}^*(\mathrm{B},\ \alpha,\ T,\ p)]$$

According to Gibbs function criterion, $\mathrm{d}G_{T,\ p} \leqslant 0 \quad \begin{matrix} <,\ \mathrm{ir} \\ =,\ \mathrm{r} \end{matrix}$

While two phases are in equilibrium, $\mathrm{d}G_{T,\ p} = 0$

Since $\mathrm{d}n_\mathrm{B} \neq 0$

So $$G_\mathrm{m}^*(\mathrm{B},\ \alpha,\ T,\ p) = G_\mathrm{m}^*(\mathrm{B},\ \beta,\ T,\ p) \tag{3.9.1}$$

(3.9.1) is the **condition for phase equilibrium of one-component system.**

That is, the molar Gibbs function of pure material in two phases must be equal when it established phase equilibrium at T, p.

3.9.2 Clapeyron equation (克拉佩龙方程)

When the one-component system established phase equilibrium at T, p:

$$B^*(\alpha,\ T,\ p) \xrightleftharpoons{eq} B^*(\beta,\ T,\ p)$$

According to the condition for phase equilibrium of one-component system,

$$G_m^*(B,\ \alpha,\ T,\ p) = G_m^*(B,\ \beta,\ T,\ p)$$

If change T and p of the equilibrium system, and establish new phase equilibrium at $T+dT$ and $p+dp$, according to the condition for phase equilibrium of one-component system, we can get: $G_m^*(B,\ \alpha,\ T,\ p) + dG_m^*(\alpha) = G_m^*(B,\ \beta,\ T,\ p) + dG_m^*(\beta)$

Therefore $$dG_m^*(\alpha) = dG_m^*(\beta)$$

According to the fundamental equation of thermodynamics, $dG=-SdT+Vdp$

We can get: $-S_m^*(\alpha)dT + V_m^*(\alpha)dp = -S_m^*(\beta)dT + V_m^*(\beta)dp$

Rewriting it, we have $\dfrac{dp}{dT} = \dfrac{S_m^*(\beta) - S_m^*(\alpha)}{V_m^*(\beta) - V_m^*(\alpha)} = \dfrac{\Delta_\alpha^\beta S_m^*}{\Delta_\alpha^\beta V_m^*}$

For a phase change at equilibrium (a reversible phase change), we have

$$\Delta_\alpha^\beta S_m^* = \frac{\Delta_\alpha^\beta H_m^*}{T}$$

Therefore, $\dfrac{dp}{dT} = \dfrac{\Delta_\alpha^\beta H_m^*}{T\Delta_\alpha^\beta V_m^*}$ (any two phase equilibriums of pure materials) (3.9.2)

Or $\dfrac{dT}{dp} = \dfrac{T\Delta_\alpha^\beta V_m^*}{\Delta_\alpha^\beta H_m^*}$ (any two phase equilibriums of pure materials) (3.9.3)

(3.9.2) and (3.9.3) are **Clapeyron equation.**

Comment on Clapeyron equation:

$\Delta_\alpha^\beta H_m^*$ and $\Delta_\alpha^\beta V_m^*$: $\alpha \rightarrow \beta$, that is, the initial and final states are same.

For instance: $H_2O(l) \rightarrow H_2O(g)$

If $\Delta_\alpha^\beta V_m^* = V_m^*(H_2O,\ g) - V_m^*(H_2O,\ l)$

Then $\Delta_\alpha^\beta H_m^* = \Delta_{vap} H_m^*$

3.9.3 Clausius-Clapeyron equation (克劳休斯-克拉佩龙方程)

(1) Two phase equilibrium of $l \xrightleftharpoons{T,\ p} g$:

Depending on Clapeyron equation, $\dfrac{dp}{dT} = \dfrac{\Delta_\alpha^\beta H_m^*}{T\Delta_\alpha^\beta V_m^*} = \dfrac{\Delta_{vap} H_m^*}{T\ [V_m^*(g) - V_m^*(l)]}$

Two approximate conditions:

① $[V_m^*(g) - V_m^*(l)] \approx V_m^*(g)$

For phase equilibrium between a gas and a liquid (or a solid), $V_m^*(g)$ is much larger than $V_m^*(l)$ unless T is near the critical temperature, in which case the vapor and liquid densities are rather close.

② We assume gas to be the perfect gas, then $pV_{m}^{*}(g)=RT$

With these two approximations, the Clapeyron equation becomes

$$\frac{dp}{dT}=\frac{\Delta_{vap}H_{m}^{*}}{RT^{2}}p$$

That is

$$\frac{d\ln\{p\}}{dT}=\frac{\Delta_{vap}H_{m}^{*}}{RT^{2}} \tag{3.9.4}$$

(3.9.4) is the differential formula of **Clausius-Clapeyron equation.**

(2) Clausius-Clapeyron integral equation

① Indefinite integral formula

Depending on the differential formula of Clausius-Clapeyron equation, we have

$$\int d\ln\{p\}=\int\frac{\Delta_{vap}H_{m}^{*}}{RT^{2}}dT$$

If the temperature interval is not large and if they are not near the critical temperature, $\Delta_{vap}H_{m}^{*}$ will vary only slightly, so that

The third approximate condition: $\Delta_{vap}H_{m}^{*}$ is constant.

Then

$$\ln\{p\}=-\frac{\Delta_{vap}H_{m}^{*}}{RT}+B \tag{3.9.5}$$

(3.9.5) is the indefinite integral formula of **Clausius-Clapeyron equation.**

According to (3.9.5), a plot of $\ln\{p\}$ dependence of $\{1/T\}$ is a straight line.

$$\text{the slope}=-\Delta_{vap}H_{m}^{*}/R$$

$$\text{the intercept}=B$$

② Definite integral formula

Depending on the differential formula of Clausius-Clapeyron equation, we have

$$\int_{p_1^*}^{p_2^*}d\ln\{p\}=\int_{T_1}^{T_2}\frac{\Delta_{vap}H_{m}^{*}}{RT^{2}}dT$$

Then, $\ln\frac{p_2^*}{p_1^*}=-\frac{\Delta_{vap}H_{m}^{*}}{R}\left(\frac{1}{T_2}-\frac{1}{T_1}\right)$ (liquid-gas or solid-gas equilibrium of pure materials)

(3.9.6)

(3.9.6) is the definite integral formula of **Clausius-Clapeyron equation**.

Conclusion:

① Comparison of Clausius-Clapeyron equation and Clapeyron equation :

the sharpness of Clausius-Clapeyron equation is low.

② Comparison of Clausius-Clapeyron equation and Clapeyron equation :

the scope of application of Clausius-Clapeyron equation is narrow.

(3) Trouton's rule (特鲁顿规则)

A useful rule of thumb for relating enthalpies and entropies of liquids and gases is **Trouton's rule**, which states that $\Delta_{vap}S_{m}^{*}$ for vaporization of a liquid at its normal boiling point is roughly $88\,J\cdot K^{-1}\cdot mol^{-1}$, that is,

For non-polar liquid, $\dfrac{\Delta_{vap}H_m^*}{T_b^*} = \Delta_{vap}S_m^* = 88\text{J} \cdot \text{K}^{-1} \cdot \text{mol}^{-1}$

Where T_b^* is normal boiling point of pure liquid.

The rule fails for highly polar liquids and for liquids boiling below 150K.

(4) The pressure-dependence of the saturated vapor pressure of liquid or solid

From the deduction of thermodynamics, we can get:

$$\frac{dp^*(g)}{dp(l)} = \frac{V_m^*(l)}{V_m^*(g)} > 0$$

Therefore, if external pressure $dp(l)$ increases, the saturated pressure $dp^*(g)$ increases, but the increment is not large.

So, the influence of external pressure to the saturated vapor pressure may not be considered.

Example 3. 12

The saturated vapor pressure of chloroform is 21. 3kPa at $t=20℃$, and 71. 4kPa at $t=50℃$. Please calculate $\Delta_{vap}H_m$ of chloroform.

Answer:

According to the definite integral formula of Clausius-Clapeyron equation,

$$\ln\frac{p_2}{p_1} = -\frac{\Delta_{vap}H_m}{R}\left(\frac{1}{T_2} - \frac{1}{T_1}\right) = \frac{\Delta_{vap}H_m(T_2 - T_1)}{RT_2T_1}$$

So,

$$\Delta_{vap}H_m = \frac{RT_1T_2}{(T_2 - T_1)}\ln\frac{p_2}{p_1} = \left(\frac{8.315 \times 293.2 \times 323.2}{30}\ln\frac{71.4}{21.3}\right)\text{J} \cdot \text{mol}^{-1} = 31.8\text{kJ} \cdot \text{mol}^{-1}$$

Example 3. 13

The saturated vapor pressure of solid benzene is 3. 27kPa at $t=0℃$, and 12. 30kPa at $t=20℃$. The saturated vapor pressure of liquid benzene is 10. 02 kPa at $t=20℃$, $\Delta_{vap}H_m$ of liquid benzene is 34. 17 kJ · mol^{-1}.

Find: (1) the saturated vapor pressure of liquid benzene at $t=30℃$;

(2) $\Delta_{sub}H_m$ of solid benzene;

(3) $\Delta_{fus}H_m$ of solid benzene.

Answer:

(1) According to the definite integral formula of Clausius-Clapeyron equation,

$$\ln\frac{p(30℃)}{p(20℃)} = -\frac{\Delta_{vap}H_m}{R}\left(\frac{1}{T_2} - \frac{1}{T_1}\right)$$

$$= \frac{\Delta_{vap}H_m}{R}\frac{T_2 - T_1}{T_1T_2} = \frac{34.17\times10^3(303.2-293.2)}{8.315\times293.2\times303.2} = 0.4623$$

Example 3. 13

$p(30℃) = 15.90\text{kPa}$

(2) According to the definite integral formula of Clausius-Clapeyron equation,

$$\ln\frac{p_2}{p_1} = -\frac{\Delta_{sub}H_m}{R}\left(\frac{1}{T_2} - \frac{1}{T_1}\right) = \frac{\Delta_{sub}H_m}{R}\frac{(T_2 - T_1)}{T_1T_2}$$

So, $\Delta_{sub}H_m = \left(\frac{8.315 \times 273.2 \times 293.2}{293.2 - 273.2}\ln\frac{12.30 \times 10^3}{3.27 \times 10^3}\right) J \cdot mol^{-1}$

$= 44.12kJ \cdot mol^{-1}$

(3) $\Delta_{sub}H_m = \Delta_{fus}H_m + \Delta_{vap}H_m$

So, $\Delta_{fus}H_m = \Delta_{sub}H_m - \Delta_{vap}H_m = (44.12-34.17)kJ \cdot mol^{-1} = 9.95kJ \cdot mol^{-1}$

Supplementary Examples of Chapter 3

EXERCISES

(习题)

1. 1mol diatomic perfect gas whose initial state is $T_1 = 300K$, $p_1 = 200kPa$ changes to the final state of 300K, 100kPa through the following different paths. Calculate Q and ΔS of each step and path.

(1) Expands isothermally and reversibly;

(2) First cooled isochoricly to decrease the pressure to 100kPa, then heated isobaricly to T_2;

(3) Expands adiabaticly and reversibly to decrease the pressure to 100kPa, then heated isobaricly to T_2.

Answer: (1) 1.729kJ, 5.76J/K; (2) $Q_1 = -3.118kJ$, $\Delta S_1 = -14.41J \cdot K^{-1}$, $Q_2 = 4.365kJ$, $\Delta S_2 = 20.17J \cdot K^{-1}$, $Q = 1.247kJ$, $\Delta S = 5.76J \cdot K^{-1}$; (3) $Q_1 = 0kJ$, $\Delta S_1 = 0J \cdot K^{-1}$, $Q_2 = Q = 1.57kJ$, $\Delta S = \Delta S_2 = 5.76J \cdot K^{-1}$

2. At 300K, 1mol perfect gas whose initial state is $p_1 = 100kPa$ changes to the final state through the following different processes. Calculate Q, ΔS and ΔS_{iso} of each process.

(1) Expands reversibly to the final pressure 50kPa;

(2) Expands irreversibly and against constant external pressure of 50 kPa to equilibrium state;

(3) Expands freely to twice of it's initial volume.

Answer: (1) 1.729kJ, $5.763J \cdot K^{-1}$, $0J \cdot K^{-1}$; (2) 1.247kJ, $5.763J \cdot K^{-1}$, $1.606J \cdot K^{-1}$; (3) 0kJ, $5.763J \cdot K^{-1}$, $5.763J \cdot K^{-1}$

3. 4mol monatomic perfect gas whose initial state is $T_1 = 750K$, $p_1 = 150kPa$ is cooled isochoricly to 50kPa, then compressed isothermally and reversibly to 100kPa. Calculate Q, W, ΔU, ΔH and ΔS of the whole process.

Answer: −30.71kJ, 5.763kJ, −24.94kJ, −41.57kJ, −77.86J/K

4. 5mol monatomic perfect gas whose initial state is $T_1 = 300K$, $p_1 = 50kPa$ is compressed adiabatically and reversibly to 100 kPa, then cooled isobaricly to $85dm^3$. Calculate Q, W, ΔU, ΔH and ΔS of the whole process.

Answer: −19.892kJ, 13.935kJ, −5.958kJ, −9.930kJ, $-68.66J \cdot K^{-1}$

5. At 101.325kPa, the normal boiling point of CH_3OH is 64.65℃, $\Delta_{vap}H_m = 35.32kJ \cdot mol^{-1}$ in the condition. Calculate Q, W, ΔU, ΔH, ΔS, ΔA and ΔG of the following process:

1kg CH_3OH (l, 64.65℃, 101.325kPa) → CH_3OH (g, 64.65℃, 101.325kPa)

Answer: 1102.65kJ, −87.59kJ, 1014.65kJ, 1102.65kJ, 3.263kJ/K, −87.59kJ, 0

6. The normal boiling point of water is 100℃, $C_{p,m}(H_2O, l) = 75.20J \cdot mol^{-1} \cdot K^{-1}$, $\Delta_{vap}H_m = 40.67kJ \cdot mol^{-1}$, $C_{p,m}(H_2O, g) = 33.57J \cdot mol^{-1} \cdot K^{-1}$. Calculate Q, W, ΔU, ΔH, ΔS, ΔA and ΔG of the following process:

1mol H_2O(l, 60℃, 101325Pa) → 1mol H_2O(g, 60℃, 101325Pa)

Answer: 42.34kJ, −2.77kJ, 39.57kJ, 42.34kJ, 113.7J/K, 1.69kJ, 4.46kJ

7. ΔS of the process 1mol C_6H_6(l, −5℃, 100kPa) →1mol C_6H_6(s, −5℃, 100kPa) is −35.5J/K, the saturated vapor pressure of solid benzene is 2280Pa at −5℃, $\Delta_{fus}H_m = 9874 J \cdot mol^{-1}$. Calculate the saturated vapor pressure of the supercooled liquid benzene at −5℃.

Answer: 2680Pa

8. 4 mol perfect gas whose initial state is $T_1 = 300K$, $p_1 = 100kPa$ is heated isobaricly to 600K. Calculate Q, W, ΔU, ΔH, ΔS, ΔA and ΔG of the process.

Given: the perfect gas $S^\ominus_m(300K) = 150.0J \cdot K^{-1} \cdot mol^{-1}$, $C_{p,m} = 30.00J \cdot K^{-1} \cdot mol^{-1}$.

Answer: 36.0kJ, −9.978kJ, 26.02kJ, 36.0kJ, 83.18J·K^{-1}, −203.9kJ, −193.9kJ

9. 8mol diatomic perfect gas whose initial state is $T_1 = 400K$, $p_1 = 0.20MPa$ changes to the appointed final state through the following three different processes, calculate Q, W, ΔU, ΔH, ΔS, ΔA and ΔG of each process.

(1) Expands isothermally and reversibly to 0.10MPa;

(2) Expands freely to 0.10MPa;

(3) Expands isothermally and against constant external pressure of 0.10MPa to equilibrium.

Answer:

Process	W/kJ	Q/kJ	ΔU/kJ	ΔH/kJ	ΔS/J·K^{-1}	ΔA/kJ	ΔG/kJ
(1)	−18.41	18.41	0	0	46.10	−18.44	−18.44
(2)	0	0	0	0	46.10	−18.44	−18.44
(3)	−13.30	−13.30	0	0	46.10	−18.44	−18.44

10. There is a small glass bottle filled with 0.1mol liquid ethyl ether in an isochoricly sealed vacuum vessel of $10dm^3$, the vacuum vessel is placed in a thermostatic bath of 35℃. 35℃ is the boiling point of ethyl ether at 101.325kPa. $\Delta_{vap}H_m$(ethyl ether) = 25.10kJ·mol^{-1} in the condition. Now the glass bottle is broken and liquid ethyl ether vaporized to equilibrium, calculate (1) the pressure of ethyl ether vapor; (2) Q, W, ΔU, ΔH, ΔS, ΔA, and ΔG of the process.

Answer: (1) 25.664kPa; (2) 2.254kJ, 0, 2.254kJ, 2.51kJ, 9.288J·K^{-1}, −0.61kJ, −0.353kJ

11. A small glass bottle soak with 0.1 mol liquid ethyl ether is placed in a thermal insulating bottle of $t = 35℃$, $p = 101325Pa$, and $V = 10dm^3$, which is filled with $N_2(g)$. After breaking the small glass bottle soak, liquid ethyl ether totally vaporized, the mixed gases can be regarded as the perfect gases. Calculate (1) the partial pressure of ethyl ether vapor in the mixed gases; (2) ΔH, ΔS, and ΔG of $N_2(g)$; (3) ΔH, ΔS, and ΔG of ethyl ether.

Given: the normal boiling point of ethyl ether at 101.325kPa is 35℃, and $\Delta_{vap}H_m$(ethyl ether) = 25.10kJ·mol^{-1} in the condition.

Answer: (1) 25.664kPa; (2) 0, 0, 0; (3) 2510kJ, 9.288J · K^{-1}, −0.353kJ

12. (1) Calculate $\Delta_r G_m^\ominus$(298K) of this process: C(graphite) →C(diamond);

(2) How to control pressure to make the above change realize.

Given: at 298K, $\Delta_c H_m^\ominus$(C, graphite) = −393.51kJ/mol, $\Delta_c H_m^\ominus$(C, diamond) = −395.407kJ · mol^{-1}, $S_m^\ominus$(C, graphite) = 5.694J · K^{-1} · mol^{-1}, $S_m^\ominus$(C, diamond) = 2.439J · K^{-1} · mol^{-1}, ρ(C, graphite) = 2.260g · cm^{-3}, ρ(C, diamond) = 3.520g · cm^{-3}.

Answer: (1) 2.867kJ/mol; (2) $p > 1.51\times10^9$Pa

13. The saturated vapor pressure of water is 41.891kPa at 77℃, the normal boiling point of water is 100℃ at 101.325kPa. Please calculate:

(1) The value of A and B in the relationship of vapor pressure and temperature of water: $\lg(p/\text{Pa}) = -A/T+B$;

(2) $\Delta_{vap}H_m(H_2O)$ in the scope of temperature;

(3) What is the pressure when the boiling point of water is 105℃.

Answer: (1) A=2179.133K, B=10.8455; (2) 41.719kJ · mol^{-1}; (3) 121.042kPa.

14. The normal boiling point of H_2O and CH_3Cl are 100℃ and 61.5℃ respectively, $\Delta_{vap}H_m(H_2O)$ = 40.668kJ · mol^{-1} and $\Delta_{vap}H_m(CHCl_3)$ = 29.50kJ · mol^{-1}. Calculate the corresponding temperature when the two liquids have the same saturated vapor pressure.

Answer: 262.9℃

Chapter 4 The thermodynamics of multi-component systems (多组分系统热力学)

Learning objectives:

(1) Define *partial molar quantity* and describe how any thermodynamic quantity depends on the composition of the system.

(2) Derive the *Gibbs-Duhem equation* connecting changes of the chemical potentials of the components of a system.

(3) State *Raoult's Law* for the partial pressure of a gas above a mixture.

(4) State *Henry's Law* and use it to derive the solubility of gases in liquids.

(5) State *mixture of ideal liquid* and *ideal dilute solution*.

(6) Explain what is meant by *colligative properties* and give examples.

Multi-component systems can be divided into mixtures and solutions. In solution, the solvent is normally the component present in greatest amount, other components are called solutes. We adopt the convention that the solvent is denoted by the letter A. Because liquid mixtures and liquid solutions are the most common, we will focus our attention on them in this chapter.

According to the conductive properties of solute, solutions can be divided into electrolyte solutions and non-electrolyte solutions, we will focus our attention on the non-electrolyte solutions in this chapter.

4.1 Mixture and solution (混合物和溶液)

4.1.1 Composition scale of mixture (混合物的通用组成标度)

(1) Mole fraction of B (B 的摩尔分数)

B is the arbitrary component of mixture.

$$x_B \overset{\text{def}}{=\!=\!=} n_B / \sum_A n_A \qquad (4.1.1)$$

Comment:

① Unit of x_B is 1.

② $\sum x_B = 1$, that is, the sum of the mole fraction of all components of a mixture must equal 1. For instance, mixture (A, B):

$$x_A = \frac{n_A}{n_A + n_B}$$

$$x_B = \frac{n_B}{n_A + n_B}$$

So $$x_A + x_B = \frac{n_A + n_B}{n_A + n_B} = 1$$

(2) Mass fraction of B (B 的质量分数)

$$w_B \overset{\text{def}}{=\!=\!=} m_B / \sum_A m_A \tag{4.1.2}$$

Comment:

① Unit of w_B is 1.

② $\sum w_B = 1$, that is, the sum of the mass fraction of all components of a mixture must equal 1.

(3) Volume fraction of B (B 的体积分数)

$$\varphi_B \overset{\text{def}}{=\!=\!=} x_B V^*_{m,B} / \sum_A x_A V^*_{m,A} \tag{4.1.3}$$

$V^*_{m,A}$ and $V^*_{m,B}$—molar volume of pure A and B in T, p of mixture.

Comment:

① Unit of φ_B is 1.

② $\sum \varphi_B = 1$, that is, the sum of the volume fraction of all components of a mixture must equal 1.

(4) Amount-of-substance concentration of B (or molarity of B) (B 的物质的量浓度)

$$c_B \overset{\text{def}}{=\!=\!=} n_B / V \tag{4.1.4}$$

Unit: $mol \cdot m^{-3}$

n_B—amount of substance of B;

V—total volume of mixture.

(5) Mass concentration of B (B 的质量浓度)

$$\rho_B \overset{\text{def}}{=\!=\!=} m_B / V \tag{4.1.5}$$

Unit: $kg \cdot m^{-3}$

(6) Molecular concentration of B (B 的分子浓度)

$$C_B \overset{\text{def}}{=\!=\!=} N_B / V \tag{4.1.6}$$

Unit: m^{-3}

4.1.2 Composition scale of solute B in solution (溶液中溶质 B 的通用组成标度)

(1) Molality of solute B (溶质 B 的质量摩尔浓度)

$$b_B\ (m_B) \overset{\text{def}}{=\!=\!=} n_B / m_A \tag{4.1.7}$$

Unit: $mol \cdot kg^{-1}$

n_B—amount of substance of solute B;

m_A—mass of solvent A.

(2) Mole fraction of solute B (溶质 B 的摩尔分数)

$$x_B \overset{\text{def}}{=\!=\!=} n_B / \sum_A n_A \tag{4.1.8}$$

Comment:

① Unit of x_B is 1.

② $\sum x_B = 1$, that is, the sum of the mole fraction of solvent and solute of a solution must equal 1.

For very dilute solution, $$b_B \approx \frac{x_B}{M_A}$$

Or $$x_B \approx b_B \cdot M_A$$

(3) Amount-of-substance concentration of solute B (溶质 B 的物质的量浓度)

$$c_B \overset{\text{def}}{=\!=\!=} n_B/V \tag{4.1.9}$$

Unit: $mol \cdot m^{-3}$

n_B—amount of substance of solute B;

V—total volume of solution.

For very dilute solution, $b_B \approx \frac{c_B}{\rho} = \frac{c_B}{\rho_A}$

Or $$c_B \approx b_B \cdot \rho_A$$

(4) Mole ratio of solute B (溶质 B 的摩尔比)

$$r_B \overset{\text{def}}{=\!=\!=} n_B/n_A \tag{4.1.10}$$

Unit: 1

4.2 Partial molar quantities (偏摩尔量)

V, U, H, S, A, G are extensive properties. There are molar quantities for one-component system (pure material), that is,

$V^*_{m,B} \overset{\text{def}}{=\!=\!=} \frac{V}{n_B}$ ——molar volume

$U^*_{m,B} \overset{\text{def}}{=\!=\!=} \frac{U}{n_B}$ ——molar thermodynamic energy

$H^*_{m,B} \overset{\text{def}}{=\!=\!=} \frac{H}{n_B}$ ——molar enthalpy

$S^*_{m,B} \overset{\text{def}}{=\!=\!=} \frac{S}{n_B}$ ——molar entropy

$G^*_{m,B} \overset{\text{def}}{=\!=\!=} \frac{G}{n_B}$ ——molar Gibbs function

$V^*_{m,B}$, $U^*_{m,B}$, $H^*_{m,B}$, $S^*_{m,B}$, $G^*_{m,B}$ are intensive properties, the star indicates a property of a pure substance or a collection of pure substance. In liquid solution or liquid mixture, V_B, U_B, H_B, S_B, A_B, G_B(n_B = 1mol) are not equal to $V^*_{m,B}$, $U^*_{m,B}$, $H^*_{m,B}$, $S^*_{m,B}$, $G^*_{m,B}$.(在液态混合物或溶液中，单位物质量的组分 B 的 V_B、U_B、H_B、S_B、A_B、G_B 与在同样温度、压力下单独存在时相应的摩尔量通常并不相等。包括理想液态混合物和理想稀溶液中的某些单位物质量的广度量，与其纯态时的广度量也不相等。)

In order to denote the above-mentioned differences, the concept of partial molar quantity(偏摩尔量) is put forward. For instance, molar volume of pure water and alcohol is 18.09 and

58. 35cm^3 · mol^{-1} respectively, however, in mixture of 0. 5mol water and 0. 5mol alcohol, partial molar volume of water and alcohol is 17. 0 and 57. 4cm^3 · mol^{-1} respectively.

4. 2. 1 Definition of partial molar quantity (偏摩尔量的定义)

In multi-component homogeneous system, every variable of thermodynamic function has not only two variables, it also has relationship with the mole of every substance which consists the system. Suppose Z stands for the extensive property of V, U, H, S, A, G and so on, therefore, for the multi-component homogeneous system, $Z = f(T, p, n_A, n_B, \cdots\cdots)$, the total differential of Z is $dZ = \left(\frac{\partial Z}{\partial T}\right)_{p,\, n(B)} dT + \left(\frac{\partial Z}{\partial p}\right)_{T,\, n(B)} dp + \left(\frac{\partial Z}{\partial n_A}\right)_{T,\, p,\, n(C,\, C\neq A)} dn_A + \left(\frac{\partial Z}{\partial n_B}\right)_{T,\, p,\, n(C,\, C\neq B)} dn_B + \cdots\cdots$

The subscript $n(B)$ in the first two partial derivatives indicates that all mole numbers are held constant; the subscript $n(C,\ C\neq B)$ indicates that all mole numbers except $n(B)$ are held constant.

We can define the **partial molar quantity** Z_B as

$$Z_B \stackrel{\text{def}}{=\!=} \left(\frac{\partial Z}{\partial n_B}\right)_{T,\, p,\, n(C,\, C\neq B)} \tag{4.2.1}$$

Implication of partial molar quantity Z_B:

Change ratio of any extensive property Z with n_B when T, p and n_C are constant.

According to the definition of partial molar quantity Z_B, we can get:

$$V_B \stackrel{\text{def}}{=\!=} \left(\frac{\partial V}{\partial n_B}\right)_{T,\, p,\, n(C,\, C\neq B)} \quad \text{——partial molar volume}$$

$$U_B \stackrel{\text{def}}{=\!=} \left(\frac{\partial U}{\partial n_B}\right)_{T,\, p,\, n(C,\, C\neq B)} \quad \text{——partial molar thermodynamic energy}$$

$$\cdots\cdots$$

$$G_B \stackrel{\text{def}}{=\!=} \left(\frac{\partial G}{\partial n_B}\right)_{T,\, p,\, n(C,\, C\neq B)} \quad \text{——partial molar Gibbs function}$$

Comment on partial molar quantity:

(1) Implication: change ratio of any extensive property Z with n_B when T, p and n_C are constant, or change of extensive property Z when adding unit amount of substance B to a large amount of mixture or solution in which composition is constant.

(2) Only extensive properties of system have partial molar quantities.

(3) Partial molar quantity is intensive property.

(4) Only change ratio of any extensive property Z with n_B when T, p and n_C are constant can be called partial molar quantity.

(5) Partial molar quantity is function of T, p and composition.

(6) Only one component, partial molar quantity = molar quantity.

(7) Partial molar quantity can be positive or negative.

4. 2. 2 Collected formula of partial molar quantity (偏摩尔量的集合公式)

For multi-component homogeneous system,

$Z = f(T, p, n_A, n_B, \cdots\cdots)$, Z is an arbitrary extensive property.

$$dZ = \left(\frac{\partial Z}{\partial T}\right)_{p,\ n(B)} dT + \left(\frac{\partial Z}{\partial p}\right)_{T,\ n(B)} dp + \left(\frac{\partial Z}{\partial n_A}\right)_{T,\ p,\ n(C,\ C\neq A)} dn_A + \left(\frac{\partial Z}{\partial n_B}\right)_{T,\ p,\ n(C,\ C\neq B)} dn_B + \cdots\cdots$$

If $dT=0$ and $dp=0$,

$$dZ = Z_A dn_A + Z_B dn_B + \cdots\cdots = \sum_B Z_B dn_B$$

Then

$$Z = \sum_B n_B Z_B \qquad (4.2.2)$$

(4.2.2) is the **collected formula of partial molar quantity**.

For instance, mixture(A, B):

According to the collected formula of partial molar quantity, we have

$Z = n_A Z_A + n_B Z_B$, Z is an arbitrary extensive property.

So, when Z stands for V, we can get, $V = n_A V_A + n_B V_B \neq n_A V^*_{m,\ A} + n_B V^*_{m,\ B}$

4.2.3 Gibbs-Duhem equation (吉布斯-杜亥姆方程)

According to the collected formula of partial molar quantity, $Z = \sum_B n_B Z_B$

If $dT=0$ and $dp=0$,

$$dZ = \sum_B n_B dZ_B + \sum_B Z_B dn_B$$

$$dZ = Z_A dn_A + Z_B dn_B + \cdots\cdots = \sum_B Z_B dn_B$$

Equating these two expressions for dZ, we get

$$\sum_B n_B dZ_B = 0 \qquad (4.2.3)$$

Or

$$\sum_B x_B dZ_B = 0 \qquad (4.2.4)$$

(4.2.3) and (4.2.4) are **Gibbs-Duhem equation**, which show that there is certain relationship of partial molar quantity and apply to any infinitesimal process. The change of other partial molar quantities can be get from the change of certain partial molar quantities.

For instance, solution or mixture (A, B):

According to Gibbs-Duhem equation, we have

$$x_A\, dZ_A + x_B\, dZ_B = 0$$

That is

$$x_A\, dZ_A = -x_B\, dZ_B$$

It means when dZ_A increases dZ_B must decrease.

4.2.4 Relations among different partial molar quantities (不同偏摩尔量之间的关系)

For many of the thermodynamic relations between extensive properties of a homogeneous system, there are corresponding relations with the extensive variables replaced by partial molar quantities. So that, relations of different partial molar quantities are same as molar quantities, that is, all the relations discussed in previous chapters are held for partial molar qualities.

For instance, from $H=U+pV$, we can get, $H_B=U_B+pV_B$

From $A=U-TS$, we can get, $A_B=U_B-TS_B$

From $dU = TdS-pdV$, we can get, $dU_B = TdS_B-pdV_B$

Similarly,

$$\left(\frac{\partial G_{\mathrm{B}}}{\partial p}\right)_{T,\ n(\mathrm{B})} = V_{\mathrm{B}}$$

$$\left[\frac{\partial\ (G_{\mathrm{B}}/T)}{\partial T}\right]_{p,\ n_{\mathrm{B}}} = -\frac{H_{\mathrm{B}}}{T^2}$$

4.3 Chemical potential(化学势)

4.3.1 Definition of chemical potential (化学势的定义)

$$\mu_{\mathrm{B}} \overset{\mathrm{def}}{=\!=\!=} G_{\mathrm{B}} = \left(\frac{\partial G}{\partial n_{\mathrm{B}}}\right)_{T,\ p,\ n(\mathrm{C},\ \mathrm{C}\neq\mathrm{B})} \tag{4.3.1}$$

4.3.2 Thermodynamic fundamental equation of multi-component homogeneous system changing of composition (多组分均相组成可变系统的热力学基本方程)

For multi-component homogeneous system (mixture or solution),

$G = f\ (T,\ p,\ n_{\mathrm{A}},\ n_{\mathrm{B}},\ \cdots\cdots)$, the total differential is

$$\mathrm{d}G = \left(\frac{\partial G}{\partial T}\right)_{p,\ n(\mathrm{B})}\mathrm{d}T + \left(\frac{\partial G}{\partial p}\right)_{T,\ n(\mathrm{B})}\mathrm{d}p + \left(\frac{\partial G}{\partial n_{\mathrm{A}}}\right)_{T,\ p,\ n(\mathrm{C},\ \mathrm{C}\neq\mathrm{A})}\mathrm{d}n_{\mathrm{A}} + \left(\frac{\partial G}{\partial n_{\mathrm{B}}}\right)_{T,\ p,\ n(\mathrm{C},\ \mathrm{C}\neq\mathrm{B})}\mathrm{d}n_{\mathrm{B}} + \cdots\cdots$$

Since $\left(\frac{\partial G}{\partial T}\right)_{p,\ n_{\mathrm{B}}} = -S$; $\left(\frac{\partial G}{\partial p}\right)_{T,\ n_{\mathrm{B}}} = V$

So

$$\mathrm{d}G = -S\mathrm{d}T + V\mathrm{d}p + \sum_{\mathrm{B}} \mu_{\mathrm{B}}\mathrm{d}n_{\mathrm{B}} \tag{4.3.2}$$

According to the definition formula of G, $G = H - TS$, we have

$$\mathrm{d}G = \mathrm{d}H - T\mathrm{d}S - S\mathrm{d}T$$

That is, $\mathrm{d}H = \mathrm{d}G + T\mathrm{d}S + S\mathrm{d}T = -S\mathrm{d}T + V\mathrm{d}p + \sum_{\mathrm{B}} \mu_{\mathrm{B}}\mathrm{d}n_{\mathrm{B}} + T\mathrm{d}S + S\mathrm{d}T$

So,

$$\mathrm{d}H = T\mathrm{d}S + V\mathrm{d}p + \sum_{\mathrm{B}} \mu_{\mathrm{B}}\mathrm{d}n_{\mathrm{B}} \tag{4.3.3}$$

Similarly,

According to the definition formula of H, $H = U + pV$, we have

$$\mathrm{d}U = T\mathrm{d}S - p\mathrm{d}V + \sum_{\mathrm{B}} \mu_{\mathrm{B}}\mathrm{d}n_{\mathrm{B}} \tag{4.3.4}$$

According to the definition formula of A, $A = U - TS$, we have

$$\mathrm{d}A = -S\mathrm{d}T - p\mathrm{d}V + \sum_{\mathrm{B}} \mu_{\mathrm{B}}\mathrm{d}n_{\mathrm{B}} \tag{4.3.5}$$

(4.3.2), (4.3.3), (4.3.4) and (4.3.5) are **the thermodynamic fundamental equation of multi-component homogeneous system changing of composition**.

Applicable condition:

Homogeneous closed system changing of composition and open system.

From the thermodynamic fundamental equation of multi-component homogeneous system changing of composition, we can get **four expressions of chemical potential**:

(1) $$\mathrm{d}S = 0,\ \mathrm{d}V = 0,\ \mathrm{d}n_{\mathrm{C}} = 0,\ \mu_{\mathrm{B}} = \left(\frac{\partial U}{\partial n_{\mathrm{B}}}\right)_{S,\ V,\ n(\mathrm{C},\ \mathrm{C}\neq\mathrm{B})} \tag{4.3.6}$$

(2) $dS=0,\ dp=0,\ dn_C=0,\ \mu_B = \left(\frac{\partial H}{\partial n_B}\right)_{S,\ p,\ n(C,\ C\neq B)}$ (4.3.7)

(3) $dT=0,\ dV=0,\ dn_C=0,\ \mu_B = \left(\frac{\partial A}{\partial n_B}\right)_{T,\ V,\ n(C,\ C\neq B)}$ (4.3.8)

(4) $dT=0,\ dp=0,\ dn_C=0,\ \mu_B = \left(\frac{\partial G}{\partial n_B}\right)_{T,\ p,\ n(C,\ C\neq B)}$ (4.3.9)

Only (4.3.9) is also chemical potential and partial molar quantity.

4.3.3 Equilibrium criterion of material (物质平衡判据)

For multi-component homogeneous system (mixture or solution),

$$dG = -SdT + Vdp + \sum_B \mu_B dn_B$$

If $dT=0$ and $dp=0$, $dG = \sum_B \mu_B dn_B$

So, for multi-component heterogeneous system, we can add the Gibbs free energy of each phase to get dG for the system because the state function G is extensive,

$$dG = dG^\alpha + dG^\beta = \sum_B \mu_B^\alpha dn_B^\alpha + \sum_B \mu_B^\beta dn_B^\beta = \sum_\alpha \sum_B \mu_B^\alpha dn_B^\alpha$$

Where α is arbitrary phase, and B is arbitrary substance of same phase.

If $W'=0$, depending on Gibbs function criterion: $dG_{T,\ p} \leqslant 0$

We can get

$$\sum_\alpha \sum_B \mu_B^\alpha dn_B^\alpha \leqslant 0 \quad \begin{matrix} <,\ \text{ir} \\ =,\ \text{r} \end{matrix} \tag{4.3.10}$$

(4.3.10) is the **equilibrium criterion of material (or chemical potential criterion)**.

There are two kinds of material equilibrium, phase equilibrium and reaction equilibrium.

(1) Condition for phase equilibrium

Consider a heterogeneous system that is in equilibrium, and suppose that dn_B moles of substance B were to flow from α phase to β phase. This process is:

$$B(\alpha) \xrightleftharpoons{T,\ p} B(\beta)$$

$$\mu_B^\alpha \qquad\qquad \mu_B^\beta$$

$$dn_B^\alpha = -dn_B^\beta \qquad dn_B^\beta > 0$$

$$\sum_\alpha \sum_B \mu_B^\alpha dn_B^\alpha = \sum_\alpha \mu_B^\alpha dn_B^\alpha = \mu_B^\alpha dn_B^\alpha + \mu_B^\beta dn_B^\beta \leqslant 0$$

That is

$$dn_B^\beta(\mu_B^\beta - \mu_B^\alpha) \leqslant 0$$

For $dn_B^\beta > 0$

So

$$(\mu_B^\beta - \mu_B^\alpha) \leqslant 0 \quad \begin{matrix} <,\ \text{ir} \\ =,\ \text{r} \end{matrix} \tag{4.3.11}$$

(4.3.11) is the **criterion of phase equilibrium**.

So, if a process is spontaneous, $\mu_B^\beta < \mu_B^\alpha$, that is, any spontaneous process always proceeds to the direction of small chemical potential, until chemical potentials of two phases are equal.

$\mu_B^\beta = \mu_B^\alpha$ or $G_B^\beta = G_B^\alpha$ is the **condition for phase equilibrium**.

According to the condition for phase equilibrium, for one-component system,

$$G_{m,B}^{*,\beta} = G_{m,B}^{*,\alpha}$$

That is, molar Gibbs function of pure material in two phases must be equal when it established phase equilibrium at T, p.

(2) Equilibrium condition of chemical reaction

For multi-component homogeneous system,

$$\sum_{\alpha}\sum_{B}\mu_B^{\alpha}dn_B^{\alpha} = \sum_{B}\mu_B^{\alpha}dn_B^{\alpha} = \sum_{B}\mu_B dn_B \leqslant 0$$

According to $d\xi = \dfrac{dn_B}{\nu_B}$, we can get, $dn_B = \nu_B d\xi$

So
$$\sum_{B}\nu_B\mu_B d\xi \leqslant 0 \quad \begin{matrix} <, \ \text{ir} \\ =, \ \text{r} \end{matrix} \tag{4.3.12}$$

(4.3.12) is the **equilibrium criterion of chemical reaction.**

$\sum_{B}\nu_B\mu_B = 0$ is the **equilibrium condition of chemical reaction.**

If definite affinity of chemical reaction A is defined as, $A \overset{\text{def}}{=\!=\!=} -\sum \nu_B\mu_B$

The equilibrium criterion of chemical reaction can also be expressed as

$$A \geqslant 0 \quad \begin{matrix} >, \ \text{spontaneous} \\ =, \ \text{equilibrium} \\ <, \ \text{backward, spontaneous} \end{matrix} \tag{4.3.13}$$

The principal conclusion of this section is: in a closed system in thermodynamic equilibrium, the chemical potential of any given component is the same in every phase in which that component is present.

4.4 Chemical potential of gas and fugacity (气体的化学势和逸度)

$\mu_B = f(T, p, \text{composition})$, it is an intensive property.

4.4.1 Expression of chemical potential of perfect gas (理想气体化学势的表达式)

(1) Expression of chemical potential of pure perfect gas

According to the definition of chemical potential, we have

$$\mu^* = G_B^* = G_m^*$$

So
$$d\mu^* = dG_m^*$$

According to the fundamental equation of thermodynamics,

$$dG_m^* = -S_m^* dT + V_m^* dp$$

We can get
$$d\mu^* = -S_m^* dT + V_m^* dp$$

If $dT=0$
$$d\mu^* = V_m^* dp$$

For pg (perfect gas)
$$V_m^* = \frac{RT}{p}$$

So $$d\mu^* = \frac{RT}{p}dp$$

Then $$\int_{\mu^\ominus}^{\mu^*} d\mu^* = \int_{p^\ominus}^{p} \frac{RT}{p}dp$$

Therefore $$\mu^*(g, T, p) = \mu^\ominus(g, T) + RT\ln(p/p^\ominus) \quad (4.4.1)$$

Where $\mu^*(g, T, p)$ is the chemical potential of arbitrary state, and $\mu^\ominus(g, T)$ is the chemical potential of standard state. Note that $\mu^\ominus(g, T)$ depends only on T because the pressure is fixed at 100 kPa for the standard state.

(2) Expression of chemical potential for component B in the perfect gas mixture

The molecular picture of a perfect gas mixture is the same as for a pure prefect gas, namely, molecules that have negligible volume and negligible intermolecular interaction energy.

So depending on (4.4.1) we have

$$\mu_B(g, T, p, y_c) = \mu_B^\ominus(g, T) + RT\ln(p_B/p^\ominus) \quad (4.4.2)$$

Or simply $$\mu_B(g) = \mu_B^\ominus(g, T) + RT\ln(p_B/p^\ominus) \quad (4.4.3)$$

Where p_B is partial pressure of component B in pg mixture.

According to Dalton's law, $\sum p_B = p$, $p_B = p\,y_B$

4.4.2 Expression of chemical potential of real gas and fugacity(真实气体化学势的表达式和逸度)

(1) Expression of chemical potential of pure real gas and fugacity

According to the definition of chemical potential, we have

$$\mu^* = G_B^* = G_m^*$$

So $$d\mu^* = dG_m^*$$

According to the fundamental equation of thermodynamics,

$$dG_m^* = -S_m^* dT + V_m^* dp$$

We can get $$d\mu^* = -S_m^* dT + V_m^* dp$$

If $dT = 0$ $$d\mu^* = V_m^* dp$$

For real gas $$V_m^* \neq \frac{RT}{p}$$

V_m of real gas obeys the van der Waals equation or the virial equation:

$$(p + a/V_m^2)(V_m - b) = RT$$

$$pV_m = RT\left(1 + \frac{B}{V_m} + \frac{C}{V_m^2} + \frac{D}{V_m^3} + \cdots\right)$$

So that the integral of the chemical potential will be difficult, therefore, Lewis puts forward the concept of fugacity to simplify the integral.

fugacity (逸度)—pressure of real gas, that is, corrected pressure of gas.

Symbol: $\tilde{p}$

$$\tilde{p} = \varphi \cdot p$$

Where φ is fugacity factor, unit is 1.

We can get

$$\mu^*(g, T, p) = \mu^\ominus(g, T) + RT\ln(\tilde{p}/p^\ominus) = \mu^\ominus(g, T) + RT\ln(\varphi \cdot p/p^\ominus) \quad (4.4.4)$$

(2) Expression of chemical potential for component B in real gas mixture

$$\mu_B(g, T, p, y_c) = \mu_B^\ominus(g, T) + RT\ln(\tilde{p}_B/p^\ominus) \quad (4.4.5)$$

Where $\tilde{p}_B$ is partial fugacity of component B in real gas mixture.

According to **Lewis-Randall fugacity rule**, $\tilde{p}_B = \tilde{p} \cdot y_B$

Lewis-Randall fugacity rule is similar to Dalton's law of partial pressures.

4.5 Raoult's Law and Henry's Law (拉乌尔定律和亨利定律)

4.5.1 Gas-liquid equilibrium of liquid mixture or solution (液态混合物或溶液的气-液平衡)

If multi-component A, B, C… constitute liquid mixture or solution, the gas-liquid equilibrium is shown in figure 4.1, according to the definition of partial pressure, we have:

$p_A = y_A p$

$p_B = y_B p$

$p_C = y_C p$

So, $p = p_A + p_B + p_C + \cdots\cdots = \sum_B p_B$

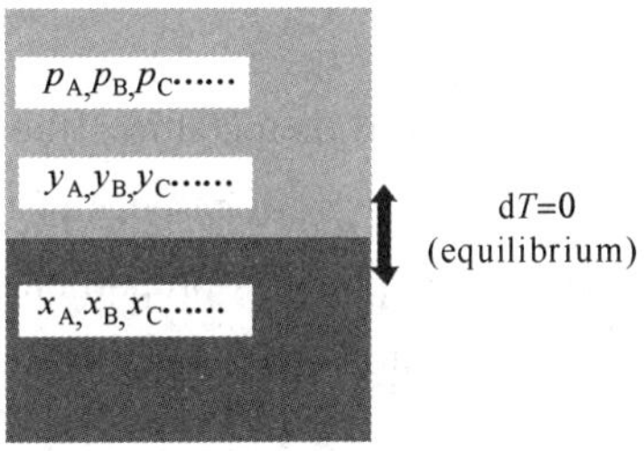

Fig. 4.1 g⇌l of dilute solution

If component B, C…are involatile, then:

$$p = p_A$$

So,

$p = f$ (T, composition and nature of component)

4.5.2 Raoult's law (拉乌尔定律)

$$p_A = p_A^* x_A \quad (4.5.1)$$

Where x_A is the mole fraction of solvent A in solution, p_A^* is the saturated vapor pressure of pure solvent A at same temperature, p_A is the equilibrium partial pressure of solvent A in gaseous phase.

François-Marie Raoult

Statement:

In dilute solution, the partial pressure of solvent A in gaseous phase equals the product of the saturated vapor pressure of pure solvent A and the mole fraction of solvent A in solution at the same temperature while two phases are in equilibrium.

For dilute solution (A, B), $x_A + x_B = 1$

So $$x_A = 1 - x_B$$

Then $$p_A = p_A^*(1 - x_B) = p_A^* - p_A^* x_B$$

That is $$p_A^* - p_A = p_A^* x_B$$

The reduction in vapor pressure, Δp_A, is $p_A^* - p_A$. Therefore, we can write this as

$$\Delta p_A = p_A^* - p_A = p_A^* x_B \quad (4.5.2)$$

Conclusion:

The vapor pressure of solvent A will be lowered with adding of solute B.

Applicable condition:

(1) Solvent of dilute solution;

(2) Arbitrary component of ideal liquid mixture.

4.5.3 Henry's law (亨利定律)

(1) Henry's law

$$p_B = k_{x,B} x_B \quad (4.5.3)$$

Where x_B is the mole fraction of volatile solute B in equilibrium liquid phase, p_B is equilibrium partial pressure of volatile solute B in equilibrium gas phase, $k_{x,B}$ is Henry's coefficient whose unit is Pa.

$$k_{x,B} = f(T)$$

Statements:

At a given temperature, the solubility of microsoluble gas B in solvent A is proportional to the partial pressure p_B in gaseous phase. Or at a given temperature, the partial pressure of volatile solute B in gaseous phase is proportional to its mole fraction in equilibrium liquid phase in dilute solution.

(2) Different formulas of Henry's law

$$p_B = k_{b,B} b_B \quad (4.5.4)$$

Where the unit of $k_{b,B}$ is Pa · (mol · kg^{-1})$^{-1}$

$$p_B = k_{c,B} c_B \quad (4.5.5)$$

Where the unit of $k_{c,B}$ is Pa · (mol · m^{-3})$^{-1}$.

Applicable condition :

Volatile solute B of dilute solution.

(3) Comment on Henry's law

① Condition of application: volatile solute B of dilute solution.

② Preferred mode: if k is known, the mode can be determined by the unit of k.

③ A certain number of gas solute can be considered respectively.

④ Comparison of Raoult's law and Henry's law gives: if solvent A obeys Raoult's law, solute B must obey Henry's law, or vice versa.

⑤ $k = k(T)$, if T is different, k is different.

⑥ We can use Henry's law as long as the molecule morphology of solute in gaseous phase is same as that in liquid phase.

For instance: HCl in hydrochloric acid aqueous solution.

g: molecule HCl

l: H^+, Cl^-

The molecule morphology of solute in gaseous phase is not same as that in liquid phase, so we cannot use Henry's law.

4.6 Mixture of ideal liquid(理想液态混合物)

4.6.1 Definition and features of mixture of ideal liquid (理想液态混合物的定义和特征)

(1) Definition of mixture of ideal liquid

Liquid mixture that obeys Raoult's law ($p_B = p_B^* x_B$) throughout its composition range (from $x_B = 0$ to $x_B = 1$) is called mixture of ideal liquid.

(2) Features of mixture of ideal liquid

① Microscopic characteristics

a. $f_{AA}^* = f_{BB}^* = f_{AB}$

Where f_{AA}^* means the interaction of pure A−A, f_{AB} means the interaction of A−B.

b. V(molecule A) = V(molecule B)

② Macroscopic characteristics

a. One more component $\xrightarrow[(T,\ p)\text{mixing}]{\Delta_{\text{mix}}H=0}$ mixture of ideal liquid.

b. One more component $\xrightarrow[(T,\ p)\text{mixing}]{\Delta_{\text{mix}}V=0}$ mixture of ideal liquid.

4.6.2 Chemical potential of arbitrary component in mixture of ideal liquid (理想液态混合物中任意组分的化学势)

The gas-liquid equilibrium of mixture of ideal liquid is shown in figure 4.2.

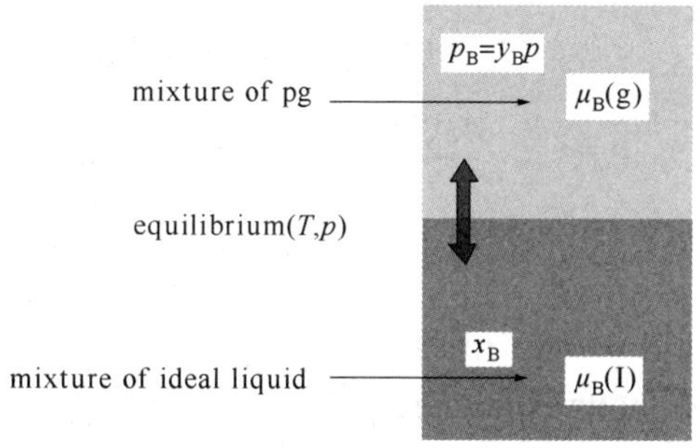

Fig. 4.2 g⇌l of mixture of ideal liquid

According to **condition for phase equilibrium**

$$\mu_B(l) = \mu_B(g)$$

For

$$\mu_B(g) = \mu_B^\ominus(g,\ T) + RT\ln(p_B/p^\ominus)$$

$$\mu_B(l) = \mu_B^\ominus(g,\ T) + RT\ln(p_B/p^\ominus)$$

According to **Raoult's law**

$$p_B = p_B^* x_B$$

$$\mu_B(l) = \mu_B^\ominus(g,\ T) + RT\ln(p_B^* x_B/p^\ominus) = \mu_B^\ominus(g,\ T) + RT\ln(p_B^*/p^\ominus) + RT\ln x_B$$

While

$$\mu_B^*(l,\ T,\ p) = \mu_B^\ominus(g,\ T) + RT\ln(p_B^*/p^\ominus) = \mu_B^*(l)$$

So

$$\mu_B(l) = \mu_B^*(l) + RT\ln x_B(\text{theoretical formula}) \qquad (4.6.1)$$

According to the standard state of liquid B[$\mu_B^\ominus(l, T)$]: T, $p^\ominus$, hypothetical state of pure liquid B.

$$\underset{\mu_B^\ominus(l)}{B(l, T, p^\ominus)} \xrightarrow{\Delta\mu_B} \underset{\mu_B^*(l)}{B(l, T, p)}$$

$$\Delta\mu_B = \Delta G_B = \int dG_m = \int_{p^\ominus}^{p} V_{m,B}^*(l)\,dp = \mu_B^*(l) - \mu_B^\ominus(l)$$

So
$$\mu_B^*(l) = \mu_B^\ominus(l) + \Delta\mu_B = \mu_B^\ominus(l) + \int_{p^\ominus}^{p} V_{m,B}^*(l)\,dp$$

Therefore

$$\mu_B(l) = \mu_B^\ominus(l) + RT\ln x_B\text{(calculation formula)} \tag{4.6.2}$$

4.6.3 Mixing properties of mixture of ideal liquid（理想液态混合物的混合性质）

$dT=0$, $dp=0$

(1) $\Delta_{mix}V=0$, that is, volume of mixing process does not change.

(2) $\Delta_{mix}H=0$, that is, enthalpy of mixing process does not change.

(3) $\Delta_{mix}S=-R\sum n_B\ln x_B>0$, that is, entropy of mixing process increases.

(4) $\Delta_{mix}G=RT\sum n_B\ln x_B<0$, that is, Gibbs function of mixing process decreases.

We can get the mixing process is spontaneous from the mixing properties of mixture of ideal liquid.

Let us prove the mixing properties of mixture of ideal liquid.

For instance, $\Delta_{mix}H=0$, according to the **theoretical formula** of chemical potential of arbitrary component in mixture of ideal liquid,

$$\mu_B(l) = \mu_B^*(l) + RT\ln x_B$$

$$\frac{\mu_B(l)}{T} = \frac{\mu_B^*(l, T)}{T} + R\ln x_B$$

$$\left\{\frac{\partial[\mu_B(l)/T]}{\partial T}\right\}_{p,x_B} = \left\{\frac{\partial[\mu_B^*(l, T)/T]}{\partial T}\right\}_p$$

According to **Gibbs-Helmholtz equation**, $\left[\dfrac{\partial(G/T)}{\partial T}\right]_p = -\dfrac{H}{T^2}$

We can get, $\left[\dfrac{\partial(G_B/T)}{\partial T}\right]_p = -\dfrac{H_B}{T^2}$ and $\left[\dfrac{\partial(G_m^*/T)}{\partial T}\right]_p = -\dfrac{H_{m,B}^*}{T^2}$

So

$$\left\{\frac{\partial[\mu_B(l)/T]}{\partial T}\right\}_{p,x_B} = -\frac{H_B}{T^2}$$

$$\left\{\frac{\partial[\mu_B^*(l)/T]}{\partial T}\right\}_p = -\frac{H_{m,B}^*}{T^2}$$

Therefore
$$H_B = H_{m,B}^*$$

So we have $\Delta_{mix}H = \sum_B n_B H_B - \sum_B n_B H_{m,B}^* = \sum_B n_B(H_B - H_{m,B}^*) = 0$

We also can prove the mixing property of mixture of ideal liquid, $\Delta_{mix}S=-R\sum n_B\ln x_B$.

According to the **theoretical formula** of chemical potential of arbitrary component in mixture of ideal liquid,

$$\mu_B(1) = \mu_B^*(1) + RT\ln x_B$$

$$\left\{\frac{\partial\,[\mu_B(1)]}{\partial T}\right\}_{p,\,x_B} = \left\{\frac{\partial\,[\mu_B^*(1,\ T)]}{\partial T}\right\}_p + R\ln x_B$$

According to $$\left[\frac{\partial G_B\ (l)}{\partial T}\right]_p = -S_B = \left\{\frac{\partial\ [\mu_B(1)]}{\partial T}\right\}_{p,\,x_B}$$

And $$\left[\frac{\partial G_m^*\ (l)}{\partial T}\right]_p = -S_{m,\,B}^* = \left\{\frac{\partial\ [\mu_B^*(1)]}{\partial T}\right\}_{p,\,x_B}$$

We can get $$-S_B = -S_{m,\,B}^* + R\ln x_B$$

That is $$S_B - S_{m,\,B}^* = -R\ln x_B$$

So we have, $\Delta_{mix}S = \sum_B n_B S_B - \sum_B n_B S_{m,\,B}^* = \sum_B n_B\,(S_B - S_{m,\,B}^*) = -R\sum_B n_B\ln x_B$

$\Delta_{mix}G = \Delta_{mix}H - T\Delta_{mix}S = RT\sum_B n_B\ln x_B$

4.6.4 Gas-liquid equilibrium of mixture of ideal liquid (理想液态混合物的气-液平衡)

For mixture of ideal liquid (A, B): A and B are volatile.

$dT=0$, $dp=0$, gas-liquid equilibrium: $p=p_A+p_B$

(1) Relations between the total vapor pressure and equilibrium liquid composition ($p\sim x_B$)

$$p = p_A + p_B = p_A^* x_A + p_B^* x_B = p_A^*\ (1 - x_B) + p_B^* x_B = p_A^* + (p_B^* - p_A^*)x_B$$

$$p_A = p_A^* x_A = p_A^*\ (1 - x_B)$$

$$p_B = p_B^* x_B$$

Volatility: if A>B, then: $p_A^* > p_B^*$

Conclusion: $p_A^* > p > p_B^*$

(2) Relations between equilibrium vapor composition and liquid composition ($y_B\sim x_B$)

Combine **Dalton's law** and **Raoult's law**, we have

$$p_A = py_A = p_A^* x_A$$

So $$\frac{y_A}{x_A} = \frac{p_A^*}{p} > 1$$

That is $$y_A > x_A$$

$$p_B = py_B = p_B^* x_B$$

So $$\frac{y_B}{x_B} = \frac{p_B^*}{p} < 1$$

That is $$y_B < x_B$$

Conclusion:

Gaseous mole fraction is large for easy volatile component; liquid mole fraction is large for hard volatile component.

(3) Relations between the total vapor pressure and equilibrium vapor composition($p \sim y_B$)

$$p = p_A + p_B = p_A^* x_A + p_B^* x_B = p_A^* (1 - x_B) + p_B^* x_B = p_A^* + (p_B^* - p_A^*) x_B$$

$$\frac{y_B}{x_B} = \frac{p_B^*}{p}$$

So
$$p = \frac{p_A^* p_B^*}{p_B^* - (p_B^* - p_A^*) y_B}$$

Example 4. 1

At 60℃, the saturated vapor pressure of methanol(A) and ethyl alcohol (B) is 83. 4kPa and 47. 0kPa respectively. A and B can form mixture of ideal liquid. Find the equilibrium vapor composition of this mixture at 60℃ expressed by mole fraction when the mass fraction of the mixture is 50%.

Answer:

$M_A = 32$, $M_B = 46$

So $$x_A = \frac{\frac{50}{32}}{\frac{50}{32} + \frac{50}{46}} = 0.5898$$

$$p = p_A + p_B = p_A^* x_A + p_B^* (1 - x_A) = 68.47\text{kPa}$$

$$y_A = \frac{p_A}{p} = \frac{p_A^* x_A}{68.47} = 0.718 \qquad y_B = 1 - y_A = 0.282$$

Example 4. 2

At 100℃, the vapor pressure of CCl_4 and $SnCl_4$ is 1.933×10^5Pa and 0.666×10^5Pa. CCl_4 and $SnCl_4$ can form mixture of ideal liquid, when external pressure $p = 1.013 \times 10^5$Pa, this mixture of ideal liquid is heated to 100℃ and then boiling, please calculate: (1) composition of this mixture of ideal liquid; (2) composition of the first air bubble while the mixture of ideal liquid is boiling.

Answer:

(1) A denotes CCl_4 and B denotes $SnCl_4$, then:

$p_A^* = 1.933 \times 10^5\text{Pa}$; $p_B^* = 0.666 \times 10^5\text{Pa}$

$$p = p_A + p_B = p_A^* x_A + p_B^* x_B = p_A^* (1 - x_B) + p_B^* x_B = p_A^* + (p_B^* - p_A^*) x_B$$

$$x_B = \frac{p - p_A^*}{p_B^* - p_A^*} = \frac{1.013 \times 10^5\text{Pa} - 1.933 \times 10^5\text{Pa}}{0.666 \times 10^5\text{Pa} - 1.933 \times 10^5\text{Pa}} = 0.726$$

$$x_A = 1 - x_B = 1 - 0.726 = 0.274$$

(2) The composition of the first bubble while boiling is the equilibrium gaseous composition of mixture, y_B, then:

$$p_B = p y_B = p_B^* x_B$$

$$y_B = \frac{x_B p_B^*}{p} = \frac{0.726 \times 0.666 \times 10^5\text{Pa}}{1.013 \times 10^5\text{Pa}} = 0.477$$

$$y_A = 1 - y_B = 0.523$$

Example 4. 3

At 85℃, 101. 3kPa, the liquid mixture of toluene (A) and benzene (B) is boiling, this mixture can be regarded as ideal liquid mixture. The vapor pressure of toluene is 46. 0kPa at 85℃, the normal boiling point of benzene is 80. 1℃. Please calculate: the liquid composition and vapor composition of this ideal liquid mixture.

Example 4. 3

Answer:

For ideal liquid mixture,

$$p = p_A + p_B = p_A^* x_A + p_B^* x_B = p_A^* (1 - x_B) + p_B^* x_B = p_A^* + (p_B^* - p_A^*) x_B$$

From Trouton's rule, we can get:

$$\Delta_{vap} H_m^* = 88 \cdot T_b^* = 88 \times (80.1 + 273.15) = 31.1 \text{kJ} \cdot \text{mol}^{-1}$$

According to Clausius-Clapeyron equation,

$$\ln \frac{p_B^* (358.15\text{K})}{p_B^* (353.25\text{K})} = -\frac{31.1 \times 10^3}{8.315} \left(\frac{1}{358.15} - \frac{1}{353.25} \right)$$

$$p_B^* (358.15\text{K}) = 117.1 \text{kPa}$$

So,

$$x_B = \frac{p - p_A^*}{p_B^* - p_A^*} = \frac{1.013 \times 10^5 - 0.46 \times 10^5}{1.171 \times 10^5 - 0.46 \times 10^5} = 0.778$$

$$x_A = 1 - x_B = 1 - 0.778 = 0.222$$

$$y_B = \frac{x_B p_B^*}{p} = \frac{0.778 \times 117.1}{101.3} = 0.899$$

$$y_A = 1 - y_B = 0.101$$

4.7 Ideal dilute solution(理想稀溶液)

4.7.1 Definition and gas-liquid equilibrium of ideal dilute solution (理想稀溶液的定义和气液平衡)

(1) Definition of ideal dilute solution

Infinite dilute solution in which solvent and solute obey Raoult's law and Henry's law respectively at given temperature is called ideal dilute solution.

(2) Gas-liquid equilibrium of ideal dilute solution

For ideal dilute solution (A, B),

If A and B are volatile, $dT=0$, $dp=0$, gas-liquid equilibrium:

$$p = p_A + p_B = p_A^* x_A + k_{x,B} x_B$$

Or

$$p = p_A + p_B = p_A^* x_A + k_{b,B} b_B$$

If B are involatile, gas-liquid equilibrium:

$$p = p_A = p_A^* x_A$$

4.7.2 Chemical potential of solvent and solute in ideal dilute solution(理想稀溶液中溶剂和溶质的化学势)

(1) Chemical potential of solvent A

$$\mu_A(1) = \mu_A^\ominus(1, T) + RT \ln x_A \quad (4.7.1)$$

The standard state of solvent A [$\mu_A^\ominus(l, T)$]: T, $p^\ominus$, hypothetical state of pure liquid A.

(2) Chemical potential of volatile solute B

The gas-liquid equilibrium of ideal dilute solution is shown in figure 4.3.

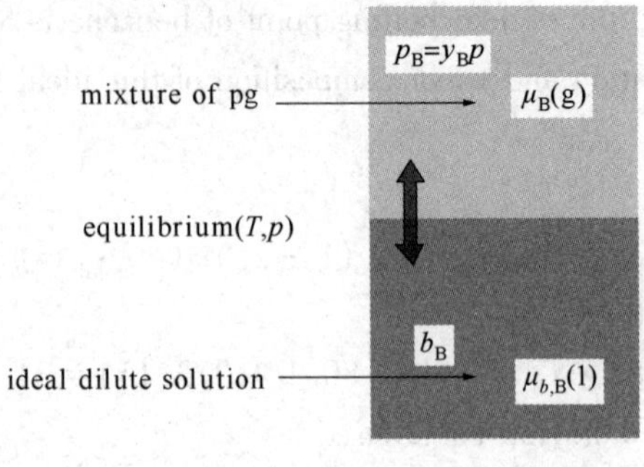

Fig. 4.3 g⇌l of ideal dilute solution

According to **condition for phase equilibrium**,

$$\mu_{b,B}(\text{solute}) = \mu_B(g) = \mu_B^\ominus(g) + RT\ln\frac{p_B}{p^\ominus}$$

According to **Henry's law**,

$$p_B = k_{b,B}b_B$$

So, $\mu_{b,B}(\text{solute}) = \mu_B^\ominus(g) + RT\ln\dfrac{k_{b,B}\cdot b_B}{p^\ominus} = \mu_B^\ominus(g) + RT\ln\dfrac{k_{b,B}\cdot b^\ominus}{p^\ominus} + RT\ln\dfrac{b_B}{b^\ominus}$

Where $b^\ominus = 1\text{mol}\cdot\text{kg}^{-1}$, it is the standard molality of solute B.

If $$\mu_{b,B}(\text{solute}, T, p, b^\ominus) = \mu_B^\ominus(g) + RT\ln\frac{k_{b,B}\cdot b^\ominus}{p^\ominus}$$

Then $$\mu_{b,B}(\text{solute}) = \mu_{b,B}(\text{solute}, T, p, b^\ominus) + RT\ln\frac{b_B}{b^\ominus}$$

The standard state of solute B [$\mu_{b,B}^\ominus(T)$]: T, $p^\ominus$, $b_B = b^\ominus = 1\text{mol}\cdot\text{kg}^{-1}$, hypothetical state of solute B which obey Henry's law.

$$\underset{\mu_{b,B}^\ominus(T)}{\text{solute B }(T, p^\ominus, b^\ominus)} \xrightarrow{\Delta\mu_B} \underset{\mu_{b,B}(T, p, b^\ominus)}{\text{solute B }(T, p, b^\ominus)}$$

$$\Delta\mu_B = \Delta G_B = \int dG_B = \int_{p^\ominus}^{p} V_B^\infty(l)\,dp = \mu_{b,B}(T, p, b^\ominus) - \mu_{b,B}^\ominus(T)$$

$$\mu_{b,B}(T, p, b^\ominus) = \mu_{b,B}^\ominus(T) + \int_{p^\ominus}^{p} V_B^\infty(l)\,dp$$

Then, $\mu_{b,B}(\text{solute}) = \mu_{b,B}(\text{solute}, T, p, b^\ominus) + RT\ln\dfrac{b_B}{b^\ominus} = \mu_{b,B}^\ominus(T) + \displaystyle\int_{p^\ominus}^{p} V_B^\infty(l)\,dp + RT\ln\dfrac{b_B}{b^\ominus}$

So $$\mu_{b,B} = \mu_{b,B}^\ominus(\text{T}) + RT\ln\frac{b_B}{b^\ominus} \tag{4.7.2}$$

Similarly $$\mu_{c,B} = \mu_{c,B}^\ominus(\text{T}) + RT\ln\frac{c_B}{c^\ominus} \tag{4.7.3}$$

4.7.3 Distribution law of ideal dilute solution(理想稀溶液的分配定律)

If ideal dilute solutions consist of solute and two liquid phases(α, β) which are immiscible and coexistent at given temperature and pressure, the molalities are b_B^α and b_B^β, the ratio of b_B^α/b_B^β must be a constant, that is, $b_B^\alpha/b_B^\beta = K$.

Let's prove it.

According to chemical potential of solute in ideal dilute solution,

$$\mu_{b,B} = \mu_{b,B}^\ominus(T) + RT\ln\frac{b_B}{b^\ominus}$$

We have

$$\mu_{b,B}^\alpha = \mu_{b,B}^{\ominus,\alpha}(T) + RT\ln\frac{b_B^\alpha}{b^\ominus}$$

$$\mu_{b,B}^\beta = \mu_{b,B}^{\ominus,\beta}(T) + RT\ln\frac{b_B^\beta}{b^\ominus}$$

When

$$\alpha \xrightarrow{\text{equilibrium}} \beta,\quad \mu_{b,B}^\alpha = \mu_{b,B}^\beta$$

Then

$$\mu_{b,B}^{\ominus,\alpha}(T) + RT\ln\frac{b_B^\alpha}{b^\ominus} = \mu_{b,B}^{\ominus,\beta}(T) + RT\ln\frac{b_B^\beta}{b^\ominus}$$

So

$$\ln\frac{b_B^\alpha}{b_B^\beta} = [\mu_{b,B}^{\ominus,\beta}(T) - \mu_{b,B}^{\ominus,\alpha}(T)]/RT$$

When $dT=0$,

$$\frac{b_B^\alpha}{b_B^\beta} = K(T)$$

Example 4.4

HCl dissolved in benzene reaches equilibrium at 20℃. When the partial pressure of HCl in gas phase is 101.325 kPa, the mole fraction of HCl in solution is 0.0425. It is known that the saturated vapor pressure of benzene is 10.0kPa at 20℃. If the total pressure of HCl and benzene vapor is 101.325kPa at 20℃, how many grams of HCl can be dissolved in 100g of benzene? ($M_{HCl}=36.5$, $M_{benzene}=78$)

***Answer*:**

In dilute solution, HCl is solute denoted by B, benzene is solvent denoted by A. Solute B obeys Henry's law. Solvent A obeys Raoult's law, therefore,

$$p_B = k_{x,B}x_B$$

$$k_{x,B} = \frac{p_B}{x_B} = \frac{101.325}{0.0425} = 2384.12\text{kPa}$$

When the total pressure is 101.325kPa,

$$p = p_A^* x_A + k_{x,B}x_B = p_A^*(1-x_B) + k_{x,B}x_B$$

$$x_B = \frac{p-p_A^*}{k_{x,B}-p_A^*} = \frac{101.325-10.0}{2384.12-10.0} = 0.0385$$

$$x_B = \frac{n_B}{n_A+n_B} = \frac{w/36.5}{100/78+w/36.5} = 0.0385$$

$$w = 1.87\text{g}$$

That is, 1.87g HCl can be dissolved in 100g benzene.

Example 4.5

At 97.11℃, alcohol aqueous solution can be regarded as ideal dilute solution, when the mass fraction of alcohol is 3%, the total vapor pressure of this solution is 101.325kPa, the vapor pressure of pure water is 91.3kPa at the same temperature. Please calculate the total vapor pressure of the solution when the mole fraction of alcohol is 2%.

***Answer*:**

A denotes solvent water and B denotes solute alcohol, alcohol aqueous solution can be regarded as ideal dilute solution, then A obeys Raoult's law and B obeys Henry's law, so:

$$p = p_A + p_B = p_A^* x_A + k_{x,B} x_B$$

$$x_B = \frac{\dfrac{w_B}{M_B}}{\dfrac{w_B}{M_B} + \dfrac{w_A}{M_A}} = 0.01195$$

Example 4.5

$$101.325\text{kPa} = 91.3\text{kPa}(1 - 0.01195) + k_{x,B} \times 0.01195$$

So, $k_{x,B} = 930\text{kPa}$

when $x_B = 0.02$

$$p = p_A + p_B = p_A^* x_A + k_{x,B} x_B = (91.3 \times 0.98 + 930 \times 0.02)\ \text{kPa} = 108.1\text{kPa}$$

4.8 Colligative properties of ideal dilute solution

(理想稀溶液的依数性)

Colligative properties(依数性)—the properties depend on the amount of solute B but not on its nature.

4.8.1 Depression of vapor pressure (vapor pressure of solvent A)(蒸气压下降)

We consider a solution of an involatile solute in a solvent. By involatile, we mean that the concentration of the solute to the vapor pressure above the solution is negligible.

$$\Delta p_A = p_A^* - p_A = p_A^* x_B \tag{4.8.1}$$

Where Δp_A is depression value of vapor pressure of solvent A, p_A^* is the saturated vapor pressure of pure solvent A, x_B is the mole fraction of solute B.

That is, after adding the involatile solute B, the vapor pressure of solvent A will depress.

Under these conditions, Δp_A is independent of the nature of B and depends only on its mole fraction in solution.

4.8.2 Depression of freezing point (precipitation of solid pure solvent)(凝固点降低)

For ideal dilute solution, see figure 4.4. Solvent A obeys Raoult's law, then:

$$p_A = p_A^* x_A$$

Since, $0<x_A<1$

So, $p_A < p_A^*$

Defining the freezing-point depression ΔT_f as

$$\Delta T_f \overset{\text{def}}{=\!=} T_f^* - T_f = k_f b_B \qquad (4.8.2)$$

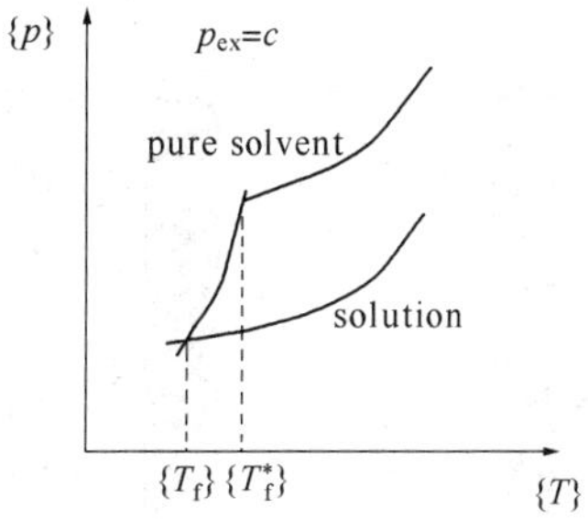

Fig. 4.4 Depression of freezing point

Where ΔT_f is the freezing-point depression for the ideal dilute solution and T_f^* is the freezing point of pure solvent A.

k_f is **freezing point depression coefficient.**

$$k_f \overset{\text{def}}{=\!=} \frac{R(T_f^*)^2 M_A}{\Delta_{fus} H_{m,A}^*}$$

k_f only depends on solvent A.

Application:

(1) Determination of the solvent activity factor.

(2) Determination of molar masses of small molecules, such as: urea.

4.8.3 Elevation of boiling point (solute B: involatile)(沸点升高)

For ideal dilute solution, see figure 4.5.

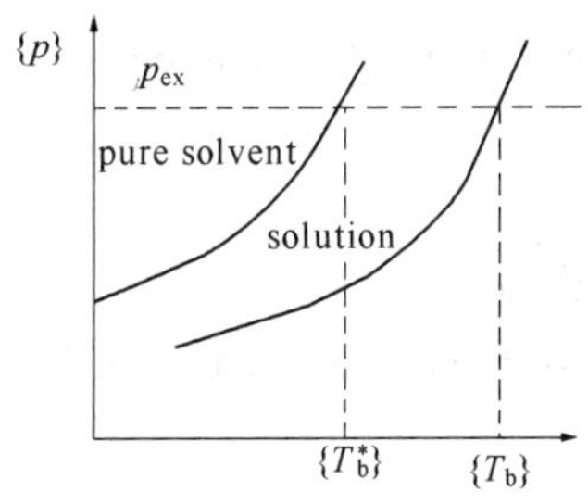

Fig. 4.5 Elevation of boiling point

Solvent A obeys Raoult's law, then:

$$p_A = p_A^* x_A$$

Since, $0<x_A<1$

So, $p_A < p_A^*$

When solute B is involatile,

$$p = p_A < p_A^*$$

Defining the boiling-point elevation ΔT_b as

$$\Delta T_b \overset{\text{def}}{=\!=} T_b - T_b^* = k_b b_B \qquad (4.8.3)$$

Where ΔT_b is the boiling-point elevation for the ideal dilute solution and T_b^* is the boiling point of pure solvent A.

k_b is **boiling point elevation coefficient.**

$$k_b \overset{\text{def}}{=\!=} \frac{R(T_b^*)^2 M_A}{\Delta_{vap} H_{m,A}^*}$$

k_b only depends on solvent A.

Boiling-point elevations can be used for molar mass determination, but the method is subject to greater experimental errors than the freezing-point depression method.

4.8.4 Osmotic pressure(渗透压)

Semipermeable membrane—membrane allows only certain chemical species (for example, the solvent not the solute) to pass through it.

In figure 4.6, the semipermeable membrane allows solvent A to pass through it but does not allow the passage of solute B. In the left tube, we put pure A, and in the right, a solution of B in A.

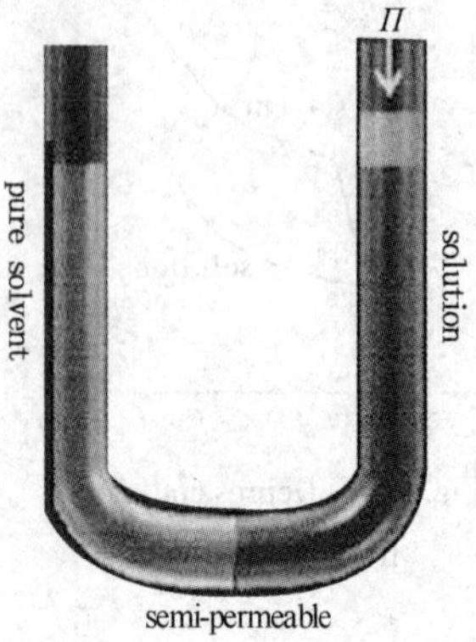

Fig. 4. 6 Osmotic pressure

Suppose that we initially fill the tubes so that the heights of the liquids in the two tubes are equal. The tubes are thus initially at equal pressure. The chemical potential of A on the left is μ_A^*, if the solution on the right is dilute enough to be considered ideally dilute, then $\mu_{A,R} = \mu_A^* + RT\ln x_A$, which is less than $\mu_{A,L} = \mu_A^*$, since x_A is less than 1. Since $\mu_{A,L} > \mu_{A,R}$, substance A will flow through the membrane from left to right, slightly diluting the solution on the right. The liquid in the right tube rises, thereby increasing the pressure in the right tube. The increase in pressure increases $\mu_{A,R}$ until eventually equilibrium is reached with $\mu_{A,L} = \mu_{A,R}$.

Let the equilibrium pressures in the left and right tube be p and $p + \Pi$, respectively. We call Π the osmotic pressure. It is the extra pressure that must be applied to the solution to make μ_A in the solution equal to μ_A^*, so as to achieve membrane equilibrium for species A between the solution and pure A.

Osmotic pressure(Π)—the pressure at osmotic equilibrium, that is, the pressure required to prevent osmosis.

$$\Pi = c_B RT \tag{4.8.4}$$

Where the concentration c_B equals n_B/V. (Note the formal resemblance to the state equation of the perfect gas, $p = cRT$, where $c = n/V$.)

Application:

Determination of molar masses of macromolecules, such as, high polymer.

The large values of Π given by dilute solution make osmotic-pressure measurements valuable in determining molar mass of substance with high molecular weights.

Example 4. 6

Osmotic pressure of one cane sugar aqueous solution is 252kPa at 30℃.

Please calculate: (1) the molality of the cane sugar in the cane sugar aqueous solution;

(2) the depression of freezing point of this solution ΔT_f;

(3) the elevation of boiling point of this solution at normal pressure ΔT_b.

Given: $k_f(H_2O) = 1.86\text{K} \cdot \text{mol}^{-1} \cdot \text{kg}^{-1}$, $k_b(H_2O) = 0.52\text{K} \cdot \text{mol}^{-1} \cdot \text{kg}^{-1}$

Answer:

(1) A denotes solvent water and B denotes solute cane sugar, then:

$$\Pi = c_B RT$$

$$c_B = \frac{\Pi}{RT} = \frac{252 \times 10^3}{8.315 \times 303.15} = 100\text{mol} \cdot \text{m}^{-3}$$

$$b_B = \frac{n_B}{m_A} = \frac{\frac{n_B}{V}}{\frac{m_A}{V}} = \frac{c_B}{\rho_A} = \frac{c_B}{\rho} = \frac{100}{1000} = 0.1\text{mol} \cdot \text{kg}^{-1}$$

(2) For $k_f(H_2O) = 1.86\text{K} \cdot \text{mol}^{-1} \cdot \text{kg}^{-1}$

$$\Delta T_f = k_f b_B = 1.86 \times 0.1 = 0.186\text{K}$$

(3) For k_b (H_2O) = 0.52K · mol^{-1} · kg^{-1}

$$\Delta T_b = k_b b_B = 0.52 \times 0.1 = 0.052\text{K}$$

4.9 Real liquid mixture, real liquid solution and activity (真实液态混合物、真实溶液和活度)

4.9.1 Positive deviation and negative deviation(正偏差与负偏差)

Any component in the real liquid mixture do not accord with Raoult's Law; in real liquid solution, solvent does not obey Raoult's Law, solute does not obey Henry's Law. There is deviation between ideal liquid mixture and real liquid mixture and there is deviation between ideal dilute solution and real liquid solution.

The deviation is shown in figure 4.7.

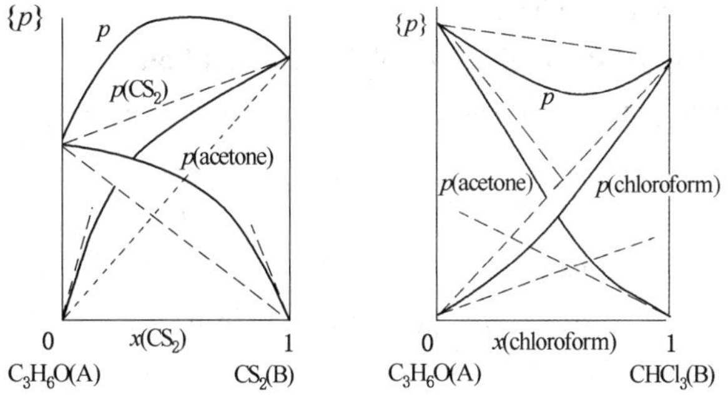

(a) 29℃ positive deviation (b) 35℃ negative deviation

Fig. 4.7 Positive deviation and negative deviation

Positive deviation—the vapor pressure in real liquid mixture or real liquid solution is larger than that calculated by Raoult's Law.

That is, $p_A > p_A^* x_A$; $p_B > p_B^* x_B$ $(0 \leqslant x_A \leqslant 1;\ 0 \leqslant x_B \leqslant 1)$

Negative deviation——the vapor pressure in real liquid mixture or real liquid solution is smaller than that calculated by Raoult's Law.

That is, $p_A < p_A^* x_A$; $p_B < p_B^* x_B$ $(0 \leqslant x_A \leqslant 1;\ 0 \leqslant x_B \leqslant 1)$

4.9.2 Activity and activity factor(活度与活度因子)

(1) Activity and activity factor of arbitrary component B in real liquid mixture

According to chemical potential of arbitrary component in mixture of ideal liquid,

$$\mu_B(\text{l}) = \mu_B^\ominus(\text{l},\ T) + RT\ln(x_B)$$

For real liquid mixture, we have

$$\mu_B(\text{l}) = \mu_B^\ominus(\text{l},\ T) + RT\ln a_B$$

Where a_B is activity, it is put forward by Lewis.

Activity(a_B)—composition of arbitrary component B in real liquid mixture, that is, the calibration to cope with deviations from ideal behavior.

$$a_B \overset{\text{def}}{=\!=\!=} f_B x_B$$

Where f_B is activity factor.

Since for mixture of ideal liquid, $p_B = p_B^* x_B$

Now for real liquid mixture, we have $p_B = p_B^* a_B$.

(2) Activity of solvent and solute in real solution

① Activity and osmotic factor of solvent A in real solution

According to chemical potential of solvent A in ideal dilute solution,

$$\mu_A(l) = \mu_A^\ominus(l, T) + RT\ln x_A$$

For real liquid solution, we have

$$\mu_A(l) = \mu_A^\ominus(l, T) + RT\ln a_A$$

Where a_A is activity of solvent A.

$$\varphi \overset{\text{def}}{=\!=\!=} -(M_A \sum_B b_B)^{-1} \ln a_A$$

Where φ is osmotic factor of solvent A.

② Activity and activity factor of solute B in real solution

According to chemical potential of solute B in ideal dilute solution,

$$\mu_{b,B} = \mu_{b,B}^\ominus(l, T) + RT\ln \frac{b_B}{b^\ominus}$$

For real liquid solution, we have

$$\mu_{b,B} = \mu_{b,B}^\ominus(l, T) + RT\ln a_{b,B}$$

Where $a_{b,B}$ is activity of solute B.

$$a_{b,B} \overset{\text{def}}{=\!=\!=} \gamma_{b,B} \frac{b_B}{b^\ominus}$$

Where $\gamma_{b,B}$ is activity factor of solute B.

Example 4.7

At 50℃, in liquid mixture ethyl alcohol (A)-H_2O(B), the liquid composition and the equilibrium vapor composition are determined to be $x_B = 0.556$, $y_B = 0.428$, the total pressure $p = 24.832\text{kPa}$, please calculate the activity and the activity factor of water. Given $\Delta_{vap} H_m(H_2O, l) = 40.668\text{kJ} \cdot \text{mol}^{-1}$ which can be regarded as a constant under the given conditions.

Answer:

$$\ln \frac{p_B^*}{101.325} = -\frac{40.668\times10^3}{8.315}\left(\frac{1}{50+273.15} - \frac{1}{373.15}\right)$$

We can get $p_B^* = 13.3\text{kPa}$

For $y_B = \dfrac{p_B^* a_B}{p}$

$$a_B = \frac{p y_B}{p_B^*} = \frac{24.832\times0.428}{13.3} = 0.799$$

For $a_B = \gamma_B x_B$, $\gamma_B = \dfrac{a_B}{x_B} = \dfrac{0.799}{0.556} = 1.437$

Supplementary Examples of Chapter 4

EXERCISES

(习题)

1. At 18℃ and 101. 325kPa, 0. 045g O_2 and 0. 02g N_2 can be dissolved in $1dm^3$ water. Now the solution of $1dm^3$ saturated by air of 202. 65kPa is heat to boil, O_2 and N_2 are drived out and dried. Please find the volume and composition of the dried gas at 18℃ and 101. 325kPa. Assumed that the air is ideal gas mixture and the composition of volume are: $\varphi(O_2)=21\%$, $\varphi(N_2)=79\%$.

Answer: $V=4.121\times10^{-5}m^3$; $y(O_2)=0.344$, $y(N_2)=0.656$

2. At 80℃, the saturated vapor pressure of benzene C_6H_6 and toluene C_7H_8 is 100kPa and 38. 7kPa respectively. C_6H_6 and C_7H_8 can form mixture of ideal liquid. The vapor composition is $y(C_6H_6)=0.300$ when the mixture of ideal liquid reaches gas-liquid equilibrium. Calculate the liquid composition and the total pressure.

Answer: $x(C_6H_6)=0.142$, $x(C_7H_8)=0.858$; $p=47.40kPa$

3. At 293K, HCl(g) is dissolved in C_6H_6(l) to form an ideal dilute solution. When the gas-liquid equilibrium is reached, the mole fraction of HCl in the liquid phase is 0. 0385, and the mole fraction of C_6H_6 in the gas phase is 0. 095. The saturated vapor pressure of C_6H_6(l) at 293K is known to be 10. 01kPa. Find: (1) The total pressure of the gas phase at gas-liquid equilibrium; (2) The Henry's coefficient $k_{x,B}$ of HCl(g) at 293K.

Answer: (1) 101. 32kPa; (2) 2381. 56kPa

4. 10g $C_6H_{12}O_6$ can be dissolved in 400g alcohol, the boiling point elevation of the solution is 0. 1428℃. The boiling point elevation of the solution is 0. 1250℃ while there is 2g organics being dissolved in 100g alcohol. Please calculate the molar mass of the organics.

Answer: $165g\cdot mol^{-1}$

5. Blood can be regarded as aqueous solution, which is freezed at −0. 56℃ and 101. 325kPa. It is known that the freezing point depression coefficient of water $k_f=1.86K\cdot mol^{-1}\cdot kg$. Please find:

(1) The osmotic pressure of blood at 37℃;

(2) At the same temperature, how much $C_{12}H_{22}O_{11}$ is needed to put into $1dm^3$ aqueous solution so that the osmotic presure will be the same as blood.

Answer: (1) 777kPa; (2) 103g

6. 0. 5455g solute was solved into 25g CCl_4. The equilibrium vapor partial pressure of CCl_4 was 11. 19kPa. While at the same temperature, the saturated vapor pressure of pure CCl_4 is 11. 4kPa. Please find:

(1) The molar mass of the solute;

(2) According to the results of element analysis, the solute is composed of 94. 34% C and 5. 66% H, try to determine the molecular formula of the solute.

Answer: $M_B=177g\cdot mol^{-1}$; $C_{14}H_{10}$

7. By putting 13. 7g $C_6H_5C_6H_5$ into 100g C_6H_6, the boiling point of the solution was 82. 4℃.

Given that the boiling point of pure C_6H_6 is 80.1℃. Please calculate:

(1) The boiling point elevation coefficient of C_6H_6;

(2) Molar vaporization enthalpy $\Delta_{vap}H_m(C_6H_6)$ of C_6H_6.

Answer: $k_b = 2.58K \cdot mol^{-1} \cdot kg$; $\Delta_{vap}H_m(C_6H_6) = 31.4kJ \cdot mol^{-1}$

8. At a certain temperature, iodine was dissolved in CCl_4. The solution can be regarded as ideal dilute solution when the mole fraction of iodine $x(I_2)$ was 0.01–0.04. Now, two groups of data of the partial pressure in vapor phase and mole fraction in liquid phase of I_2 have been showed as below when I_2 is in liquid-gas equilibrium:

$p(I_2, g)$/kPa	1.638	16.72
$x(I_2)$	0.03	0.5

Please calculate the activity and activity factor of iodine when $x(I_2)= 0.5$.

Answer: 0.306; 0.612

Chapter 5　Chemical equilibrium(化学平衡)

Learning objectives:

(1) Define the *Gibbs function of a reaction* and calculate it from tables of Gibbs functions of formation.

(2) Express the *equilibrium constant* of a reaction in terms of the standard Gibbs functions of the reactants and products.

(3) Derive and use *van't Hoff isothermal equation*.

(4) State how the temperature affects the position of equilibrium of exothermic and endothermic reactions.

(5) State how the pressure affects the equilibrium constant.

Chemical reactions can proceed in both positive and negative directions at the same times. How to control the reaction conditions to make the reaction proceed in the direction we need and what is the maximum extent to which the reaction can proceed under the given conditions? By applying the basic principles and laws of thermodynamics to chemical reactions, we can determine the direction of reaction, the conditions of equilibrium and the maximum limit that can be reached by reaction. We can also derive the quantitative relationship of substances in equilibrium and express it in terms of the equilibrium constant. This chapter deals with chemical equilibria in the light of some of the consequences of the second law of thermodynamics. The thermodynamic calculation of standard equilibrium constants of chemical reactions and the influence of some factors on chemical equilibrium are discussed.

5.1　Standard equilibrium constant of chemical reaction (化学反应的标准平衡常数)

5.1.1　Molar Gibbs function change of chemical reaction (摩尔反应吉布斯函数变)

For multi-component homogeneous system,

$$dG = -SdT + Vdp + \sum_B \mu_B dn_B$$

If $dT=0$, $dp=0$, then, $dG = \sum_B \mu_B dn_B$

From $d\xi = \dfrac{dn_B}{\nu_B}$, we have $dn_B = \nu_B d\xi$

So
$$dG = \sum_B \nu_B \mu_B d\xi$$

Therefore
$$(\partial G/\partial \xi)_{T,p} = \sum_B \nu_B \mu_B$$

Definite $$\Delta_r G_m \xlongequal{\text{def}} \sum_B \nu_B \mu_B \tag{5.1.1}$$

Where $\Delta_r G_m$ is molar Gibbs function change of chemical reaction.

$$A \xlongequal{\text{def}} -\sum \nu_B \mu_B = -\Delta_r G_m$$

Where A is affinity of chemical reaction.

So, we can get the equilibrium criterion of chemical reaction,

$$\Delta_r G_m = \sum_B \nu_B \mu_B \leqslant 0 = \begin{cases} <, & \text{to right} \\ =, & \text{equilibrium} \\ >, & \text{to left} \end{cases} \tag{5.1.2}$$

Or

$$A \geqslant 0 = \begin{cases} >, & \text{to right} \\ =, & \text{equilibrium} \\ <, & \text{to left} \end{cases} \tag{5.1.3}$$

Similarly, $$\Delta_r G_m^\ominus(T) \xlongequal{\text{def}} \sum_B \nu_B \mu_B^\ominus(T) \tag{5.1.4}$$

Where $\Delta_r G_m^\ominus$ is standard molar Gibbs function change of chemical reaction, here, every component is in its standard state.

5.1.2 Definition of standard equilibrium constant of chemical reaction(化学反应标准平衡常数的定义)

For arbitrary reaction, $0 = \sum_B \nu_B B$.

$$K^\ominus(T) \xlongequal{\text{def}} \exp\left\{\left[-\sum_B \nu_B \mu_B^\ominus(T)\right]/RT\right\} = \exp\{[-\Delta_r G_m^\ominus(T)]/RT\} \tag{5.1.5}$$

That is, $\Delta_r G_m^\ominus(T) = -RT\ln K^\ominus(T)$

Comment on $K^\ominus(T)$:

(1) $K^\ominus$ is only function of temperature, $K^\ominus(T)$.

(2) Unit of $K^\ominus$ is 1.

(3) $K^\ominus$ is related to the reaction stoichiometric equation.

For instance, $SO_2+1/2O_2 = SO_3$, $\Delta_r G_{m,1}^\ominus(T) = -RT\ln K_1^\ominus(T)$

$2SO_2+O_2 = 2SO_3$, $\Delta_r G_{m,2}^\ominus(T) = -RT\ln K_2^\ominus(T)$

Since, $\Delta_r G_{m,2}^\ominus(T) = 2\Delta_r G_{m,1}^\ominus(T)$

So we have $K_2^\ominus(T) = [K_1^\ominus(T)]^2$

5.2 Thermodynamic calculation of standard equilibrium constant(标准平衡常数的热力学计算)

According to the definition of standard equilibrium constant of chemical reaction,

$$K^\ominus(T) \xlongequal{\text{def}} \exp\left\{\left[-\sum_B \nu_B \mu_B^\ominus(T)\right]/RT\right\} = \exp\{[-\Delta_r G_m^\ominus(T)]/RT\}$$

We have $\Delta_r G_m^\ominus(T) = -RT\ln K^\ominus(T)$, so if we want to calculated $K^\ominus(T)$, we can calculate $\Delta_r G_m^\ominus(T)$ firstly.

5.2.1 Calculate $\Delta_r G_m^\ominus (T)$ from $\Delta_f G_m^\ominus$ (B, β, T)

(1) Definition of standard molar Gibbs function change of formation

$\Delta_f G_m^\ominus$ (B, β, T) —the Gibbs function changes that occur when unit amount of the compound in its standard state is formed from its elements in their standard states, that is, standard molar Gibbs function change of formation reaction is the standard molar enthalpy of formation of product B.

Formation reaction:

① Reactant : steady state simple substance.

② Product B: $\nu_B = +1$.

Comment:

$$\Delta_f G_m^\ominus(\text{steady elementary substance},\ T) = 0 \tag{5.2.1}$$

(2) Calculate $\Delta_r G_m^\ominus (T)$ from $\Delta_f G_m^\ominus$ (B, β, T)

$$\Delta_r G_m^\ominus (T) = \sum_B \nu_B \Delta_f G_m^\ominus (B,\ \beta,\ T) \tag{5.2.2}$$

When T=298.15K, $\Delta_r G_m^\ominus (298.15\text{K}) = \sum_B \nu_B \Delta_f G_m^\ominus (B,\ \beta,\ 298.15\text{K})$

We can get $\Delta_f G_m^\ominus$ (B, β, 298.15K) from appendix IX.

5.2.2 Calculate $\Delta_r G_m^\ominus (T)$ from $\Delta_f H_m^\ominus$ (B, β, T), $\Delta_c H_m^\ominus$ (B, β, T), $S_m^\ominus$ (B, β, T) and $C_{p,m}$ (B, β, T)

(1) Calculate $\Delta_r H_m^\ominus (T)$

$$\Delta_r H_m^\ominus (298.15\text{K}) = \sum_B \nu_B \Delta_f H_m^\ominus (B,\ \beta,\ 298.15\text{K})$$

Or

$$\Delta_r H_m^\ominus (298.15\text{K}) = -\sum_B \nu_B \Delta_c H_m^\ominus (B,\ \beta,\ 298.15\text{K})$$

According to **Kirchhoff formula**, we can calculate $\Delta_r H_m^\ominus (T)$

$$\Delta_r H_m^\ominus(T) = \Delta_r H_m^\ominus(298.15\text{K}) + \int_{298.15\text{K}}^{T} \sum \nu_B C_{p,m}(B)\,dT$$

(2) Calculate $\Delta_r S_m^\ominus (T)$

$$\Delta_r S_m^\ominus (298.15\text{K}) = \sum_B \nu_B S_m^\ominus (B,\ \beta,\ 298.15\text{K})$$

$$\Delta_r S_m^\ominus(T) = \Delta_r S_m^\ominus(298.15\text{K}) + \int_{298.15\text{K}}^{T} \frac{\sum \nu_B C_{p,m}(B)\,dT}{T}$$

(3) Calculate $\Delta_r G_m^\ominus (T)$

$$\Delta_r G_m^\ominus (T) = \Delta_r H_m^\ominus (T) - T\Delta_r S_m^\ominus (T) \tag{5.2.3}$$

5.2.3 Calculate $\Delta_r G_m^\ominus (T)$ from relative reactions

Theoretical basis: G is a state function.

(1) C(graphite)+O_2(g) → CO_2(g), $K_1^\ominus (T)$

(2) CO(g)+1/2 O_2(g)→ CO_2(g), $K_2^\ominus (T)$

(3) C(graphite)+1/2 O_2(g)→ CO(g), $K_3^\ominus (T)$ = ?

Since, (3) = (1) −(2)

So $$\Delta_r G^{\ominus}_{m,3}(T) = \Delta_r G^{\ominus}_{m,1}(T) - \Delta_r G^{\ominus}_{m,2}(T)$$

$$-RT\ln K^{\ominus}_3 = -RT\ln K^{\ominus}_1 + RT\ln K^{\ominus}_2 = -RT\ln \frac{K^{\ominus}_1}{K^{\ominus}_2}$$

Therefore $$K^{\ominus}_3 = \frac{K^{\ominus}_1}{K^{\ominus}_2}$$

Example 5.1

According to following standard thermodynamic data, find $\Delta_r G^{\ominus}_m$ and $K^{\ominus}$ of reaction at 25℃: $SO_2(g) + (1/2)O_2(g) = SO_3(g)$

Substance	$SO_2(g)$	$SO_3(g)$	$O_2(g)$
$\Delta_f H^{\ominus}_m$ (298.15K)/kJ · mol^{-1}	−296.9	−359.2	0
$S^{\ominus}_m$ (298.15K)/J · K^{-1} · mol^{-1}	248.1	256.23	205.03

***Answer*:**

$\Delta_r H^{\ominus}_m(298.15K) = \sum_B \nu_B \Delta_f H^{\ominus}_m(B, \beta, 298.15K) = -62.3kJ \cdot mol^{-1}$

$\Delta_r S^{\ominus}_m(298.15K) = \sum_B \nu_B S^{\ominus}_m(B, \beta, 298.15K) = -94.39J \cdot K^{-1} \cdot mol^{-1}$

$\Delta_r G^{\ominus}_m(298.15K) = \Delta_r H^{\ominus}_m(298.15K) - 298.15 \times \Delta_r S^{\ominus}_m(298.15K) = -35.06kJ \cdot mol^{-1}$

Since, $\Delta_r G^{\ominus}_m(298.15K) = -RT\ln K^{\ominus}(298.15K)$

$$K^{\ominus}(298.15K) = 7.2 \times 10^{-7}$$

5.3 Relations between $K^{\ominus}(T)$ and T (标准平衡常数与温度的关系)

5.3.1 Relations between $K^{\ominus}(T)$ and T (标准平衡常数与温度的关系)

From $\Delta_r G^{\ominus}_m(T) = -RT\ln K^{\ominus}(T)$, we have:

$$\ln K^{\ominus}(T) = -\frac{\Delta_r G^{\ominus}_m(T)}{RT}$$

Differentiation with respect to T gives

$$\frac{d\ln K^{\ominus}(T)}{dT} = -\frac{1}{R}\frac{d}{dT}\left[\frac{\Delta_r G^{\ominus}_m(T)}{T}\right]$$

According to Gibbs-Helmholtz equation, $\left[\frac{\partial}{\partial T}\left(\frac{G}{T}\right)\right]_p = -\frac{H}{T^2}$

We can get $$\frac{d}{dT}\left[\frac{\Delta_r G^{\ominus}_m(T)}{T}\right] = -\frac{\Delta_r H^{\ominus}_m(T)}{T^2}$$

So $$\frac{d\ln K^{\ominus}(T)}{dT} = \frac{\Delta_r H^{\ominus}_m(T)}{RT^2} \qquad (5.3.1)$$

(5.3.1) is **Van't Hoff equation**. $\Delta_r H_m^{\ominus}(T)$ is the standard enthalpy change for the reaction at temperature T. The greater the value of $\Delta_r H_m^{\ominus}(T)$, the faster the equilibrium constant $K^{\ominus}(T)$ changes with temperature.

5.3.2 Integral formula of Van't Hoff equation (范特霍夫方程的积分式)

(1) Indefinite integral formula (不定积分式)

Jacobus H. van't Hoff

Depending on **Van't Hoff equation**, $\dfrac{\mathrm{d}\ln K^{\ominus}(T)}{\mathrm{d}T}=\dfrac{\Delta_r H_m^{\ominus}(T)}{RT^2}$, we have:

$$\int \mathrm{d}\ln K^{\ominus}(T)=\int \frac{\Delta_r H_m^{\ominus}(T)}{R}\frac{\mathrm{d}T}{T^2}$$

Then, $\ln K^{\ominus}(T)=-\dfrac{\Delta_r H_m^{\ominus}(T)}{R}\cdot\dfrac{1}{T}+B$ [$\Delta_r H_m^{\ominus}(T)$ is constant] (5.3.2)

A plot of $\{\ln K^{\ominus}\}$ dependent on $\left\{\dfrac{1}{T}\right\}$ is a straight line,

$$\text{the slope}=-\Delta_r H_m^{\ominus}(T)/R$$

From the slop we can calculate $\Delta_r H_m^{\ominus}(T)$.

And the intercept $=B=\Delta_r S_m^{\ominus}(T)/R$

From the intercept we can calculate $\Delta_r S_m^{\ominus}(T)$.

According to $-RT\ln K^{\ominus}(T)=\Delta_r G_m^{\ominus}(T)=\Delta_r H_m^{\ominus}(T)-T\Delta_r S_m^{\ominus}(T)$

We have

$$\ln K^{\ominus}(T)=-\frac{\Delta_r H_m^{\ominus}(T)}{R}\cdot\frac{1}{T}+\frac{\Delta_r S_m^{\ominus}(T)}{R} \quad (5.3.3)$$

Comparison (5.3.3) with (5.3.2), we can get: $B=\Delta_r S_m^{\ominus}(T)/R$

(2) Definite integral formula (定积分式)

Depending on **Van't Hoff equation**, $\dfrac{\mathrm{d}\ln K^{\ominus}(T)}{\mathrm{d}T}=\dfrac{\Delta_r H_m^{\ominus}(T)}{RT^2}$, we have:

$$\int_{K_1^{\ominus}}^{K_2^{\ominus}} \mathrm{d}\ln K^{\ominus}(T)=\int_{T_1}^{T_2}\frac{\Delta_r H_m^{\ominus}(T)}{R}\frac{\mathrm{d}T}{T^2}$$

$\Delta_r H_m^{\ominus}(T)$ usually varies rather slowly with T, so if the temperature interval T_2-T_1 is reasonably small, it's generally a good approximation to neglect the temperature dependence of $\Delta_r H_m^{\ominus}(T)$. Moving $\Delta_r H_m^{\ominus}(T)$ outside the integral sign and integrating, we get

$$\ln\frac{K_2^{\ominus}(T_2)}{K_1^{\ominus}(T_1)}=\frac{\Delta_r H_m^{\ominus}}{R}\left(\frac{1}{T_1}-\frac{1}{T_2}\right)\ [\ \Delta_r H_m^{\ominus}(T)\ \text{is constant}] \quad (5.3.4)$$

If $\Delta_r H_m^{\ominus}(T)$ is essentially constant over the temperature range, one can use (5.3.4) to find $\Delta_r H_m^{\ominus}(T)$ from only two values of $K^{\ominus}$ at different temperatures.

Example 5.2

In the scope of temperature 454~475K, $2C_2H_5OH(g) \rightleftharpoons CH_3COOC_2H_5(g)+2H_2(g)$, the relationship of $K^\ominus$ and T is $\lg K^\ominus = -\frac{2100}{T/K} + 4.67$. At 473K, the standard molar enthalpy of formation of ethyl alcohol is $\Delta_f H_m^\ominus = -235.34\,kJ \cdot mol^{-1}$. Calculate $\Delta_f H_m^\ominus(473K)$ of ethyl acetate.

Answer:

$$\lg K^\ominus = -\frac{2100}{T/K} + 4.67$$

Then $\ln K^\ominus = -\frac{2100}{T/K} \times \ln 10 + 4.67 \times \ln 10$

According to the indefinite integral formula of Van't Hoff equation,

$\ln K^\ominus = -\frac{\Delta_r H_m^\ominus}{RT} + C$

We can get $\Delta_r H_m^\ominus = R \times 2100 \times \ln 10 = 40.2\,kJ \cdot mol^{-1}$

$\Delta_r H_m^\ominus = \sum v_B \Delta_f H_m^\ominus(B)$, therefore,

$\Delta_f H_m^\ominus(CH_3COOC_2H_5) = \Delta_r H_m^\ominus + 2\Delta_f H_m^\ominus(C_2H_5OH) = -430.5\,kJ \cdot mol^{-1}$

5.4 Chemical equilibrium of perfect gas mixture reaction
(理想气体混合物反应的化学平衡)

5.4.1 Expression of standard equilibrium constant (标准平衡常数的表达式)

For perfect gas mixture reaction: $0 = \sum_B \nu_B B$

$$\Delta_r G_m(T) = \sum_B \nu_B \mu_B$$

According to expression of chemical potential for component B in pg mixture,

$$\mu_B(g) = \mu_B^\ominus(g, T) + RT\ln(p_B/p^\ominus)$$

We have $\Delta_r G_m(T) = \sum_B \nu_B \mu_B^\ominus(g, T) + \sum_B \nu_B RT\ln(p_B/p^\ominus)$

Since $\Delta_r G_m^\ominus(T) \stackrel{\text{def}}{=\!=} \sum_B \nu_B \mu_B^\ominus(T)$

So $\Delta_r G_m(T) = \Delta_r G_m^\ominus(T) + RT\sum_B \ln(p_B/p^\ominus)^{\nu_B}$

At chemical equilibrium, $0 = \Delta_r G_m(T) = \Delta_r G_m^\ominus(T) + RT\ln\prod_B (p_B^{eq}/p^\ominus)^{\nu_B}$

So that $\Delta_r G_m^\ominus(T) = -RT\ln\prod_B (p_B^{eq}/p^\ominus)^{\nu_B}$

According to the definition of standard equilibrium constant of reaction

$$\Delta_r G_m^\ominus(T) = -RT\ln K^\ominus(T)$$

We can get the expression of standard equilibrium constant

$$K^\ominus(pgm, T) = \prod_B (p_B^{eq}/p^\ominus)^{\nu_B} \qquad (5.4.1)$$

For instance, pg mixture reaction: $aA + bB \rightarrow yY + zZ$ (pg)

$$K^\ominus(pgm, T) = \frac{(p_Y^{eq}/p^\ominus)^y (p_Z^{eq}/p^\ominus)^z}{(p_A^{eq}/p^\ominus)^a (p_B^{eq}/p^\ominus)^b}$$

Since $$p_B^{eq} = p^{eq} y_B^{eq}$$

So $$K^\ominus(\text{pgm}, T) = \prod_B (y_B^{eq} p^{eq}/p^\ominus)^{\nu_B} \quad (5.4.2)$$

Depending on the expression of standard equilibrium constant, we can calculate equilibrium composition y_B^{eq} according to standard equilibrium constant.

5.4.2 Other expression of equilibrium constant(平衡常数的其他表达式)

$$K_p = \prod_B (p_B^{eq})^{\nu_B}$$

So $$K^\ominus = K_p / (p^\ominus)^{\sum_B \nu_B}$$

$$K_n = \prod_B (n_B)^{\nu_B}$$

$$K_y = \prod_B (y_B)^{\nu_B}$$

If $$\sum_B \nu_B = 0$$

Then $$K^\ominus = K_p = K_n = K_y$$

5.5 Chemical equilibrium of real gas mixture reaction (真实气体混合物反应的化学平衡)

For real gas mixture reaction: $0 = \sum_B \nu_B B$

$$\Delta_r G_m(T) = \sum_B \nu_B \mu_B$$

According to expression of chemical potential for component B in real gas mixture,

$$\mu_B(g) = \mu_B^\ominus(g, T) + RT\ln(\tilde{p}_B/p^\ominus)$$

We have, $\Delta_r G_m(T) = \sum_B \nu_B \mu_B^\ominus(g, T) + \sum_B \nu_B RT\ln(\tilde{p}_B/p^\ominus)$

Since $$\Delta_r G_m^\ominus(T) \overset{\text{def}}{=\!=} \sum_B \nu_B \mu_B^\ominus(T)$$

So $$\Delta_r G_m(T) = \Delta_r G_m^\ominus(T) + RT\sum_B \ln(\tilde{p}_B/p^\ominus)^{\nu_B}$$

At chemical equilibrium, $0 = \Delta_r G_m(T) = \Delta_r G_m^\ominus(T) + RT\ln\prod_B (\tilde{p}_B^{eq}/p^\ominus)^{\nu_B}$

So, $$\Delta_r G_m^\ominus(T) = -RT\ln\prod_B (\tilde{p}_B^{eq}/p^\ominus)^{\nu_B}$$

According to the definition of standard equilibrium constant of reaction

$$\Delta_r G_m^\ominus(T) = -RT\ln K^\ominus(T)$$

We can get the expression of standard equilibrium constant

$$K^\ominus(\text{gm}, T) = \prod_B (\tilde{p}_B^{eq}/p^\ominus)^{\nu_B} \quad (5.5.1)$$

According to the formula of fugacity, $\tilde{p}_B^{eq} = \varphi_B^{eq} p_B^{eq}$

So, $K^\ominus(\text{gm}, T) = \prod_B (\varphi_B^{eq} p_B^{eq}/p^\ominus)^{\nu_B} = \prod_B (\varphi_B^{eq})^{\nu_B} \cdot \prod_B (p_B^{eq}/p^\ominus)^{\nu_B}$ (5.5.2)

5.6 Van't Hoff isothermal equation and determination of direction of chemical reaction (范特霍夫等温方程与化学反应方向的判定)

For perfect gas mixture reaction: $0=\sum_B \nu_B B$

$$\Delta_r G_m(T)=\sum_B \nu_B \mu_B$$

According to expression of chemical potential for component B in pg mixture,

$$\mu_B(g)=\mu_B^\ominus(g,\ T)+RT\ln(p_B/p^\ominus)$$

We have, $\Delta_r G_m(T)=\sum_B \nu_B \mu_B^\ominus(g,\ T)+\sum_B \nu_B RT\ln(p_B/p^\ominus)$

Since $$\Delta_r G_m^\ominus(T) = \sum_B \nu_B \mu_B^\ominus(T)$$

So $$\Delta_r G_m(T)=\Delta_r G_m^\ominus(T)+RT\sum_B \ln(p_B/p^\ominus)^{\nu_B} \qquad (5.6.1)$$

Where p_B is partial pressure of arbitrary component B in arbitrary state, (5.6.1) is **Van't Hoff isothermal equation of pg reaction.**

Definite $$J_p(\text{pgm},\ T)\overset{\text{def}}{=\!=\!=}\prod_B (p_B/p^\ominus)^{\nu_B}$$

Where J_p (pgm, T) is partial pressure ratio of perfect gas mixture reaction.

For real gas mixture reaction: $0=\sum_B \nu_B B$

$$\Delta_r G_m(T)=\sum_B \nu_B \mu_B$$

According to expression of chemical potential for component B in real gas mixture,

$$\mu_B(g)=\mu_B^\ominus(g,\ T)+RT\ln(\tilde{p}_B/p^\ominus)$$

We have, $\Delta_r G_m(T)=\sum_B \nu_B \mu_B^\ominus(g,\ T)+\sum_B \nu_B RT\ln(\tilde{p}_B/p^\ominus)$

Since $$\Delta_r G_m^\ominus(T) = \sum_B \nu_B \mu_B^\ominus(T)$$

So $$\Delta_r G_m(T)=\Delta_r G_m^\ominus(T)+RT\sum_B \ln(\tilde{p}_B/p^\ominus)^{\nu_B} \qquad (5.6.2)$$

(5.6.2) is **Van't Hoff isothermal equation of real gas reaction.**

Definite $$J_p(\text{gm},\ T)\overset{\text{def}}{=\!=\!=}\prod_B (\tilde{p}_B/p^\ominus)^{\nu_B}$$

Where J_p (gm, T) is partial fugacity ratio of real gas mixture reaction.

Combine (5.6.1) with (5.6.2), we have:

$$\Delta_r G_m(T)=\Delta_r G_m^\ominus(T)+RT\ln J_p(T)=-RT\ln K^\ominus(T)+RT\ln J_p(T)=RT\ln\frac{J_p(T)}{K^\ominus(T)}$$

According to Gibbs function criterion, $\Delta_r G_m \leqslant 0$
- $<$, to right
- $=$, equilibrium
- $>$, to left

We can get a new **equilibrium criterion of chemical reaction**:

$J_p(T) < K^\ominus(T)$, $\Delta_r G_m < 0$, to right

$J_p(T) = K^\ominus(T)$, $\Delta_r G_m = 0$, at equilibrium

$J_p(T) > K^\ominus(T)$, $\Delta_r G_m > 0$, to left

When we determine the direction of reaction using the new equilibrium criterion of chemical reaction, we must calculate $J_p(T)$ and $K^\ominus(T)$ firstly. $J_p(T)$ can be calculated by the definition of $J_p(\text{pgm}, T) \stackrel{\text{def}}{=\!=} \prod_B (p_B/p^\ominus)^{\nu_B} = \prod_B (py_B/p^\ominus)^{\nu_B}$, and $K^\ominus(T)$ can be calculated by $\Delta_r G_m^\ominus(T) = -RT\ln K^\ominus(T)$.

Example 5.3

Relative data of pg reaction $A(g) + 2B(g) = Y(g) + 4Z(g)$ are:

Substance	$\Delta_f H_m^\ominus$ (298.15K) / kJ · mol^{-1}	$S_m^\ominus$ (298.15K) / J · K^{-1} · mol^{-1}	$C_{p,m}$ (298.15K) / J · K^{-1} · mol^{-1}
A(g)	−74.84	186.0	3
B(g)	−241.84	188.0	14
Y(g)	−393.42	214.0	11
Z(g)	0	130.0	5

Please calculate and determine: (1) the direction of chemical reaction while $T = 800$K, $p = 0.1$MPa and the mole fraction of A, B, Y and Z are 0.3, 0.2, 0.3 and 0.2 respectively; (2) if the condition is same as (1), how to change temperature to make the direction backward.

Example 5.3

Answer:

$$(1)\ \Delta_r H_m^\ominus(298.15\text{K}) = \sum_B \nu_B \Delta_f H_m^\ominus(\text{B}, \beta, 298.15\text{K})$$

$$= (-393.42) - (-241.84 \times 2) - (-74.84) = 165.1\text{kJ} \cdot \text{mol}^{-1}$$

$$\Delta_r S_m^\ominus(298.15\text{K}) = \sum_B \nu_B S_m^\ominus(\text{B}, \beta, 298.15\text{K})$$

$$= (214.0 + 4 \times 130.0 - 186.0 - 2 \times 188.0) = 172\text{J} \cdot \text{K}^{-1} \cdot \text{mol}^{-1}$$

$$\sum \nu_B C_{p,m}(\text{B}) = (20 + 11 - 28 - 3) = 0$$

$$\Delta_r G_m^\ominus(T) = \Delta_r H_m^\ominus(T) - T\Delta_r S_m^\ominus(T) = \Delta_r H_m^\ominus(298.15\text{K}) - T\Delta_r S_m^\ominus(298.15\text{K})$$

$$= 165100\text{J} \cdot \text{mol}^{-1} - 172T\ \text{J} \cdot \text{mol}^{-1}$$

Since, $\Delta_r G_m^\ominus(T) = -RT\ln K^\ominus(T)$

$$K^\ominus(800\text{K}) = \exp[-(165100 - 172 \times 800) / 8.315 \times 800] = 0.0180 \approx 0.0200$$

$$J_p(800\text{K}) \stackrel{\text{def}}{=\!=} \prod_B (p_B/p^\ominus)^{\nu_B} = \prod_B (py_B/p^\ominus)^{\nu_B}$$

$$= \frac{(0.2 \times 10^5\text{Pa}/1 \times 10^5\text{Pa})^4 (0.3 \times 10^5\text{Pa}/10^5\text{Pa})}{(0.3 \times 10^5\text{Pa}/10^5\text{Pa})(0.2 \times 10^5\text{Pa}/10^5\text{Pa})^2} = 0.04$$

For $J_p(T) > K^\ominus(T)$, that is $\Delta_r G_m > 0$

So the direction of chemical reaction is to left while $T = 800$K.

(2) If the direction of chemical reaction is to right, the condition is:

$$\Delta_r G_m(T) = \Delta_r G_m^\ominus(T) + RT\ln J_p(T) < 0$$

That is, $165100-172T<-8.315\ T\ \ln 0.04$

So, $T>831\text{K}$

Example 5.4

Relative data of pg reaction $2\text{A(g)} \Longrightarrow \text{Y(g)}$ are:

Substance	$\Delta_f H_m^\ominus$ (298.15K) / $\text{kJ} \cdot \text{mol}^{-1}$	$S_m^\ominus$ (298.15K) / $\text{J} \cdot \text{K}^{-1} \cdot \text{mol}^{-1}$	$C_{p,m}$ (298.15K) / $\text{J} \cdot \text{K}^{-1} \cdot \text{mol}^{-1}$
A(g)	35	250	38.0
Y(g)	10	300	76.0

Please calculate and determine: (1) the direction of chemical reaction while $T=310\text{K}$, $p=100\text{kPa}$ and the mole fraction of A and Y are 0.5 respectively; (2) if the other conditions do not change, how to change ① pressure p ② temperature T ③ composition y_A to make the direction backward.

Answer:

(1) $\Delta_r H_m^\ominus(298.15\text{K}) = \sum_B \nu_B \Delta_f H_m^\ominus(\text{B}, \beta, 298.15\text{K})$

$$= (10 - 2 \times 35) = -60\text{kJ} \cdot \text{mol}^{-1}$$

$$\Delta_r S_m^\ominus(298.15\text{K}) = \sum_B \nu_B S_m^\ominus(\text{B}, \beta, 298.15\text{K})$$

$$= (300 - 2 \times 250) = -200\text{J} \cdot \text{K}^{-1} \cdot \text{mol}^{-1}$$

$$\sum \nu_B C_{p,m}(\text{B}) = (76 - 2 \times 38) = 0$$

$$\Delta_r G_m^\ominus(T) = \Delta_r H_m^\ominus(T) - T\Delta_r S_m^\ominus(T) = \Delta_r H_m^\ominus(298.15\text{K}) - T\Delta_r S_m^\ominus(298.15\text{K})$$

$$= -60000 + 310 \times 200 = 2000\text{J} \cdot \text{mol}^{-1}$$

Since, $\Delta_r G_m^\ominus(T) = -RT\ln K^\ominus(T)$

$K^\ominus(310\text{K}) = \exp[-2000/8.315 \times 310] = 0.46$

$$J_p(310\text{K}) \overset{\text{def}}{=\!=\!=} \prod_B (p_B/p^\ominus)^{\nu_B} = \prod_B (py_B/p^\ominus)^{\nu_B} = \frac{0.5p/p^\ominus}{(0.5p/p^\ominus)^2} = 2.0$$

For $J_p(T) > K^\ominus(T)$, that is $\Delta_r G_m > 0$

So the direction of chemical reaction is to left.

(2) If the direction of chemical reaction is to right, the condition is:

$$J_p(T) < K^\ominus(T)$$

① $J_p = \dfrac{p^\ominus}{0.5p} < 0.46$, we have, $p>434.8\text{kPa}$

② $K^\ominus(T) > J_p(T) = 2.0$, that is, $\ln K^\ominus(T) > \ln 2.0$

So, $\ln K^\ominus = \dfrac{-\Delta_r G_m^\ominus}{RT} = \dfrac{-\Delta_r H_m^\ominus}{RT} + \dfrac{\Delta_r S_m^\ominus}{R} = \dfrac{60000}{8.315T} + \dfrac{(-200)}{8.315} > \ln 2.0$

We can get, $T<291.6\text{K}$

③ $J_p = \dfrac{1-y_A}{y_A^2} < 0.46$, we have, $y_A>0.745$

5.7 Calculation of equilibrium conversion of reactant and equilibrium composition of system (反应物的平衡转化率与系统平衡组成的计算)

5.7.1 Definition of equilibrium conversion of reactant and equilibrium composition of system (反应物的平衡转化率与系统平衡组成的定义)

We definite equilibrium conversion = $\dfrac{\text{consumption of any reactant at equilibrium}}{\text{the initial amounts of the same reactant}}$

That is
$$x_A^{eq} \overset{\text{def}}{=\!=} \frac{n_{A,0} - n_A^{eq}}{n_{A,0}} \tag{5.7.1}$$

Where A is principal reactant, x_A^{eq} is the conversion of A at equilibrium, it is the maximum conversion of A at the given conditions.

Equilibrium rate of production
$$= \frac{\text{the amounts of any reactant that is converted into appointed products}}{\text{the initial amounts of the same reactant}}$$

Equilibrium composition (y_B^{eq})—mole fraction of any component in equilibrium system.

5.7.2 Calculation of equilibrium conversion of reactant and equilibrium composition of system (反应物的平衡转化率与系统平衡组成的计算)

We can calculate x_A^{eq} and y_B^{eq} using $K^{\ominus}(T)$.

Example 5.5

The reaction of preparation $C_6H_5CH_3(g)$ using $CH_4(g)$ and $C_6H_6(g)$ is

$CH_4(g) + C_6H_6(g) = C_6H_5CH_3(g) + H_2(g)$, known $\Delta_f G_m^{\ominus}(500K)$ of $CH_4(g)$, $C_6H_6(g)$ and $C_6H_5CH_3(g)$ are $-33.68 kJ \cdot mol^{-1}$, $161.92 kJ \cdot mol^{-1}$, $172.38 kJ \cdot mol^{-1}$ respectively. Now suitable catalysts are added to mixture of reactants with $n(CH_4, g) : n(C_6H_6, g) = 1 : 1$, please find the maximum rate of production while $p = 10^5 Pa$.

Answer:

$$\Delta_r G_m^{\ominus}(500K) = \sum_B \nu_B \Delta_f G_m^{\ominus}(B, \beta, 500K)$$
$$= [172.38 - (-33.68) - 161.92] kJ \cdot mol^{-1} = 44.14 kJ \cdot mol^{-1}$$

Since, $\Delta_r G_m^{\ominus}(T) = -RT \ln K^{\ominus}(T)$

$$\ln K^{\ominus}(500K) = -\frac{\Delta_r G_m^{\ominus}(500K)}{RT}$$
$$= \frac{-44.14 kJ \cdot mol^{-1}}{8.315 J \cdot K^{-1} \cdot mol^{-1} \times 500K} = -10.62$$

$$K^{\ominus}(500K) = 2.44 \times 10^{-5}$$

Suppose there are 1mol $CH_4(g)$ and 1mol $C_6H_6(g)$ at the beginning of the reaction, and the equilibrium conversion of reactant is x^{eq}, then:

$$CH_4(g) + C_6H_6(g) \Longrightarrow C_6H_5CH_3(g) + H_2(g)$$

$t=0$	1	1	0	0	
$t=t$	$(1-x^{eq})$	$(1-x^{eq})$	x^{eq}	x^{eq}	total: 2mol

$$K^{\ominus} = \frac{\left(\frac{x^{eq}}{2} \cdot \frac{p^{eq}}{p^{\ominus}}\right)^2}{\left(\frac{(1-x^{eq})}{2} \cdot \frac{p^{eq}}{p^{\ominus}}\right)^2} = \frac{(x^{eq})^2}{(1-x^{eq})^2} = 2.44 \times 10^{-5}$$

So we can calculate $x^{eq} = 4.9 \times 10^{-3} = 0.49\%$, that is, the maximum rate of production is 0.49%.

Example 5.6

At 100℃, $COCl_2(g) = CO(g) + Cl_2(g)$, $\Delta_r S_m^{\ominus}(373K) = 125.6 J \cdot mol^{-1} \cdot K^{-1}$, $K^{\ominus} = 8.1 \times 10^{-9}$, calculate(1)the degree of dissociation of $COCl_2$ at 100℃ when the total pressure is 200kPa (2)$\Delta_r H_m^{\ominus}$ of the above reaction at 100℃(3) temperature when the degree of dissociation of $COCl_2$ is 0.1% and the total pressure is 200kPa, assume $\Delta_r C_{p,m} = 0$.

Answer:

(1) Let the degree of dissociation of $COCl_2$ be α, then

$$COCl_2(g) = CO(g) + Cl_2(g)$$

mole number at equilibrium $\quad 1-\alpha \quad \alpha \quad \alpha$

The total amount of substance is $n = 1+\alpha$, $\sum v_B = 1$

$$K^{\ominus} = K_n\left(\frac{p}{p^{\ominus} n}\right)^{\Sigma v_B} = \frac{\alpha^2}{1-\alpha} \times \frac{(200/100)}{1+\alpha} = 8.1 \times 10^{-9}$$

We can get $\alpha = 6.37 \times 10^{-5}$

(2) $\Delta_r G_m^{\ominus} = -RT \ln K^{\ominus} = -8.315 \times 373.15 \times \ln(8.1 \times 10^{-9}) = 57840 J \cdot mol^{-1}$

$\Delta_r H_m = \Delta_r G_m + T\Delta_r S_m = 104.7 kJ \cdot mol^{-1}$

(3) Assume the equilibrium constant is $K_2^{\ominus}$ when $\alpha = 0.1\%$, the corresponding temperature is T_2.

$$K_2^{\ominus} = \frac{\alpha^2}{1-\alpha^2} \times (p/p^{\ominus}) = 2 \times 10^{-6}$$

$K_1^{\ominus} = 8.1 \times 10^{-9}$, $\Delta C_{p,m} = 0$, $\Delta_r H_m$ is a constant, bring them to

$\ln(K_2^{\ominus}/K_1^{\ominus}) = \Delta_r H_m^{\ominus}(T_2 - T_1)/RT_2T_1$

We can get $T_2 = 446K$

5.8 The response of equilibrium to the conditions (各种因素对平衡的影响)

5.8.1 Temperature (温度)

Suppose we change T, keeping p constant.

According to **Van't Hoff equation**, $\dfrac{d\ln K^{\ominus}(T)}{dT} = \dfrac{\Delta_r H_m^{\ominus}(T)}{RT^2}$

Since $K^{\ominus}(T)$ and RT^2 are positive, the sign of $\frac{\mathrm{dln}\,K^{\ominus}(T)}{\mathrm{d}T}$ is the same as the sign of $\Delta_r H_m^{\ominus}(T)$.

(1) Endothermic reactions (吸热反应): $\Delta_r H_m^{\ominus}(T) > 0$

If increase temperature, $\mathrm{d}T > 0$, then $\mathrm{dln}\,K^{\ominus}(T) > 0$, that is, $K^{\ominus}(T)$ increases. Since product partial pressures are in the numerator of $K^{\ominus}(T)$, an increase in $K^{\ominus}(T)$ means an increase in the equilibrium values of the product partial pressure and a decrease in reactant partial pressure. Since $p_B = py_B$, and p is held fixed, the mole fraction undergo changes proportional to the changes in the partial pressure. Thus an increase in temperature at constant pressure will shift the equilibrium to the right.

If decrease temperature, $\mathrm{d}T < 0$, then $\mathrm{dln}\,K^{\ominus}(T) < 0$, that is, $K^{\ominus}(T)$ decreases, the equilibrium shift to the left.

(2) Exothermic reactions (放热反应): $\Delta_r H_m^{\ominus}(T) < 0$

If increase temperature, $\mathrm{d}T > 0$, then $\mathrm{dln}\,K^{\ominus}(T) < 0$, that is, $K^{\ominus}(T)$ decreases, the equilibrium shift to the left;

If decrease temperature, $\mathrm{d}T < 0$, then $\mathrm{dln}\,K^{\ominus}(T) > 0$, that is, $K^{\ominus}(T)$ increases, the equilibrium shift to the right.

Conclusion:

Endothermic reactions: a rise in T favors the reaction.

Exothermic reactions: a decline in T favors the reaction.

5.8.2 Pressure (压力)

Suppose we isothermally change the volume of the system, thereby changing the total pressure and changing of the partial pressure of each gas. Since T is constant, $K^{\ominus}(T)$ is constant.

According to the expression of standard equilibrium constant,

$$K^{\ominus}(T) = \prod_B (p_B^{eq}/p^{\ominus})^{\nu_B} = \prod_B (p^{eq} \cdot y_B^{eq}/p^{\ominus})^{\nu_B} = (p^{eq}/p^{\ominus})^{\sum\nu_B} \cdot \prod_B (y_B^{eq})^{\nu_B}$$

(1) $\sum \nu_B > 0$

If increase pressure, $(p^{eq}/p^{\ominus})^{\sum\nu_B}$ increases, then $\prod_B (y_B^{eq})^{\nu_B}$ decreases, that is, the equilibrium shift to the left;

If decrease pressure, $(p^{eq}/p^{\ominus})^{\sum\nu_B}$ decreases, then $\prod_B (y_B^{eq})^{\nu_B}$ increases, that is, the equilibrium shift to the right.

(2) $\sum \nu_B < 0$

If increase pressure, $(p^{eq}/p^{\ominus})^{\sum\nu_B}$ decreases, then $\prod_B (y_B^{eq})^{\nu_B}$ increases, that is, the equilibrium shift to the right;

If decrease pressure, $(p^{eq}/p^{\ominus})^{\sum\nu_B}$ increases, then $\prod_B (y_B^{eq})^{\nu_B}$ decreases, that is, the equilibrium shift to the left.

Conclusion:

A rise in p makes equilibrium shift to the direction of $\sum \nu_B < 0$.

5.8.3 Inert gas (惰性气体)

Inert gas—gases that exist in system and do not participate in reaction.

We know inert gases affect $\sum n_B$.

$\sum n_B$ —sum of amount-of-substance of total gases including inert gases in system.

Suppose we add some inert gas to an equilibrium mixture, holding T and p constant.

Since T is constant, $K^\ominus(T)$ is constant.

According to the expression of standard equilibrium constant,

$$K^\ominus(T) = \prod_B (p_B^{eq}/p^\ominus)^{\nu_B} = \prod_B (p^{eq}\cdot y_B^{eq}/p^\ominus)^{\nu_B} = \prod_B \left(\frac{p^{eq}\cdot n_B^{eq}}{p^\ominus\cdot \sum_B n_B^{eq}}\right)^{\nu_B}$$

$$= \left(\frac{p^{eq}}{p^\ominus\cdot \sum_B n_B^{eq}}\right)^{\sum \nu_B}\cdot \prod_B (n_B^{eq})^{\nu_B}$$

(1) $\sum \nu_B > 0$

If adding inert gas, $\sum_B n_B$ increases, $[p^{eq}/(p^\ominus \sum_B n_B^{eq})]$ decreases, then $\prod_B (n_B^{eq})^{\nu_B}$ increases, that is, the equilibrium shift to the right;

If decreasing inert gas, $\sum_B n_B$ decreases, $[p^{eq}/(p^\ominus \sum_B n_B^{eq})]$ increases, then $\prod_B (n_B^{eq})^{\nu_B}$ decreases, that is, the equilibrium shift to the left.

(2) $\sum \nu_B < 0$

If adding inert gas, $\sum_B n_B$ increases, $[p^{eq}/(p^\ominus \sum_B n_B^{eq})]$ increases, then $\prod_B (n_B^{eq})^{\nu_B}$ decreases, that is, the equilibrium shift to the left;

If decreasing inert gas, $\sum_B n_B$ decreases, $[p^{eq}/(p^\ominus \sum_B n_B^{eq})]$ decreases, then $\prod_B (n_B^{eq})^{\nu_B}$ increases, that is, the equilibrium shift to the right.

(3) $\sum \nu_B = 0$

Suppose we add some inert gas to an equilibrium mixture, holding T and V constant. Since $p_B = \dfrac{n_B RT}{V}$, the partial pressure of each gas taking part in the reaction is unaffected by the addition of an inert gas. Thus, there is no shift in equilibrium for isochoric, isothermal addition of an inert gas.

Conclusion:

$\sum \nu_B > 0$, adding inert gas favors the reaction.

$\sum \nu_B < 0$, decreasing inert gas favors the reaction.

$\sum \nu_B = 0$, equilibrium does not shift.

For instance : $N_2 + 3H_2 \rightarrow 2NH_3$

Since $\sum \nu_B < 0$, if adding inert gas, $\sum_B n_B$ increases, the equilibrium shifts to the left, it is not favorable for formation of NH_3.

Suitable method: removing continuously the CH_4 gas which does not participate in reaction in system.

5.8.4 Input material ratio (原料配比)

For perfect gas reaction: $a\mathrm{A(g)} + b\mathrm{B(g)} \rightarrow y\mathrm{Y(g)} + z\mathrm{Z(g)}$

If there were no product Y and Z at the beginning of the reaction, then we can prove:

When input material ratio $r = \dfrac{n_B}{n_A} = \dfrac{b}{a}$, the equilibrium conversion x_A^{eq} is maximum, r favors the reaction. Here, r is optimum ratio(最佳配比).

5.9 Chemical equilibrium of reaction of perfect gas and pure condensed phase (理想气体和纯凝聚相反应的化学平衡)

5.9.1 Expression of standard equilibrium constant (标准平衡常数的表达式)

For reaction of perfect gas and pure condensed phase: $a\mathrm{A(g)} + b\mathrm{B(s)} \rightarrow y\mathrm{Y(g)} + z\mathrm{Z(s)}$

$$\Delta_r G_m(T) = \sum_B \nu_B \mu_B$$

According to expression of chemical potential,

$$\mu_A(g) = \mu_A^\ominus(g, T) + RT\ln(p_A/p^\ominus)$$

$$\mu_Y(g) = \mu_Y^\ominus(g, T) + RT\ln(p_Y/p^\ominus)$$

$$\mu_B(s) = \mu_B^\ominus(s, T) + \int_{p^\ominus}^{p} V_{m,B}^* dp$$

$$\mu_Z(s) = \mu_Z^\ominus(s, T) + \int_{p^\ominus}^{p} V_{m,Z}^* dp$$

We have,

$$\Delta_r G_m(T) = (-a\mu_A^\ominus - b\mu_B^\ominus + y\mu_Y^\ominus + z\mu_Z^\ominus) + RT\ln\frac{(p_Y/p^\ominus)^y}{(p_A/p^\ominus)^a}$$

$$= \Delta_r G_m^\ominus(T) + RT\ln\frac{(p_Y/p^\ominus)^y}{(p_A/p^\ominus)^a}$$

At chemical equilibrium, $\Delta_r G_m(T) = 0$

So
$$\Delta_r G_m^\ominus(T) = -RT\ln\frac{(p_Y^{eq}/p^\ominus)^y}{(p_A^{eq}/p^\ominus)^a}$$

According to the definition of standard equilibrium constant of reaction,

$$\Delta_r G_m^\ominus(T) = -RT\ln K^\ominus(T)$$

We can get the expression of standard equilibrium constant,

$$K^{\ominus}(T)=\frac{(p_{Y}^{eq}/p^{\ominus})^{y}}{(p_{A}^{eq}/p^{\ominus})^{\alpha}}=\prod_{B}(p_{B}^{eq}/p^{\ominus})^{\nu_B}(\text{omit condensed phase}) \quad (5.9.1)$$

5.9.2 Dissociation pressure of pure solid compound（纯固体化合物的分解压）

Dissociation pressure—the pressure of gas at equilibrium of dissociation reaction at given temperature is called the dissociation pressure of this solid compound.

For instance：$CaCO_3(s)=CaO(s)+CO_2(g)$

Dissociation pressure of $CaCO_3(s)$ is $p_{CO_2}^{eq}$.

Since $$K^{\ominus}(T)=\prod_{B}(p_{B}^{eq}/p^{\ominus})^{\nu_B}=(p_{CO_2}^{eq}/p^{\ominus})^{1}=\frac{p_{CO_2}^{eq}}{p^{\ominus}}$$

If $dT=0$, $K^{\ominus}(T)$ is constant, so $p_{CO_2}^{eq}$ is constant, that is, T is constant, dissociation pressure of pure solid compound is constant.

Example 5.6

Relative data of reaction $MCO_3(s)=MO(s)+CO_2(g)$ (M is metal) are:

Substance	$\Delta_f H_m^{\ominus}$ (298.15K) / $kJ\cdot mol^{-1}$	$S_m^{\ominus}$ (298.15K) / $J\cdot K^{-1}\cdot mol^{-1}$	$C_{p,m}$ (298.15K) / $J\cdot K^{-1}\cdot mol^{-1}$
$MCO_3(s)$	−500	167.4	108.6
$MO(s)$	−29.0	121.4	68.4
$CO_2(g)$	−393.5	213.0	40.2

Please calculate and determine: (1) relations between $\Delta_r G_m^{\ominus}(T)$ and T; (2) if $t=127$ ℃, $p=101.325$kPa and the mole fraction of CO_2 is 0.01, can $MCO_3(s)$ dissociate? (3) how to change temperature to prevent $MCO_3(s)$ from dissociating?

Answer:

(1) $$\Delta_r H_m^{\ominus}(298.15K)=\sum_B \nu_B \Delta_f H_m^{\ominus}(B,\beta,298.15K)$$

$$=-393.5-29.0+500=77.5kJ\cdot mol^{-1}=77500J\cdot mol^{-1}$$

$$\Delta_r S_m^{\ominus}(298.15K)=\sum_B \nu_B S_m^{\ominus}(B,\beta,298.15K)$$

$$=213.0+121.4-167.4=167J\cdot K^{-1}\cdot mol^{-1}$$

$$\sum \nu_B C_{p,m}(B)=40.2+68.4-108.6=0$$

$$\Delta_r G_m^{\ominus}(T)=\Delta_r H_m^{\ominus}(T)-T\Delta_r S_m^{\ominus}(T)=\Delta_r H_m^{\ominus}(298.15K)-T\Delta_r S_m^{\ominus}(298.15K)$$

$$=77500J\cdot mol^{-1}-167T\ J\cdot mol^{-1}$$

(2) Since, $\Delta_r G_m^{\ominus}(T)=-RT\ln K^{\ominus}(T)$

$$K^{\ominus}(400.15K)=\exp[-(77500-167\times400.15)/8.315\times400.15]=0.04$$

$$J_p(400.15K)\overset{def}{=\!=}\prod_B(p_B/p^{\ominus})^{\nu_B}=(p_{CO_2}/p^{\ominus})^{1}=\frac{101.325\times0.01}{100}\approx0.01$$

For $J_p(T)<K^{\ominus}(T)$, that is $\Delta_r G_m<0$

So the direction of chemical reaction is to right, that is, $MCO_3(s)$ can dissociate.

(3) In order to prevent $MCO_3(s)$ from dissociating, the condition is:

$$J_p(T) > K^\ominus(T)$$

That is, $\ln K^\ominus(T) < \ln J_p(T) = \ln 0.01$

So, $\ln K^\ominus = \dfrac{-\Delta_r G_m^\ominus}{RT} = \dfrac{-\Delta_r H_m^\ominus}{RT} + \dfrac{\Delta_r S_m^\ominus}{R} = \dfrac{-77500}{8.315T} + \dfrac{167}{8.315} < \ln 0.01$

We can get, $T<377\text{K}$

Supplementary Examples of Chapter 5

EXERCISES

(习题)

1. At $T=1000\text{K}$, $\Delta_r G_m^\ominus$ of reaction $C(s)+2H_2(g)=CH_4(g)$ is $\Delta_r G_m^\ominus = 19.397\text{kJ}\cdot\text{mol}^{-1}$. The composition of gas mixture reacts with carbon is $y(CH_4)=0.1$, $y(H_2)=0.8$, $y(N_2)=0.10$, try to answer:

(1) What is $\Delta_r G_m$ when $T=1000\text{K}$ and $p=100\text{kPa}$? Can CH_4 be produced?

(2) What is the pressure to make the reaction forward when $T=1000\text{K}$?

Answer: (1) $\Delta_r G_m = 3.963\text{kJ}\cdot\text{mol}^{-1}$, CH_4 can not be produced;

(2) $p>161.06\text{kPa}$

2. The conversion reaction of CH_4 is: $CH_4(g)+H_2O(g)=CO(g)+3H_2(g)$, $K^\ominus = 1.280$ at 900K. If the moles of $CH_4(g)$ and $H_2O(g)$ is 1 : 1, calculate the equilibrium composition of the system at 900K and 101.325 kPa.

Answer: $y(CH_4) = y(H_2O) = 0.146$, $y(CO) = 0.177$, $y(H_2) = 0.531$

3. Calculate $K^\ominus(T)$ of the reaction: $CO(g)+H_2O(g)=CO_2(g)+H_2(g)$

Given: at the same temperature, two reaction equations and their standard equilibrium constants are as below:

$$C(\text{graphite})+H_2O(g)=CO(g)+H_2(g) \qquad K_1^\ominus(T)$$

$$C(\text{graphite})+2H_2O(g)=CO_2(g)+2H_2(g) \qquad K_2^\ominus(T)$$

Answer: $K^\ominus = K_2^\ominus/K_1^\ominus$

4. The solid NH_4HS was put into a vacuum vessel, and was decomposed to $NH_3(g)$ and $H_2S(g)$ at 25℃, the pressure was 66.66kPa when the reaction reached equilibrium.

(1) The pressure of $H_2S(g)$ was 39.99kPa when $NH_4HS(s)$ was put into the vessel, calculate the equilibrium pressure in the vessel;

(2) Beginning with $NH_3(g)$ of 6.666kPa in the vessel, what pressure of $H_2S(g)$ was needed to form solid NH_4HS.

Answer: (1) 77.7kPa; (2) $p(H_2S) > 166.65\text{kPa}$

5. There is a reaction of perfect gas: $A(g) + B(g) = C(g) + D(g)$. At the beginning, both A and B are 1mol, when the reaction reaches equilibrium at 25℃, the amount of substance of A and B are 1/3 mol respectively.

(1) Calculate $K^\ominus$ of reaction;

(2) At the beginning, A is 1mol, B is 2mol;

(3) At the beginning, A is 1mol, B is 2mol, and C is 0.5mol;

(4) At the beginning, C is 1mol, D is 2mol.

Please calculate the amount of substance of C respectively when the reaction reaches equilibrium.

Answer: (1) 4; (2) 0.845mol; (3) 1.290mol; (4) 0.543mol

6. The steam goes through heated coal layer at high temperature, and produces water coal gas as the reaction:

$$C(graphite) + H_2O(g) = CO(g) + H_2(g)$$

If $K^\ominus$ is 2.472 and 37.58 at 1000K and 1200K respectively, calculate $K^\ominus$ (1100K).

Answer: 11.0

7. For the gas phase reaction $CO_2+H_2S = COS+H_2O$, 4.4g of CO_2 was added to an empty bottle with a volume of 2.5dm^3 at 610K, and then H_2S was charged to bring the total pressure to 1000kPa. The mole fraction of H_2O in the system after reaction equilibrium is 0.02. The temperature is increased to 620K, and the mole fraction of H_2O in the system after equilibrium is 0.03. Each gas is treated as a perfect gas. (1) Calculate $K^\ominus$ and $\Delta_r G_m^\ominus$ at 610K; (2) Calculate $\Delta_r H_m^\ominus$ which does not change with temperature; (3) At 610K, add inert gas to the bottle to double the pressure, how does the amount of COS change?

Answer: (1) 2.8×10^{-3}, 29.81$kJ \cdot mol^{-1}$; (2) 273.88$kJ \cdot mol^{-1}$; (3) The amount of COS does not change

8. The relationship between $\Delta_r G_m^\ominus$ and T of the reaction $3CuCl(g) = Cu_3Cl_3(g)$ is as follow:

$$\Delta_r G_m^\ominus/(J \cdot mol^{-1}) = -528858 - 52.34\,(T/K)\lg(T/K) + 438.2\,(T/K)$$

Calculate: (1) $\Delta_r H_m^\ominus$ and $\Delta_r S_m^\ominus$ of the reaction at 2000K;

(2) The total pressure of the system when the mole fraction of Cu_3Cl_3 in the equilibrium mixture is 0.5 at 2000K.

Answer: (1) $\Delta_r H_m^\ominus$ (2000K) = $-483.4 kJ \cdot mol^{-1}$, $\Delta_r S_m^\ominus$ (2000K) = $-242.6 J \cdot K^{-1} \cdot mol^{-1}$;

(2) 212.3kPa

Chapter 6 Phase equilibrium(相平衡)

Learning objectives:

(1) State and derive the *phase rule*.

(2) Define the term *phase*, *component* and *degree of freedom*.

(3) Apply the phase rule to one-component systems.

(4) Use the phase rule to interpret *liquid-vapor composition diagrams* of two-component systems.

(5) State the *lever rule*.

(6) Construct and interpret *liquid-liquid phase diagrams* and explain the term *consolute temperature*.

(7) Construct and interpret *liquid-solid phase diagrams* and explain the term *eutectic* and *eutectic halt*.

Phase rule and phase diagram are two main parts of phase equilibrium. The phase rule discusses the relationship between the number of phases, the number of independent components in an equilibrium system and the variables describing the equilibrium system. Phase diagram is used to show the relationship between properties and conditions in phase equilibrium system. Phase rule is the theoretical basis of phase diagram. In this chapter, we will introduce some basic typical phase diagram. Through the study of these phase diagrams, we should master the analysis of phase diagram, lever rule and plotting of cooling curve.

6.1 Phase rule (相律)

6.1.1 Basic concepts(基本概念)

(1) Number of species(物种数)

Number of species—the number of chemical substance present in the equilibrium system. Symbol: S.

(2) Number of components(组分数)

Number of components—the minimum number of independent species necessary to define the composition of all the phases present in the system. Symbol: C.

$$C \overset{\text{def}}{=\!=\!=} S - R - R'$$

Where R is number of independent reaction stoichiometric equation(独立的化学计量方程式个数), R' is number of independent additional restrictions among other compositions in a phase except the relation of $\sum x_B = 1$(独立的限制条件个数).

(3) Phase and number of phase(相与相数)

① Phase(相)

Phase—a state of matter that is uniform throughout, not only in chemical composition, but also in physical state.

② Number of phase(相数)

Gas: a single phase, that is, a gas or a gaseous mixture is a single phase.

Liquid: full miscible, a single phase;

full immiscible, number of phase = number of substance;

partially miscible, number of phase can be determined by degree of mutual solubility.

Solid: number of phase = number of substance (except the solid solution is one phase.)

Solid solution—solid substances molecules are mixed at the molecular level, and the composition of the solution can be varied continuously over a certain range, such as alloy.

Symbol: P.

(4) Degree of freedom and number of degrees of freedom(自由度与自由度数)

Degree of freedom—independent intensive variables needed to specify its state and maintain original number of phase at phase equilibrium. Such as T, p, etc.

Number of degrees of freedom:

f—number of intensive degrees of freedom;

F—sum of numbers of intensive and extensive degrees of freedom. In this chapter, number of degrees of freedom means f.

6.1.2 Phase rule(相律)

Gibbs put forward the mathematical expression of phase rule, that is

$$f = C - P + 2 \qquad (6.1.1)$$

Where f is number of degrees of freedom, C is number of components, P is number of phase.

Comment:

(1) S kinds of chemical substances coexisting in a phase are not demanded.

(2) 2 in the mathematical expression of phase rule means two effecting factors of temperature and pressure. The mathematical expression of phase rule should be $f = C-P+n$, if we consider other factors.

(3) The mathematical expression of phase rule is usually $f = C-P+1$ for condensed phase system with no gas phase because we do not consider the effect of pressure on phase equilibrium.

$$f' = C - P + 1 \ (T \text{ or } p \text{ is fixed}) \qquad (6.1.2)$$

Where f' is conditional number of degrees of freedom, C is number of components, P is number of phase.

$$f'' = C - P \ (T \text{ and } p \text{ are fixed}) \qquad (6.1.3)$$

Where f'' is biconditional number of degrees of freedom, C is number of components, P is number of phase.

6.1.3 Application of phase rule(相律的应用)

(1) To determine f (number of effecting factors to phase equilibrium systems)

Example 6.1

Find C, P and f in the following equilibrium systems:

① $NH_4Cl(s)$ is at equilibrium with $NH_3(g)$ and $HCl(g)$ in a vacuum vessel;

② $NH_4Cl(s)$ is at equilibrium with arbitrary $NH_3(g)$ and $HCl(g)$;

③ $NH_4HCO_3(s)$ is at equilibrium with $NH_3(g)$, $H_2O(g)$ and $CO_2(g)$ in a vacuum vessel.

Answer:

① The equilibrium reaction is: $NH_4Cl(s) = NH_3(g) + HCl(g)$

$$S = 3,\quad R = 1,\quad P = 2$$

Since in vacuum vessel, $n(NH_3, g) : n(HCl, g) = 1 : 1$, $R' = 1$

So, $C = S - R - R' = 3 - 1 - 1 = 1$

For T and p are not fixed, we use the formula of phase rule $f = C - P + 2$ to calculate f, that is, $f = 1 - 2 + 2 = 1$.

② The equilibrium reaction is: $NH_4Cl(s) = NH_3(g) + HCl(g)$

$$S = 3,\ R = 1,\ P = 2$$

Since $NH_3(g)$ and $HCl(g)$ are arbitrary, $n(NH_3, g) : n(HCl, g) \neq 1 : 1$, $R' = 0$

So, $C = S - R - R' = 3 - 1 - 0 = 2$

For T and p are not fixed, we use the formula of phase rule $f = C - P + 2$ to calculate f, that is, $f = 2 - 2 + 2 = 2$.

③ The equilibrium reaction is: $NH_4HCO_3(s) = NH_3(g) + CO_2(g) + H_2O(g)$

$$S = 4,\ R = 1,\ P = 2$$

Since in vacuum vessel, $n(NH_3, g) : n(CO_2, g) = 1 : 1$, $n(NH_3, g) : n(H_2O, g) = 1 : 1$ and $n(CO_2, g) : n(H_2O, g) = 1 : 1$, number of relations is 3, but number of independent relations is 2, so $R' = 2$.

Therefore, $C = S - R - R' = 4 - 1 - 2 = 1$

For T and p are not fixed, we use the formula of phase rule $f = C - P + 2$ to calculate f, that is, $f = 1 - 2 + 2 = 1$.

Example 6.2

Find C and f in the following equilibrium systems:

① $CaCO_3(s)$ is partially decomposed at equilibrium with $CaO(s)$ and $CO_2(g)$;

② $CaCO_3(s)$ is at equilibrium with arbitrary $CaO(s)$ and $CO_2(g)$.

Answer:

① The equilibrium reaction is: $CaCO_3(s) = CaO(s) + CO_2(g)$

$$S = 3,\ R = 1,\ P = 3$$

Since $CaCO_3(s)$ is partially decomposed, $n(CaO, s) : n(CO_2, g) = 1 : 1$, but $CaO(s)$ and $CO_2(g)$ are not in the same phase, $R' = 0$

So, $C = S - R - R' = 3 - 1 - 0 = 2$

For T and p are not fixed, we use the formula of phase rule $f = C - P + 2$ to calculate f, that is, $f = 2 - 3 + 2 = 1$.

② The equilibrium reaction is: $CaCO_3(s) = CaO(s) + CO_2(g)$

$$S = 3,\ R = 1,\ P = 3$$

Since $CaO(s)$ and $CO_2(g)$ are arbitrary, $n(CaO, s) : n(CO_2, g) \neq 1 : 1$, $R' = 0$

So, $C = S - R - R' = 3 - 1 - 0 = 2$

For T and p are not fixed, we use the formula of phase rule $f=C-P+2$ to calculate f, that is, $f=2-3+2=1$.

(2) To determine P_{max} (the maximum number of phase in phase equilibrium systems under certain conditions)

Example 6.3

Na_2CO_3 and H_2O can form hydrates: $Na_2CO_3 \cdot H_2O(s)$, $Na_2CO_3 \cdot 7H_2O(s)$ and $Na_2CO_3 \cdot 10H_2O(s)$. ① Find the maximum kinds of hydrates that coexist with ice $H_2O(s)$ and Na_2CO_3 aqueous solution at $p^\ominus$; ② Find the maximum kinds of hydrates that coexist with steam $H_2O(g)$ at $t=30℃$.

Answer:

The equilibrium reaction is: $Na_2CO_3(s) + H_2O(l) = Na_2CO_3 \cdot H_2O(s)$

$$Na_2CO_3(s) + 7H_2O(l) = Na_2CO_3 \cdot 7H_2O(s)$$

$$Na_2CO_3(s) + 10H_2O(l) = Na_2CO_3 \cdot 10H_2O(s)$$

$$S = 5,\ R = 3,\ R' = 0$$

So,
$$C = S-R-R' = 5-3-0=2$$

① At $p^\ominus$, p is fixed, we can use the formula of phase rule $f=C-P+1$ to calculate f, that is, $f=2-P+1=3-P$.

When $f_{min}=0$, $P_{max}=3$, for ice $H_2O(s)$ and Na_2CO_3 aqueous solution are 2 phase, while every hydrate is 1 phase, so the maximum kind of hydrates that coexists with ice $H_2O(s)$ and Na_2CO_3 aqueous solution at $p^\ominus$ is 1.

② At $t=30$ ℃, T is fixed, we can use the formula of phase rule $f=C-P+1$ to calculate f, that is, $f=2-P+1=3-P$.

When $f_{min}=0$, $P_{max}=3$, for steam $H_2O(g)$ is 1 phase, while every hydrate is 1 phase, so the maximum kinds of hydrates that coexist with steam $H_2O(g)$ at $t=30℃$ is 2.

6.2 p–T graph of one-component systems (单组分系统的 *p-T* 图)

Phase diagram (相图) —diagram describes the relations between properties of phase equilibrium system (melting point, boiling point, etc.) and conditions (T, p etc.).

The phase rule for a system containing only one component:

$$f = C - P + 2 = 1 - P + 2 = 3 - P$$

When $P_{max}=3$, $f_{min}=0$ (there are no independent intensive variables, T and p are fixed), the system is invariant system(无变量系统);

When $P=2$, $f=1$ (there is only one independent intensive variable, that is T or p), the system is univariant system(单变量系统);

When $P_{min}=1$, $f_{max}=2$ (there are two independent intensive variables, that are T and p), the system is bivariant system(双变量系统).

For one-component system, the maximum number of phase is 3, the maximum number of degrees of freedom is 2, so there are two independent intensive variables, that is, T and p, so the phase diagram of one-component system can be denoted by p-T graph.

6.2.1 p-T graph for water(水的 p-T 图)

Data of phase equilibrium of water are list in table 6.1.

Tab. 6.1 Data of phase equilibrium of water

t/℃	2-Phase equilibrium			3-Phase equilibrium
	Saturated vapor pressure/Pa		Equilibrium pressure/MPa	Equilibrium pressure /Pa
	l ⇌ g	s ⇌ g	s ⇌ l	s ⇌ l ⇌ g
-20	—	103.4	199.6	—
-15	(190.5)	165.2	161.1	—
-10	285.8	295.4	115.0	—
-5	421.0	410.3	61.8	—
0.01	610.0	610.0	610.0×10^{-6}	610.0
20	2337.8	—	—	—
60	19920.5	—	—	—
99.65	100000	—	—	—
100	101325	—	—	—
374.2	22119247	—	—	—

We can see: $P_{max}=3$, ice ⇌ water ⇌ steam, $f_{min}=0$, **triple point**

$P=2$, ice ⇌ water, ice ⇌ steam, water ⇌ steam, $f=1$, **2-phase line**

$P_{min}=1$, ice, water, steam, $f_{max}=2$, **1-phase region**

We can get the p-T graph for water depending on the data of phase equilibrium of water, the p-T graph for water is shown in figure 6.1.

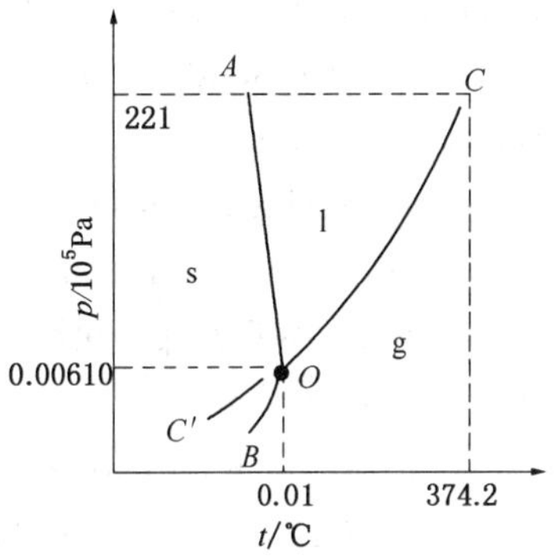

Fig. 6.1 p-T graph for water

Analysis of p-T graph for water:

(1) Triple point: $P=3$, $f=0$, s, l and g are in equilibrium, point O (0.01℃, 610.0Pa).

(2) 2-phase line:

OC line: liquid-vapor line of water, critical point C(374.2℃, 22.1MPa).

OC' line: liquid-vapor line of supercooled water, supercooled water is in metas-table state.

黄子卿
与水的三相点

OB line: solid-vapor line of ice.

OA line: solid-liquid line of ice.

Notice that: the slope of OA line is negative.

(3) 1-phase region:

solid phase region(BOA).

liquid phase region(COA).

gas phase region(BOC).

Comment on phase diagram of water:

① Triple point O (0.01℃, 610.0Pa).

② The slope of OA line is negative.

③ Distinguish triple point and ice point of water.

From figure 6.2 we can get, triple point is the self-character of substance which can not be changed. Triple point is the temperature three phases of water, ice and steam coexist, such as triple point of H_2O is $T=273.16K$, the corresponding pressure is $p=610.0Pa$.

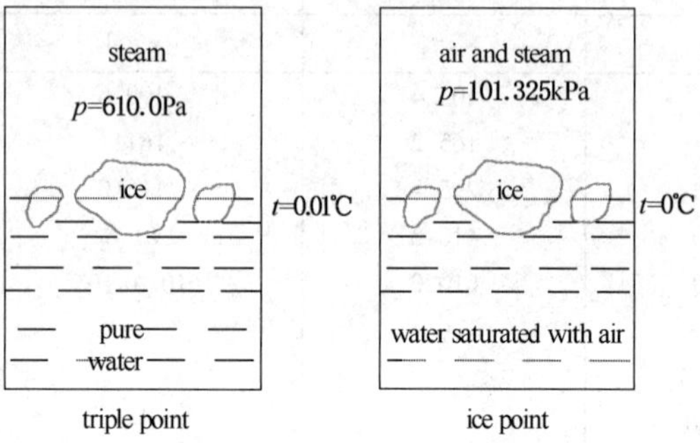

Fig. 6.2 Triple point and ice point of water

Ice point is the temperature two phases of water and ice coexist under the atmospheric pressure, ice point of H_2O is $T=273.15K$, the corresponding pressure is $p=101.325kPa$.

Ice point of water is lower 0.01K than triple point of water which is caused by two kinds of factors:

① Increasing of outside pressure making the freezing point decrease 0.00748K;

② Dissolving of air in the water making the freezing point decrease 0.00241K.

6.2.2 p-T graph for carbon dioxide and supercritical CO_2 fluid (CO_2 的 p-T 图和超临界 CO_2 流体)

(1) p-T graph for carbon dioxide

The p-T graph for carbon dioxide is shown in figure 6.3.

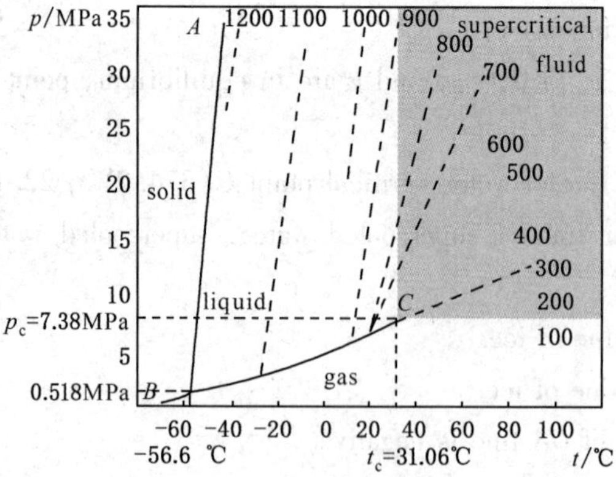

Fig. 6.3 p-T graph for CO_2

Analysis of p-T graph for CO_2:

① Triple point: $P=3$, $f=0$, s, l and g are in equilibrium, point O (−56.6℃, 0.518MPa)

② 2-phase line:

OC line: liquid-vapor line of carbon dioxide, critical point *C* (31.06℃, 7.38MPa)

OB line: solid-vapor line of carbon dioxide

OA line: solid-liquid line of carbon dioxide

Notice that: the slope of *OA* line is positive.

③ 1-phase region:

solid phase region(*BOA*).

liquid phase region(*COA*).

gas phase region(*BOC*).

Supercritical CO_2 fluid—CO_2 above point *C*.

(2) Application of supercritical CO_2 fluid

Extraction fractionation: a new technique.

Five advantages of supercritical CO_2 fluid:

① ρ of supercritical CO_2 fluid is large, capability of miscibility is strong, so efficiency of extraction is high.

② Cheap and easy to be made.

③ Atoxic, inert and easy to be separate.

④ Critical temperature (31.06℃) is near to room temperature, so fit to separate natural product.

⑤ Critical pressure (7.38MPa) is suitable and so easy to bring about industrialization.

6.2.3 *p-T* graph for sulfur(硫的 *p-T* 图)

The *p-T* graph for sulfur is shown in figure 6.4.

Two crystal forms of solid sulfur:

① Monoclinic sulfur(β-sulfur).

② Rhombic sulfur.

Three triple points:

① Triple point *B*: β-sulfur, rhombic sulfur and gas sulfur are in equilibrium.

② Triple point *C*: β-sulfur, liquid sulfur and gas sulfur are in equilibrium.

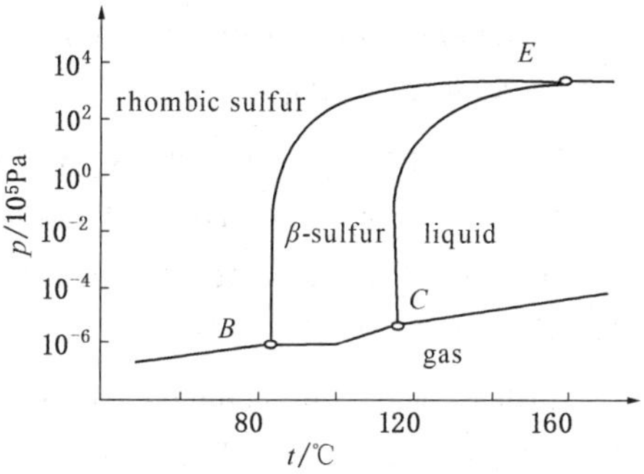

Fig. 6.4 *p-T* graph for sulfur

③ Triple point *E*: β-sulfur, rhombic sulfur and liquid sulfur are in equilibrium.

6.3 Two-component liquid-gas phase diagram of liquid full miscible system (二组分液态完全互溶系统的液-气相图)

According to mutual dissolubility of two liquid components, liquid-gas phase diagram of two-component systems can be divided into three kinds, that is,

① liquid-gas phase diagram of two-component liquid full miscible system;

② liquid-gas phase diagram of two-component liquid partially miscible system;

③ liquid-gas phase diagram of two-component liquid full immiscible system.

Three kinds of two-component systems are shown in figure 6. 5.

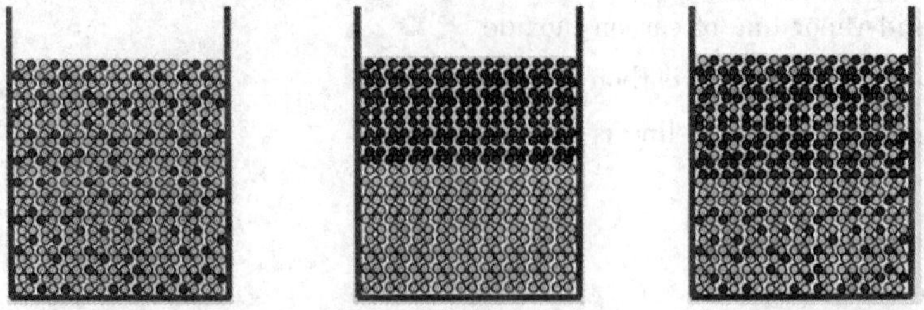

(a) liquid full miscible system (b) liquid full immiscible system (c) liquid partially miscible system

Fig. 6. 5 Three kinds of two-component systems

For two-component system, $C=2$, the phase rule is:

$$f=C-P+2=2-P+2=4-P$$

When
$$P_{min}=1,\ f_{max}=3$$
$$P=2,\ f=2$$
$$P=3,\ f=1$$
$$P_{max}=4,\ f_{min}=0$$

P is 1 at least, so f is 3 at most, there are three independent intensive variables, T, P and composition x.

For convenience, we usually keep one of these variables constant and plot a two-dimensional phase diagram. This amounts to taking a cross section of a three-dimensional plot.

① Keep temperature constant, we can get ***p-x*** diagram, quite in common use.

② Keep pressure constant, we can get ***T-x*** diagram, in common use.

③ Keep composition constant, we can get ***T-p*** diagram, not in common use.

6. 3. 1 Two-component liquid-gas phase diagram of ideal liquid mixture (二组分理想液态混合物的液-气相图)

Instead of plotting complete phase diagram, we shall usually consider only one portion of the phase diagram at a time. This section deals with the liquid-gas part of the phase diagram of a two-component ideal liquid mixture system.

(1) Two-component vapor pressure-composition diagram of ideal liquid mixture (二组分理想液态混合物的蒸气压-组成图, *p-x* 图)

For ideal liquid mixture (A: $C_6H_5CH_3$, B: C_6H_6), the vapor pressure-composition data of $C_6H_5CH_3(A)-C_6H_6(B)$ at 79. 7℃ are list in table 6. 2.

Tab. 6. 2 The vapor pressure-composition data of $C_6H_5CH_3(A)-C_6H_6(B)$ at 79. 7 ℃

x_B	y_B	$p/10^2$kPa	x_B	y_B	$p/10^2$kPa
0	0	0. 3846	0. 6344	0. 8179	0. 7722
0. 1161	0. 2530	0. 4553	0. 7327	0. 8782	0. 8331
0. 2271	0. 4295	0. 5225	0. 8243	0. 9240	0. 8907
0. 3383	0. 5667	0. 5907	0. 9189	0. 9672	0. 9845
0. 4532	0. 6656	0. 66499	0. 9565	0. 9827	0. 9179
0. 5451	0. 7574	0. 7166	1. 000	1. 000	0. 9982

In terms of data of table 6.2, we can get the two-component vapor pressure-composition diagram of ideal liquid mixture, which is shown in figure 6.6.

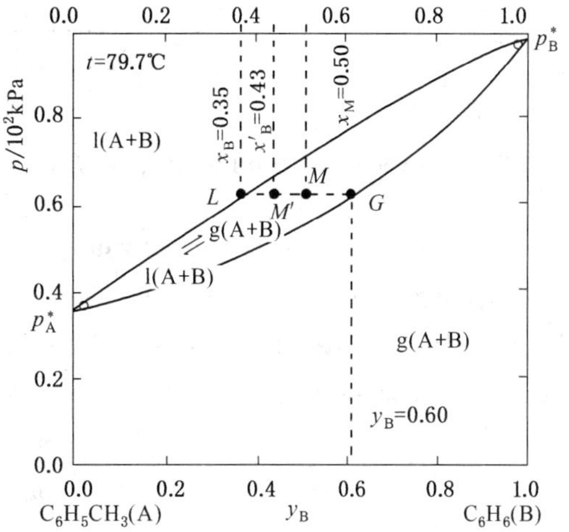

Fig. 6.6　Vapor pressure-composition diagram of $C_6H_5CH_3(A)-C_6H_6(B)$

Characteristics of phase diagram:

① Straight line: p-x_B.

② Curve line: p-y_B.

③ $p_B^* > p > p_A^*$, $p_B = py_B = p_B^* x_B$, $\frac{y_B}{x_B} = \frac{p_B^*}{p} > 1$, that is, $y_B > x_B$.

So line of p-y_B is below line of p-x_B.

Analysis of phase diagram:

① Region

1-phase region: liquid phase region, high pressure region, l(A+B); gas phase region, low pressure region, g(A+B).

$$f' = C - P + 1 = 2 - 1 + 1 = 2$$

There are two independent intensive variables, that is, p and composition x_B.

2-phase region: $l(A+B) \rightleftharpoons g(A+B)$

$$f' = C - P + 1 = 2 - 2 + 1 = 1$$

There is one independent intensive variables, that is, p or composition x_B.

② Line (2-phase line)

Line of liquid phase: p-x_B, straight line.

Line of gas phase: p-y_B, curve line.

③ Point

System point M: if system point is fixed, the composition of system x_M must be fixed.

Phase point:

Liquid phase point: $L(x_B)$;

Gas phase point: $G(y_B)$;

Isobaric tie line: line LG.

(2) Lever rule(杠杆规则)

In 2-phase equilibrium region: $l(A+B) \rightleftharpoons g(A+B)$, depending on material balance of component B, we can get lever rule.

$$n \cdot x_M = n_B = n_L x_B + n_G y_B \quad (6.3.1)$$

$$n = n_L + n_G \quad (6.3.2)$$

So, $(n_L + n_G) x_M = n_L x_B + n_G y_B$, that is,

$$\frac{n_L}{n_G} = \frac{y_B - x_M}{x_M - x_B} \quad (6.3.3)$$

Combine (6.3.2) and (6.3.3), we can calculate n_L and n_G.

Formula (6.3.3) is called **lever rule**, which can be deduced from lever.

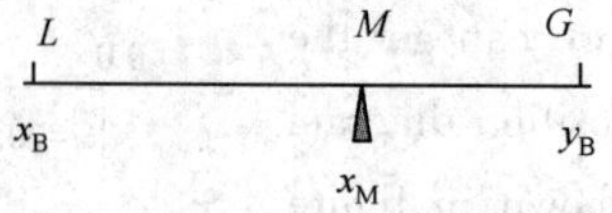

When lever is in equilibrium, $n_L\overline{LM} = n_G\overline{MG}$, so $\frac{n_L}{n_G} = \frac{\overline{MG}}{\overline{LM}} = \frac{y_B - x_M}{x_M - x_B}$, it is (6.3.3), therefore, (6.3.3) is called **lever rule**.

Frequently, the overall mass fraction of B (instead of x_B) is used as the abscissa of the phase diagram. In this case, the masses replace the numbers of moles in the above derivation, and the lever rule becomes

$$\frac{m_L}{m_G} = \frac{w_B^g - w_M}{w_M - w_B^l} \tag{6.3.4}$$

Applicable condition of lever rule: arbitrary 2-phase equilibrium region.

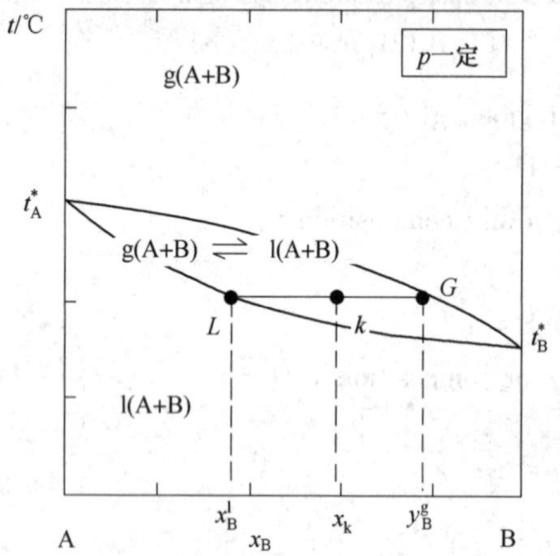

Fig. 6.7 Boiling point-composition diagram of $C_6H_5CH_3$(A)-C_6H_6(B)

(3) Two-component boiling point-composition diagram of ideal liquid mixture (二组分理想液态混合物的沸点-组成图, *t-x* 图)

For ideal liquid mixture (A: $C_6H_5CH_3$, B: C_6H_6), the boiling point-composition diagram is shown in figure 6.7.

Analysis of phase diagram:

① Region

1-phase region: liquid phase region, low temperature region, l(A+B); gas phase region, high temperature region, g(A+B).

$$f' = C - P + 1 = 2 - 1 + 1 = 2$$

There are two degrees of freedom, that is, T and composition x_B.

2-phase region: l(A+B) $\rightleftharpoons$ g(A+B)

$$f' = C - P + 1 = 2 - 2 + 1 = 1$$

There is one degree of freedom, that is, T or composition x_B.

② Line (2-phase line)

Line of liquid phase: T-x_B, bubble point line(泡点线);

Line of gas phase: T-y_B, curve line, dew point line(露点线).

③ Point

System point k: if system point is fixed, the composition of system x_k must be fixed.

Phase point:

Liquid phase point : $L(x_B^l)$;

Gas phase point : $G(y_B^g)$;

Isothermal tie line: line LG.

Example 6.4

$C_6H_5CH_3$(A) and C_6H_6(B) can form ideal liquid mixture. Given that p^* ($C_6H_5CH_3$, l, 90℃) = 54.22kPa, p^* (C_6H_6, l, 90℃) = 136.12kPa.

(1) Calculate the liquid composition (x_B) and gas composition (y_B) of the $C_6H_5CH_3$(A)-C_6H_6(B) system when it reaches l-g equilibrium under 90℃ and 101.325 kPa.

(2) If the system was formed by 100.0g $C_6H_5CH_3$ and 200.0g C_6H_6, calculate the mass of gas phase and liquid phase under 90℃ and 101.325kPa.

Answer:

(1) $C_6H_5CH_3$(A) and C_6H_6(B) can form ideal liquid mixture, A and B obey Raoult' law, then:

$$p = p_A + p_B = p_A^* x_A + p_B^* x_B = p_A^* (1 - x_B) + p_B^* x_B$$

$$x_B = \frac{p - p_A^*}{p_B^* - p_A^*} = \frac{101.325 - 54.22}{136.12 - 54.22} = 0.5752$$

$$x_A = 1 - x_B = 1 - 0.5752 = 0.4248$$

$$y_B = \frac{p_B}{p} = \frac{p_B^* x_B}{p} = \frac{136.12 \times 0.5752}{101.325} = 0.7727$$

$$y_A = 1 - y_B = 1 - 0.7727 = 0.2273$$

(2) The system was formed by 100.0g $C_6H_5CH_3$(A) and 200.0g C_6H_6(B)

So, $$n = \frac{m_A}{M_A} + \frac{m_B}{M_B} = \frac{100.0}{92.14} + \frac{200.0}{78.11} = 3.645\text{mol}$$

The system composition $x_M = \frac{n_B}{n} = \frac{200.0/78.11}{3.645} = 0.7025$

So, $$n(\text{l}) = \frac{y_B - x_M}{y_B - x_B} n = \frac{0.7727 - 0.7025}{0.7727 - 0.5752} \times 3.645 = 1.296\text{mol}$$

$$n(\text{g}) = n - n(\text{l}) = 3.645 - 1.296 = 2.349\text{mol}$$

Therefore,

$$m(\text{l}) = n(\text{l})(x_B M_B + x_A M_A) = 1.296 \times (0.5752 \times 78.11 + 0.4248 \times 92.14) = 109.0\text{g}$$

$$m(\text{g}) = m - m(\text{l}) = 100.0 + 200.0 - 109.0 = 191.0\text{g}$$

6.3.2 Two-component liquid-gas phase diagram of real liquid mixture (二组分真实液态混合物的液-气相图)

Having dealt with liquid-gas equilibrium for ideal liquid mixture, we now consider real liquid mixture. Liquid-gas phase diagram for real liquid mixture systems are obtained by measurement of the pressure and composition of the vapor in equilibrium with liquid of known composition. So there will be deviation, including positive deviation and negative deviation. If the real liquid mixture is only slightly nonideal, the deviation is general deviation. If, however, the real liquid mixture has a great enough deviation from ideality to give a maximum deviation, a new phenomenon appears.

Positive deviation(正偏差)—real vapor pressure is larger than theoretical value calculated by Raoult's law.

Negative deviation(负偏差)—real vapor pressure is smaller than theoretical value calculated by Raoult's law.

General positive deviation(一般正偏差)—real vapor pressure is larger than theoretical value calculated by Raoult's law, but in total composition, $p_B^* > p > p_A^*$.

General negative deviation(一般负偏差)—real vapor pressure is smaller than theoretical value calculated by Raoult's law, but in total composition, $p_B^* > p > p_A^*$.

Maximum positive deviation(最大正偏差)—real vapor pressure is larger than theoretical value calculated by Raoult's law, and in certain composition $p > p_B^*$, so there is a maximum value point in phase diagram.

Maximum negative deviation(最大负偏差)—real vapor pressure is smaller than theoretical value calculated by Raoult's law, and in certain composition $p < p_A^*$, so there is a minimum value point in phase diagram.

(1) Vapor pressure-composition diagram of general positive deviation system (一般正偏差系统的蒸气压-组成图)

Vapor pressure-composition diagram of general positive deviation system is shown in figure 6.8.

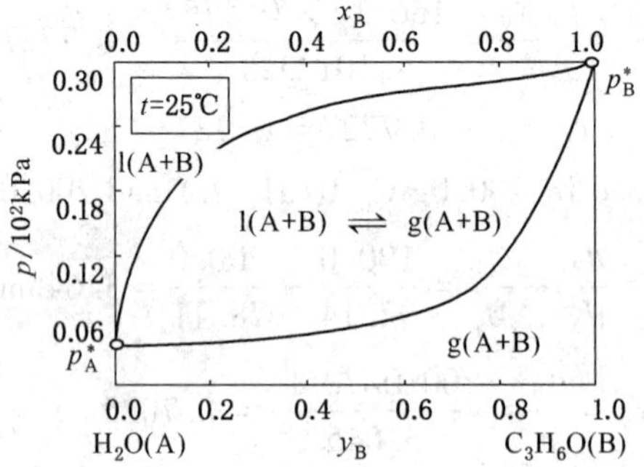

Fig. 6.8 Vapor pressure-composition diagram of $H_2O(A)-C_3H_6O(B)$

Characteristics of vapor pressure-composition diagram of general positive deviation system:

① In total composition, $p_B^* > p > p_A^*$.

② Curve line of p-x_B is up straight line of $p_A^* p_B^*$.

(2) Vapor pressure-composition diagram of maximum positive deviation system (最大正偏差系统的蒸气压-组成图)

Vapor pressure-composition diagram of maximum positive deviation system is shown in figure. 6.9.

Characteristics of vapor pressure-composition diagram of maximum positive deviation system:

① Maximum value point: $x_B = y_B$.

② Left side of maximum value point: $y_B > x_B$;

Right side of maximum value point: $y_B < x_B$.

(3) Boiling point-composition diagram of maximum positive deviation system (最大正偏差系统的沸点-组成图)

Boiling point-composition diagram of maximum positive deviation system is shown in figure 6.10.

Characteristics of boiling point-composition diagram of maximum positive deviation system:

① Minimum value point: $x_B = y_B$.

Comment:

Minimum value point in boiling point-composition diagram is called as minimum azeotropic point(最低恒沸点), the mixture corresponding to the minimum azeotropic point is called as minimum azeotropic mixture(最低恒沸混合物), not minimum azeotropic compound.

Minimum azeotropic mixture is mixture not compound, its composition in the fixed pressure is fixed value. When pressure changing, the temperature of the minimum azeotropic point changes, its composition changes following.

② Left side of minimum value point: $y_B > x_B$;

Right side of minimum value point: $y_B < x_B$.

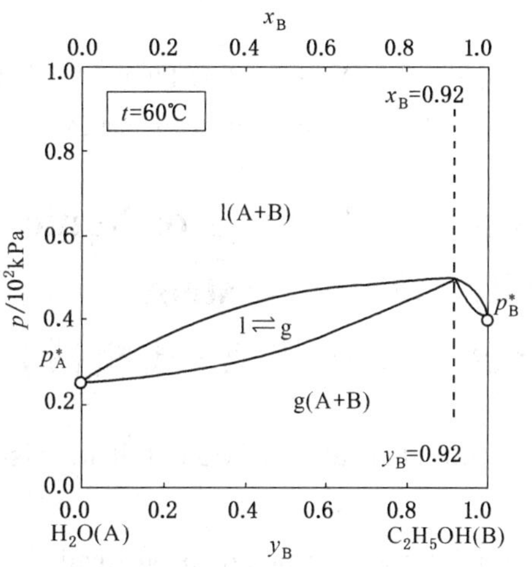

Fig. 6. 9 Vapor pressure-composition diagram of $H_2O(A)-C_2H_5OH(B)$

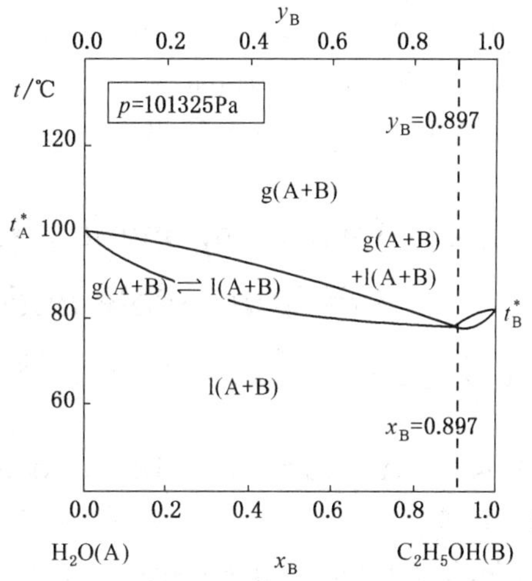

Fig. 6. 10 Boiling point-composition diagram of $H_2O(A)-C_2H_5OH(B)$

(4) Vapor pressure-composition diagram of maximum negative deviation system(最大负偏差系统的蒸气压-组成图)

Vapor pressure-composition diagram of maximum negative deviation system is shown in figure 6. 11.

Characteristics of vapor pressure-composition diagram of maximum negative deviation system:

① Minimum value point: $x_B = y_B$.

② Left side of minimum value point: $x_B > y_B$;

Right side of minimum value point: $x_B < y_B$.

(5) Boiling point-composition diagram of maximum negative deviation system (最大负偏差系统的沸点-组成图)

Boiling point-composition diagram of maximum negative

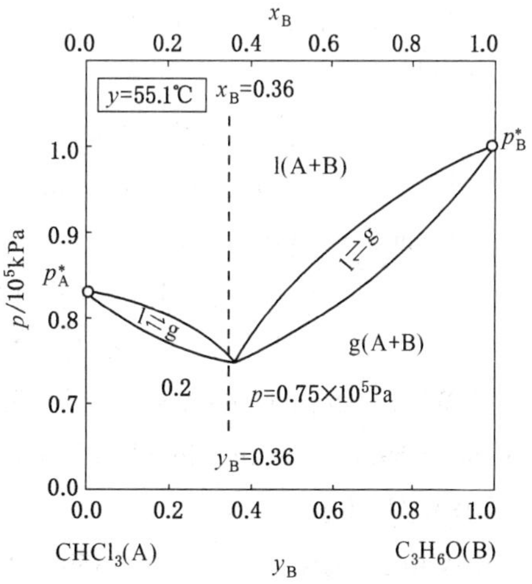

Fig. 6. 11 Vapor pressure-composition diagram of $CHCl_3(A)-C_3H_6O(B)$

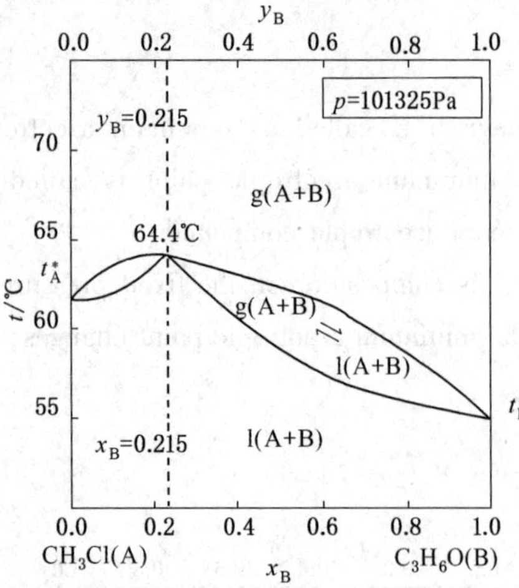

Fig. 6.12 Boiling point-composition diagram of $CHCl_3$(A)-C_3H_6O(B)

deviation system is shown in figure 6.12.

Characteristics of boiling point-composition diagram of maximum negative deviation system:

① Maximum value point: $x_B = y_B$.

Comment:

Maximum value point in boiling point-composition diagram is called as maximum azeotropic point(最高恒沸点), the mixture corresponding to the maximum azeotropic point is called as maximum azeotropic mixture(最高恒沸混合物), not maximum azeotropic compound.

② Left side of maximum value point: $x_B > y_B$;

Right side of maximum value point: $x_B < y_B$.

6.4 Two-component liquid-gas phase diagram of liquid full immiscible and partially miscible system

(二组分液态完全不互溶及部分互溶系统的液-气相图)

6.4.1 Two-component boiling point-composition diagram of liquid full immiscible system(二组分液态完全不互溶系统的沸点-组成图，*t-x* 图)

Liquid full immiscible system—system in which solubility of two liquids is so small that can be omitted when they are mixed. Such as: H_2O-paraffin, H_2O-aromatic hydrocarbon.

Consider two liquids H_2O(A) and C_6H_6(B) that form a liquid full immiscible system. The vapor pressure curve is shown in figure 6.13.

For two liquids H_2O (A) and C_6H_6 (B) are full immiscible,

$$p = p_A + p_B = p_A^* + p_B^*$$

So $$p > p_A^*, \quad p > p_B^*$$

From the figure we can see,

$$T_b(A+B) < T_b^*(A)$$

$$T_b(A+B) < T_b^*(B)$$

When external pressure $p = 101.325\text{kPa}$, the boiling point $t = 69.9℃$ for the liquid full immiscible system H_2O(A)-C_6H_6(B).

When $t = 69.9℃$, $p_B^* = 73359.3\text{Pa}$,

$$p_A^* = 27965\text{Pa}$$

Then the vapor composition

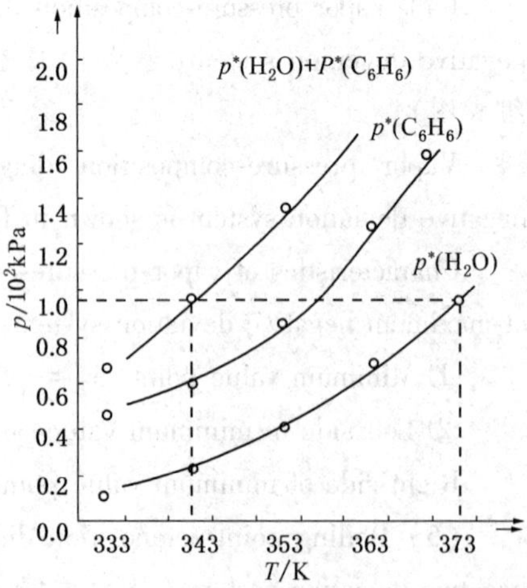

Fig. 6.13 vapor pressure curve of H_2O(A)-C_6H_6(B)

$$y_B = \frac{p_B^*}{p} = \frac{p_B^*}{p_A^* + p_B^*} = \frac{73359.3}{101325} = 0.724$$

We can get the azeotropic point E (69.9℃, 0.724).

The boiling point-composition diagram of liquid full immiscible system H_2O (A)-C_6H_6 (B) is shown in figure 6.14.

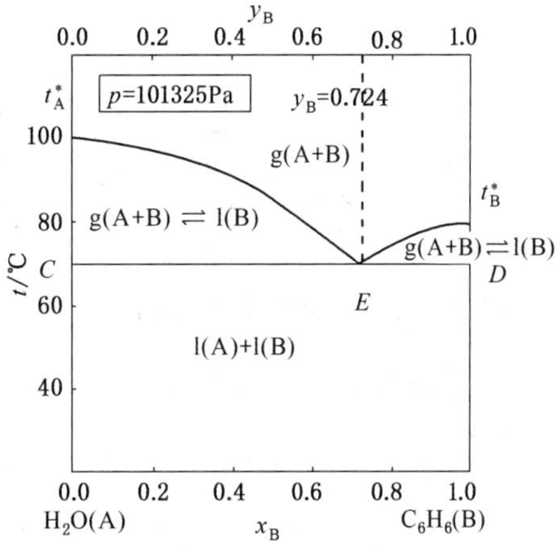

Fig. 6.14 Boiling point-composition diagram of H_2O(A)-C_6H_6(B)

Analysis of phase diagram:

(1) Point

t_A^* —boiling point of pure A

t_B^* —boiling point of pure B

E—azeotropic point

Line

$t_A^* C$ —liquid phase line of pure A

$t_B^* D$ —liquid phase line of pure B

$t_A^* E$ and $t_B^* E$ —gaseous phase line

CED—triple line, that is, l(A), l(B) and g(A+B) coexist, l(A) + l(B)→g(A+B)

$$f' = C - P + 1 = 2 - 3 + 1 = 0$$

Region

1-phase region: gas phase region, high temperature region, g(A+B)

$$f' = 2 - 1 + 1 = 2$$

2-phase region: g(A+B) + l(A), g(A+B) + l(B), l(A) + l(B)

$$f' = 2 - 2 + 1 = 1$$

(2) 2-phase region

Lever rule can be used to calculate the masses or the numbers of moles of two phases coexist in equilibrium.

(3) Triple line

The proportion is fixed, that is, $\frac{n_A}{n_B} = \frac{\overline{DE}}{\overline{CE}}$

6.4.2 Two-component boiling point-composition diagram of liquid partially miscible system(二组分液态部分互溶系统的沸点-组成图, *t-x* 图)

Liquid partially miscible system—two components are mutual soluble in the scope of fixed proportion and temperature, and form two liquid phases because of partially mutual solubility in other range. Such as: H_2O-$C_6H_5NH_2$.

Consider two liquids H_2O(A) and $C_6H_5NH_2$(B) that form a liquid partially miscible system.

(1) Two-component solubility diagram of liquid partially miscible system(二组分液态部分互溶系统的溶解度图)

Conjugate solution(共轭溶液)—solution in which solvent and solute can be exchanged. Such as: $H_2O(A)-C_6H_5NH_2(B)$.

l_α ⟶ H_2O phase: H_2O is solvent, $C_6H_5NH_2$ is solute

l_β ⟶ amine phase: $C_6H_5NH_2$ is solvent, H_2O is solute

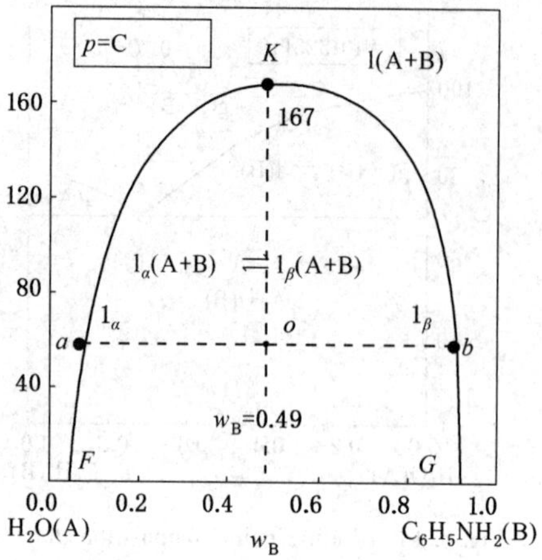

Fig. 6.15 Solubility diagram of $H_2O(A)-C_6H_5NH_2(B)$

Where ($l_\alpha + l_\beta$) form the conjugate solution. The solubility diagram of liquid partially miscible system $H_2O(A)-C_6H_5NH_2(B)$ is shown in figure 6.15.

Analysis of phase diagram:

① Point

K—critical consolute point(临界会溶点)

t_K —critical consolute temperature(临界会溶温度)

When the temperature is higher than t_K, $C_6H_5NH_2$ and H_2O can dissolve infinitely, the interface disappears, which form a single liquid phase.

Line

FK curve—solubility curve of aminobenzene dissolved in water

KG curve—solubility curve of water dissolved in aminobenzene

Region

1-phase region: liquid phase region, l(A+B)

$$f' = 2 - 1 + 1 = 2$$

2-phase region: $l_\alpha(A+B) + l_\beta(A+B)$

$$f' = 2 - 2 + 1 = 1$$

② 2-phase region

Lever rule can be used to calculate the masses or the numbers of moles of two liquid phases coexist in equilibrium.

(2) Two-component boiling point-composition diagram of liquid partially miscible system(二组分液态部分互溶系统的沸点-组成图, *t-x* 图)

① The vapor composition is between the liquid composition of l_α at *C* and l_β at *D*(气相组成介于两液相组成之间的系统)

Consider two liquids $H_2O(A)$ and n-C_4H_9OH (B) that form a liquid partially miscible system. The boiling point-composition diagram is shown in figure 6.16.

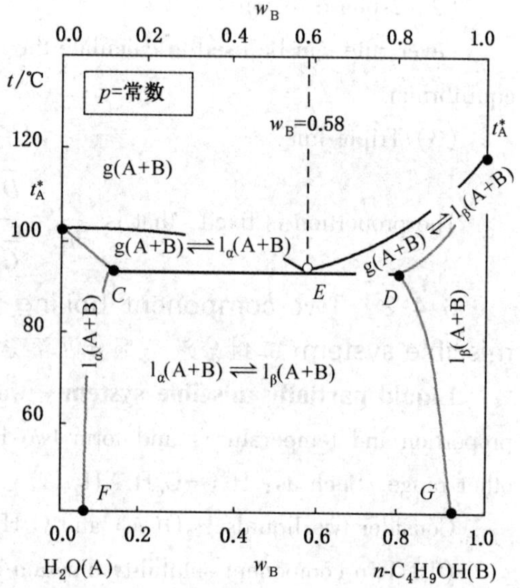

Fig. 6.16 Boiling point-composition diagram of $H_2O(A)-n$-$C_4H_9OH(B)$

Analysis of phase diagram:

i) Point

t_A^* —boiling point of pure A

t_B^* —boiling point of pure B

E—azeotropic point

Line

$t_A^* C$ and $t_B^* D$ —liquid phase line

$t_A^* E$ and $t_B^* E$ —gaseous phase line

CED—triple line, that is, $l_\alpha(A+B)$, $l_\beta(A+B)$ and $g(A+B)$ coexist,

$$l_\beta(A+B) + l_\beta(A+B) \rightarrow g(A+B)$$

$$f' = C - P + 1 = 2 - 3 + 1 = 0$$

Region

1-phase region: $g(A+B)$, $l_\alpha(A+B)$, $l_\beta(A+B)$

$$f' = 2 - 1 + 1 = 2$$

2-phase region: $g(A+B) + l_\alpha(A+B)$, $g(A+B) + l_\beta(A+B)$, $l_\alpha(A+B) + l_\beta(A+B)$

$$f' = 2 - 2 + 1 = 1$$

ii) 2-phase region

Lever rule can be used to calculate the masses or the numbers of moles of two phases coexist in equilibrium.

iii) Triple line

The proportion is fixed, that is, $\dfrac{n(l_\alpha)}{n(l_\beta)} = \dfrac{\overline{DE}}{\overline{CE}}$

② The vapor composition is on the same side of two liquid compositions of l_1 at L_1 and l_2 at L_2 (气相组成位于两液相组成同一侧的系统)

The boiling point-composition diagram of two-component liquid partially miscible system whose vapor composition is on the same side of two liquid compositions is shown in figure 6. 17.

Analysis of phase diagram:

Point

P—boiling point of pure A

Q—boiling point of pure B

Line

PL_1 and QL_2—liquid phase line

PG and QG—gaseous phase line

GL_1L_2—triple line, that is, $l_1(A+B)$, $l_2(A+B)$ and $g(A+B)$ coexist,

$$l_1(A+B) \longrightarrow g(A+B) + l_2(A+B)$$

$$f' = C - P + 1 = 2 - 3 + 1 = 0$$

Region

1-phase region: $g(A+B)$, $l_1(A+B)$, $l_2(A+B)$

$$f' = 2 - 1 + 1 = 2$$

2-phase region: g(A+B) + l_1(A+B), g(A+B) + l_2(A+B), l_1(A+B) + l_2(A+B)

$$f' = 2 - 2 + 1 = 1$$

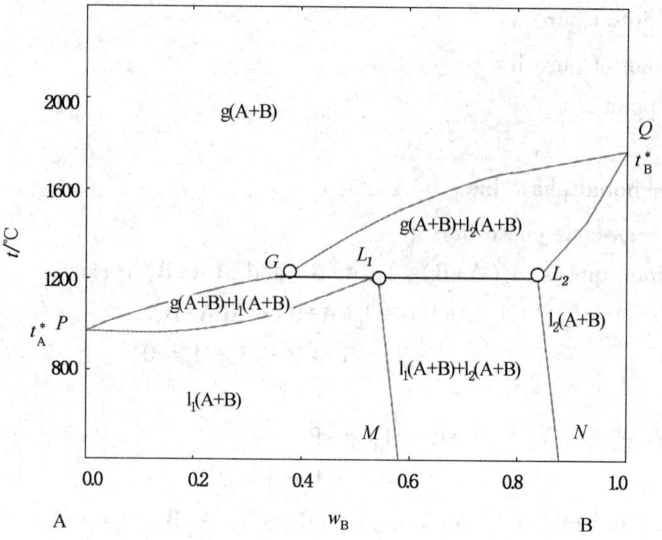

Fig. 6. 17 Boiling point-composition diagram of two-component system A-B

Example 6. 5

A and B are partially miscible in liquid phase, at $p = 100$kPa, the boiling point are120℃ and 100℃ respectively. In the gas-liquid equilibrium phase diagram, the composition of phase point C, E and D are $x_{B,C} = 0.05$, $y_{B,E} = 0.60$ and $x_{B,D} = 0.97$ respectively.

Example 6. 5

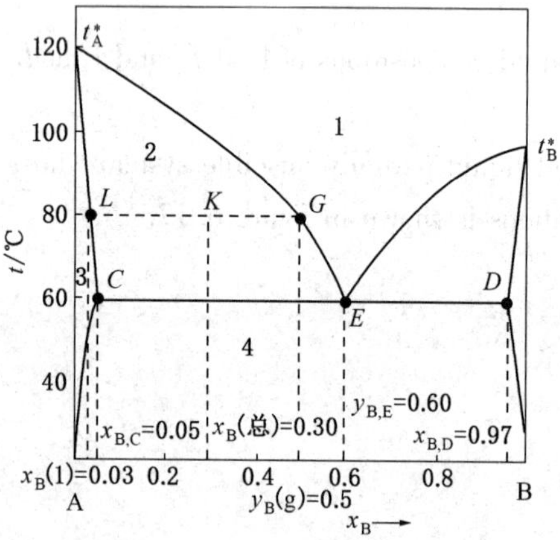

(1) List table and write P, state of aggregation (g, l or s) and ingredient (A, B or A+B) of phase and f' of all regions and CED line.

(2) Mixture forming of 3mol B and 7mol A reaches gas-liquid equilibrium at 100kPa, 80℃. Please calculate n(l) and n(g).

(3) Assume two solutions presented with equilibrium phase point C and D can be seen as ideal dilute solution. Please calculate the saturated vapor pressure of pure liquid A and B, and Henry's coefficient of solute in two solutions (composition scale is mole fraction).

Answer:

(1)

Phase region	P	Phase state and ingredient	f'
1	1	g(A+B)	2
2	2	g(A+B) + l_1(A+B)	1
3	1	l_1(A+B)	2
4	2	l_1(A+B) + l_2(A+B)	1
CED line	3	g(A+B) + l_1(A+B) + l_2(A+B)	0

(2) Mixture forming of 3mol B and 7mol A reaches gas-liquid equilibrium at 100kPa, 80℃, the system point is K, $x_K = 0.3$, it is in the 2-phase equilibrium region. Liquid phase point is L, $x_B = 0.03$, gas phase point is G, $y_B = 0.5$.

According to lever rule, we have,

$$n = n(\mathrm{l}) + n(\mathrm{g}) = 3 + 7 = 10\mathrm{mol}$$

$$\frac{n(\mathrm{l})}{n(\mathrm{g})} = \frac{y_B - x_K}{x_K - x_B} = \frac{0.5 - 0.3}{0.3 - 0.03} = \frac{0.2}{0.27}$$

We can get: $n(\mathrm{l}) = 4.26\mathrm{mol}$

$n(\mathrm{g}) = 5.74\mathrm{mol}$

(3) Two solutions presented with equilibrium phase point C and D can be seen as ideal dilute solution.

For equilibrium phase point C, A is solvent and B is solute, so A obeys Raoult's law, and B obeys Henry's law, known $x_{B,C} = 0.05$, $y_{B,E} = 0.6$

Then,

$$p_A^* x_A = p y_A$$

That is, $p_A^* = \dfrac{p y_A}{x_A} = \dfrac{p(1 - y_B)}{1 - x_B} = \dfrac{100 \times (1 - 0.6)}{1 - 0.05} = 42.1\mathrm{kPa}$

$$k_{x,B} x_B = p y_B$$

That is, $k_{x,B} = \dfrac{p y_B}{x_B} = \dfrac{100 \times 0.6}{0.05} = 1200\mathrm{kPa}$

For equilibrium phase point D, B is solvent and A is solute, so B obeys Raoult's law, and A obeys Henry's law, known $x_{B,D} = 0.97$, $y_{B,E} = 0.6$

Then,

$$p_B^* x_B = p y_B$$

That is, $p_B^* = \dfrac{p y_B}{x_B} = \dfrac{100 \times 0.6}{0.97} = 61.9\mathrm{kPa}$

$$k_{x,A} x_A = p y_A$$

That is, $k_{x,A} = \dfrac{p y_A}{x_A} = \dfrac{p(1 - y_B)}{1 - x_B} = \dfrac{100 \times (1 - 0.6)}{1 - 0.97} = 1333.3\mathrm{kPa}$

6.5 Solid-liquid phase diagram of two-component system (二组分系统的固-液相图)

6.5.1 Two-component solid-liquid phase diagram of solid full immiscible system (二组分固态完全不互溶系统的固-液相图)

We now discuss binary solid-liquid diagrams. The effect of pressure on condensed phases (solids and liquids) is slight, and unless one is interested in high-pressure phenomena, one generally holds p fixed and examines the T-x_B solid-liquid phase diagram.

Let substance A and B be miscible in all proportions in the liquid phase and completely immiscible in the solid phase. Mixing any amounts of liquids A and B will produce a single-phase system that is a solution of A plus B. Since solids A and B are completely insoluble in each other, cooling a liquid solution of A and B will cause either pure A or pure B to freeze out of the solution.

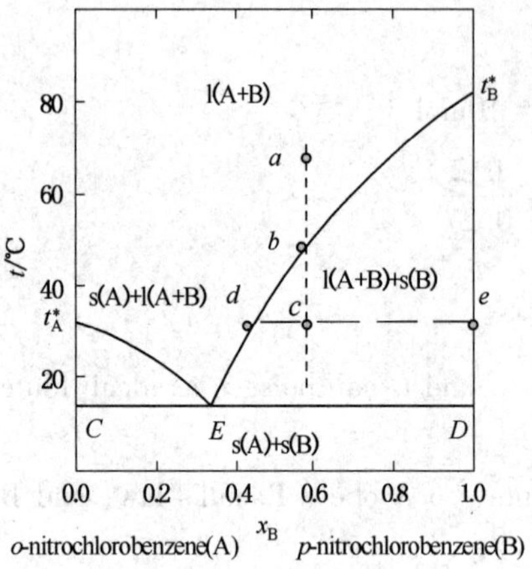

Fig. 6. 18 Melting point-composition diagram of *o*-nitrochlorobenzene(A) *p*-nitrochlorobenzene(B)

The typical appearance of the solid-liquid phase diagram for this case is shown in figure 6. 18.

Analysis of phase diagram:

(1) Point

t_A^* —melting point of pure A

t_B^* —melting point of pure B

E—eutectic point (共熔点)

Line

$t_A^* C$ —solid phase line of pure A

$t_B^* D$ —solid phase line of pure B

$t_A^* E$ and $t_B^* E$ —liquid phase line

CED— triple line, that is, s(A), s(B) and l(A+B) coexist, s(A) + s(B)→l(A+B), which is called cocrystallization line

$$f' = C - P + 1 = 2 - 3 + 1 = 0$$

Region

1-phase region: liquid phase region, high temperature region, l(A+B)

$$f' = 2 - 1 + 1 = 2$$

2-phase region: l(A+B) + s(A), l(A+B) + s(B), s(A) + s(B)

$$f' = 2 - 2 + 1 = 1$$

(2) 2-phase region

Lever rule can be used to calculate the masses or the numbers of moles of two phases coexist in equilibrium.

(3) Triple line

The proportion is fixed, that is, $\frac{n_A}{n_B} = \frac{\overline{DE}}{\overline{CE}}$

(4) Changing of system point

For a solution with system point a to the right of E, solid B will freeze out as T is lowered; for system point to the left of E, solid A will freeze out as T is lowered. At the values of T and x_B corresponding to point E, the chemical potentials of solid A, solid B, and the solution are all equal, and both A and B freeze out.

6. 5. 2 Thermal analysis method (热分析法)

Method of drawing phase diagram:

① Thermal analysis method;

② Solubility method: H_2O-salt system, such as $H_2O-(NH_4)_2SO_4$.

(1) The basic principle of thermal analysis method (热分析法基本原理)

Here, a sample of the two components is melted to a liquid solution, then one allows it to cool and measures the system's temperature as a function of time; this is repeated for several different liquid compositions to give a set of cooling curves, which are shown in figure 6.19.

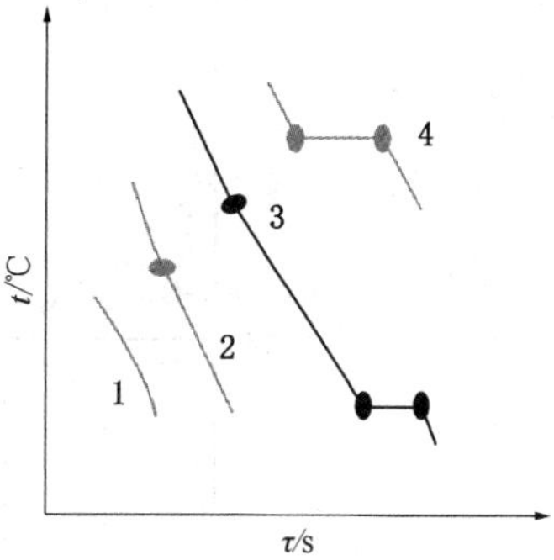

Fig. 6.19 Cooling curves

Cooling curve(步冷曲线)—the curve of temperature t changing with time τ.

Idle point (停歇点, 平台)—terrace that does not change with time.

Break point(转折点, 拐点)—point before and after which depressing rates of temperature (slop change) are different.

(2) Drawing phase diagram using thermal analysis method (利用热分析法绘制相图)

One way to determine a solid-liquid phase diagram experimentally is by thermal analysis method. We can get melting point-composition diagram through a series of cooling curve. It is shown in figure 6.20.

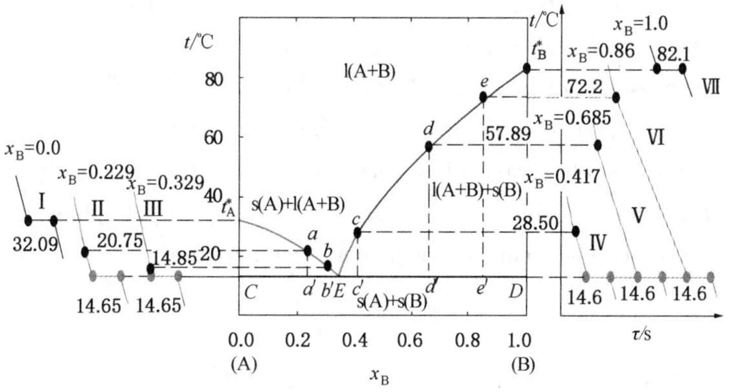

Fig. 6.20 Melting point-composition diagram of A−B system determined by thermal analysis method

(3) Drawing cooling curve according to phase diagram (根据相图绘制步冷曲线)

Drawing procedures:

① Mark abscissa(横坐标) and longitudinal coordinate(纵坐标). Longitudinal coordinate is temperature t (℃) and abscissa is time τ(s).

② Draw plumb line crossing the system point in order to get intersection point. Intersection point of curve is break point; intersection point of straight line is idle point.

③ Draw curve under break point. The slopes of lines of both sides are different.

We can get the cooling curve according to melting point-composition diagram using the above drawing procedures. It is shown in figure 6.21.

6.5.3 Two-component solid-liquid phase diagram of condensed system forming compound (生成化合物的二组分凝聚系统的固–液相图)

A fairly common occurrence is for substance A and B to form a solid compound that can exist in

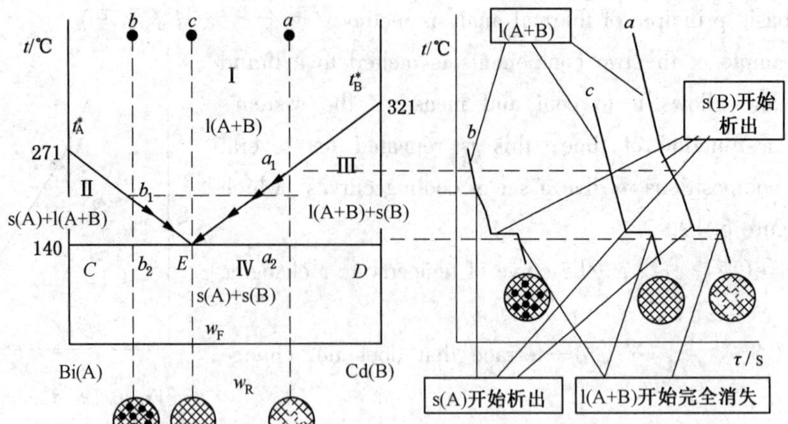

Fig. 6. 21 Drawing cooling curve according to melting point-composition diagram of A-B system

equilibrium with the liquid. Although the system has $C=3$ (instead of 2), the number of degrees of freedom is unchanged by compound formation, since we now have the equilibrium restriction A(s)+B(s) = C(s), thus, $C=S-R-R'$ = 3−1−0 = 2, C is still 2, and the system is binary.

Compound of congruent melting point(相合熔点化合物，稳定化合物)—existing forms of molecules of compound C in liquid and in solid are same.

Compound of incongruent melting point(不相合熔点化合物，不稳定化合物)—existing forms of molecules of compound C in liquid and in solid are different.

Figure 6. 22 shows the solid-liquid phase diagram for Mg(A) plus Si(B), which form the compound of congruent melting point; Figure 6. 23 shows the solid-liquid phase diagram for Na(A) plus K(B), which form the compound of incongruent melting point.

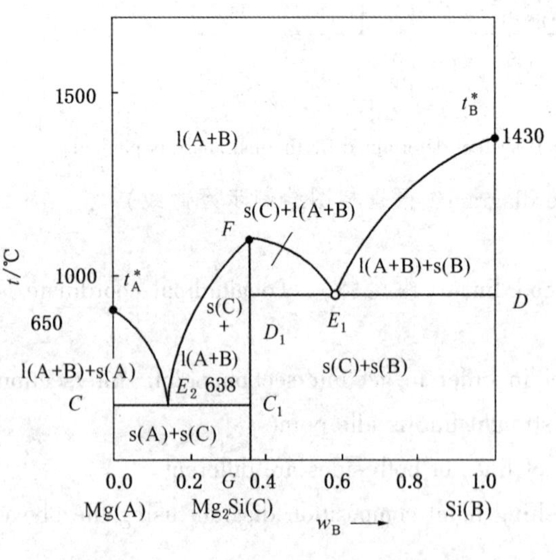

Fig. 6. 22 Melting point-composition diagram of Mg(A)−Si(B) forming compound of congruent melting point

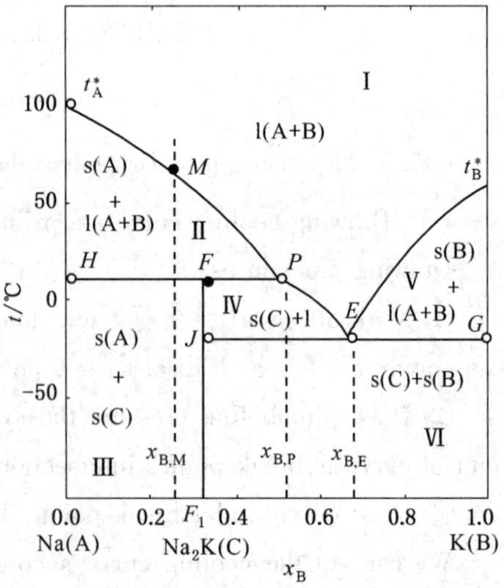

Fig. 6. 23 Melting point-composition diagram of Na(A)−K(B) forming compound of incongruent melting point

(1) Analysis of phase diagram of Mg(A)-Si(B) forming the compound of congruent melting point

① Point

t_A^* —melting point of pure A

t_B^* —melting point of pure B

E_1 and E_2—eutectic point (共熔点)

Line

$t_A^* C$ —solid phase line of pure A

$t_B^* D$ —solid phase line of pure B

FG—solid phase line of compound C

$t_A^* E_2$, FE_2, $t_B^* E_1$ and FE_1—liquid phase line

CE_2C_1—triple line, that is, s(A), s(C) and l(A+B) coexist, s(A) + s(C)→l(A+B),

$$f' = C - P + 1 = 2 - 3 + 1 = 0$$

DE_1D_1—triple line, that is, s(B), s(C) and l(A+B) coexist, s(B) + s(C)→l(A+B),

$$f' = C - P + 1 = 2 - 3 + 1 = 0$$

Region

1-phase region: liquid phase region, l(A+B)

$$f' = 2 - 1 + 1 = 2$$

2-phase region: l(A+B) + s(A), l(A+B) + s(C), s(A) + s(C), l(A+B) + s(B), l(A+B) + s(C), s(B) + s(C)

$$f' = 2 - 2 + 1 = 1$$

② 2-phase region

Lever rule can be used to calculate the masses or the numbers of moles of two phases coexist in equilibrium.

③ Triple line

The proportion is fixed, that is, $\frac{n_A}{n_C} = \frac{\overline{C_1E_2}}{\overline{CE_2}}$ and $\frac{n_C}{n_B} = \frac{\overline{DE_1}}{\overline{D_1E_1}}$

(2) Analysis of phase diagram of Na(A)-K(B) forming the compound of incongruent melting point

① Point

t_A^* —melting point of pure A

t_B^* —melting point of pure B

E—eutectic point (共熔点)

Line

$t_A^* H$ —solid phase line of pure A

$t_B^* G$ —solid phase line of pure B

FF_1—solid phase line of compound C

$t_A^* P$, PE and $t_B^* E$ —liquid phase line

HFP—triple line, that is, s(A), s(C) and l(A+B) coexist, s(C)→ l(A+B)+s(A),

$$f' = C - P + 1 = 2 - 3 + 1 = 0$$

JEG—triple line, that is, s(B), s(C) and l(A+B) coexist, s(B) + s(C)→ l(A+B),

$$f' = C - P + 1 = 2 - 3 + 1 = 0$$

Region

1-phase region: liquid phase region, l(A+B)

$$f' = 2 - 1 + 1 = 2$$

2-phase region: l(A+B) + s(A), l(A+B) + s(C), s(A) + s(C), l(A+B) + s(B), s(B) + s(C)

$$f' = 2 - 2 + 1 = 1$$

② 2-phase region

Lever rule can be used to calculate the masses or the numbers of moles of two phases coexist in equilibrium.

③ Triple line

The proportion is fixed, that is, $\frac{n_A}{n_l} = \frac{\overline{FP}}{\overline{HF}}$ and $\frac{n_C}{n_B} = \frac{\overline{GE}}{\overline{JE}}$

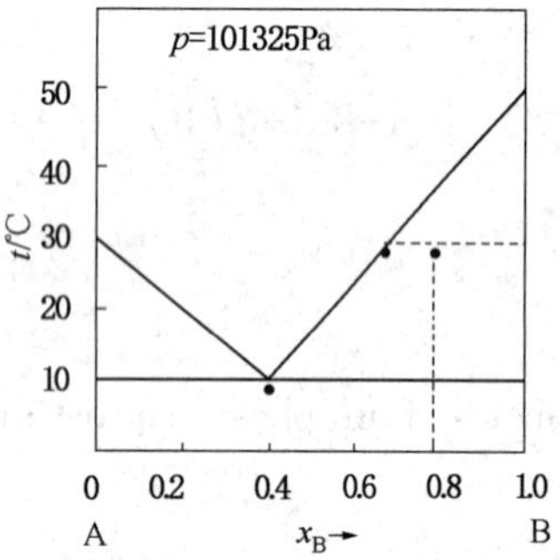

Example 6.6

Substance A and B are completely immiscible in the solid phase. The melting point of A(s) is 30℃ and B(s) is 50℃ when p = 101325Pa. At t = 10℃, A and B have the eutectic point E, the corresponding composition is $x_{B,E} = 0.4$. Suppose the solubility lines of A and B are straight line, (1) draw the melting point-composition diagram of A-B two-component system; (2) 2mol A and 8mol B form the system, please list table and write P, state of aggregation(g, l or s) and ingredient(A, B or A+B), number of moles of every phase and f' of all regions when t = 5℃, 30℃, and 50℃ according to the melting point-composition diagram of A-B two-component system.

Answer:

(1) The melting point-composition diagram of A-B two-component system is shown in the above figure.

(2)

t/℃	P	Phase state and ingredient	Number of moles	f'
5	2	s(A) + s(B)	$n_{s(A)}$ = 2mol $n_{s(B)}$ = 8mol	1
30	2	l(A+B) + s(B)	$n_{l(A+B)}$ = 6.67mol $n_{s(B)}$ = 3.33mol	1
50	1	l(A+B)	$n_{l(A+B)}$ = 10mol	2

Example 6. 7

Phase diagram of condensed system is showed in the figure.

(1) List table and write P, state of aggregation(g, l, s) and ingredient(A, B or A+B) of every phase and f' of all phase regions.

(2) Draw the cooling curve of system point a.

(3) Point out the changing of phase state at break point and idle point in the cooling curve.

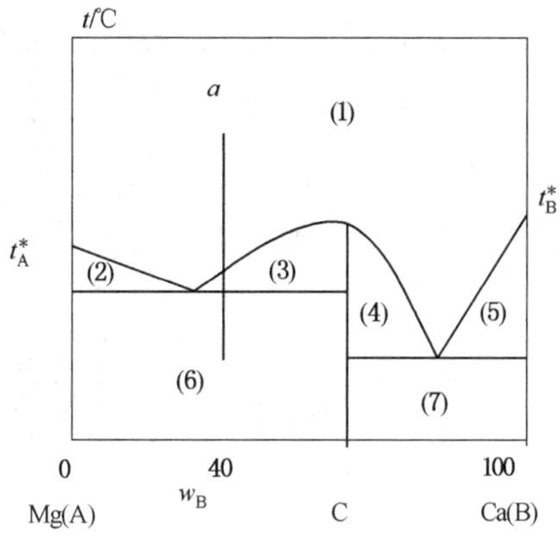

Example 6. 7

***Answer*:**

(1)

Phase region	P	Phase state and ingredient	f'	Phase region	P	Phase state and ingredient	f'
1	1	l(A+B)	2	5	2	l(A+B) + s(B)	1
2	2	l(A+B) + s(A)	1	6	2	s(A) + s(C)	1
3	2	l(A+B) + s(C)	1	7	2	s(B) + s(C)	1
4	2	l(A+B) + s(C)	1				

(2) and (3)

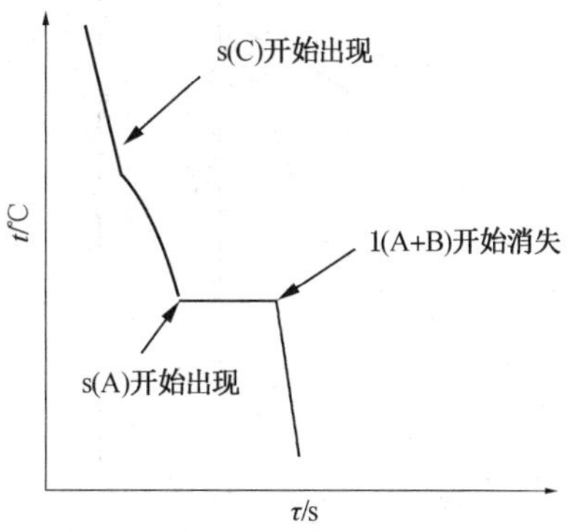

Example 6. 8

Phase diagram of condensed system forming compound of incongruent melting point is showed in the figure. Draw the cooling curve of system point a, b, c, d, e, f, g.

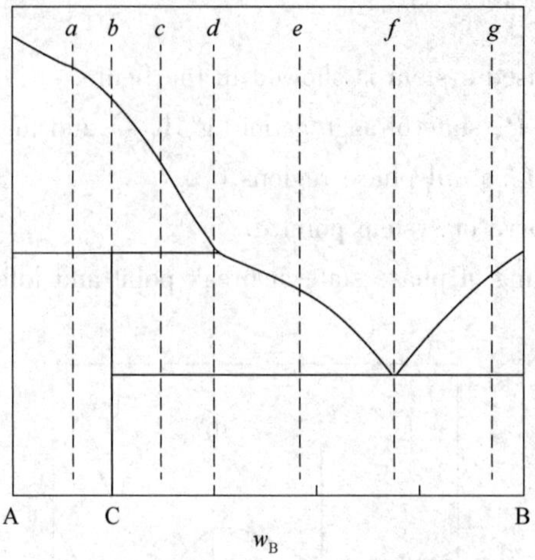

Answer:

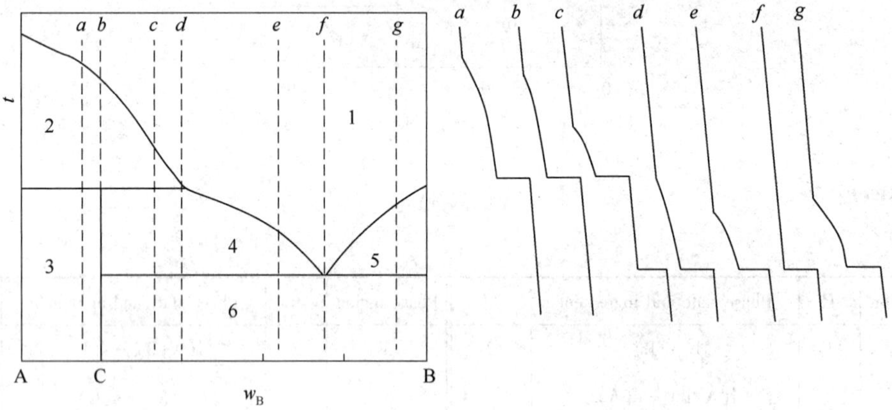

Example 6.9

The figure below is the melting point-composition phase diagram for A and B. Please answer the following questions based on the phase diagram.

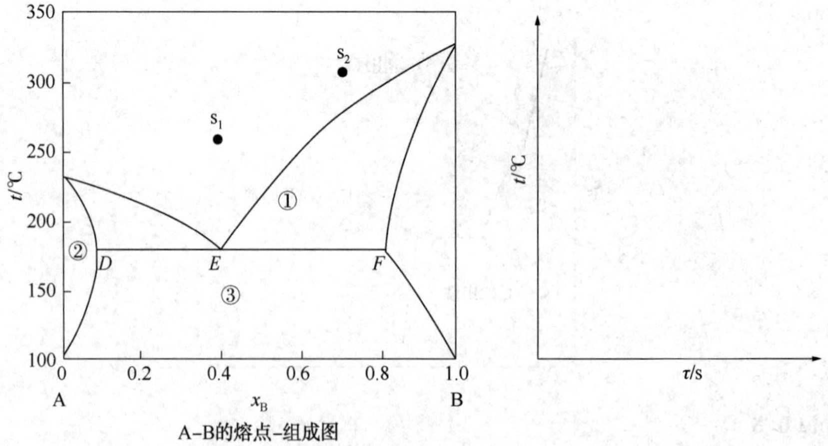

A–B的熔点-组成图

(1) Draw the cooling curve of the system points s_1 and s_2 in the right $t-\tau$ plane coordinate system.

(2) Fill in the phase number, composition and degree of freedom of the marked phase areas in table.

Phase region	phase number(P)	Phase state(composition)	degree of freedom(f')
①			
②			
③			
Line $\overline{DEF}$			

(3) How much solid matter can be precipitated from the eutectic liquid consisting of 2mol A and 3mol B when cooling down? What is this solid matter? What is the amount of substance containing A?

***Answer*:**

(1)

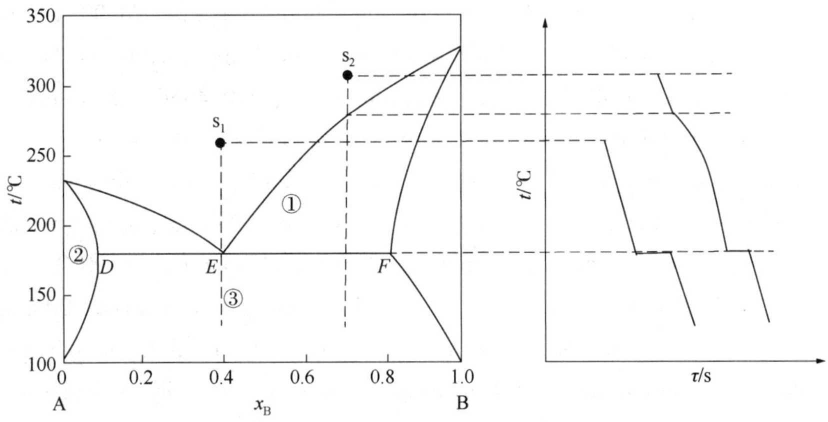

A-B的熔点-组成图

(2)

Phase region	Phase number(P)	Phase state(composition)	Degree of freedom(f')
①	2	l(A+B)+s_2(A+B)	1
②	1	s_1(A+B)	2
③	2	s_1(A+B)+s_2(A+B)	1
Line $\overline{DEF}$	3	s_1(A+B)+l(A+B)+s_2(A+B)	0

(3) Eutectic liquid consisting of 2mol A and 3mol B, $x_B = 0.6$. When the temperature drops to dT from the three-phase line, pure solid s_2(A + B) can be precipitated at most, which can be known from the lever rule:

$$n(s_2)\times(0.8-0.6) = n(l)\times(0.6-0.4)$$

$n(s_2)+n(l) = 5\text{mol}$

The solution of the above two equations is: $n(s_2) = 2.5$ mol; $n(l) = 2.5$ mol

The precipitated solid substance is a solid solution of A and B, the amount of the substance containing A is $2.5\text{mol}\times(1-0.8) = 0.5\text{mol}$.

Supplementary Examples of Chapter 6

EXERCISES
(习题)

1. Please find C, P and f in the following equilibrium systems.

(1) $I_2(s)$ is in equilibrium with it's vapor;

(2) $CaCO_3(s)$ is partially decomposed in equilibrium with $CaO(s)$ and $CO_2(g)$;

(3) $NH_4HS(s)$ is in equilibrium with $NH_3(g)$ and $H_2S(g)$ in a vacuum vessel;

(4) $NH_4HS(s)$ is in equilibrium with arbitrary $NH_3(g)$ and $H_2S(g)$;

(5) I_2 is dissolved in two immiscible liquids H_2O and CCl_4, and reaching distribution equilibrium (condensed system).

Answer: (1)1, 2, 1; (2) 2, 3, 1; (3) 1, 2, 1; (4)2, 2, 2; (5)3, 2, 2

2. Toluene (A) and benzene (B) can form ideal liquid mixture. At 90 ℃, the saturated vapor pressure is 54.22kPa and 136.12kPa respectively. Now 5 mol mixture of A−B with $x_{B,0}$ = 0.3 reaches g−l equilibrium at 90℃, the gas composition is y_B = 0.4556, please calculate :

(1) The equilibrium liquid composition x_B and the total pressure of system p;

(2) The amount of substance n(l), and n(g) when reaching g-l equilibrium.

Answer: (1)x_B = 0.2500, p = 74.70kPa; (2)n (l) = 3.784mol, n (g) = 1.216mol

3. Toluene (A) and benzene (B) can form ideal liquid mixture. At 90 ℃, the saturated vapor pressure is 54.22kPa and 136.12kPa respectively. 200.0g C_7H_8(A) and 200.0g C_6H_6(B) were put into a heat conduction vessel with a piston to form the ideal liquid mixture at 90 ℃. Keeping the temperature be constant and decreasing the pressure gradually, calculate:

(1) The corresponding pressure and the gas composition when the gas begins to produce;

(2) The corresponding pressure and the liquid composition of the last droplet when the liquid begins to disappear;

(3) The liquid composition (x_B), gas composition (y_B) of the system and n(l), n(g) when reaching g−l equilibrium at 92kPa.

Answer: (1)p = 98.54kPa, y_B = 0.7476; (2)p = 80.40kPa, x_B = 0.3197; (3)x_B = 0.4613, y_B = 0.6825, n (l) = 3.022mol, n (g) = 1.709mol

4. C_6H_5OH and H_2O reach liquid-liquid equilibrium at 30℃, the corresponding compositions of the conjugate solution(l_1−l_2) of the system are $w_1(C_6H_5OH, l_1)$ = 8.75%, $w_2(C_6H_5OH, l_2)$ = 69.9% respectively.

(1) At 30℃, the system formed by 100g C_6H_5OH and 200g H_2O reaches liquid-liquid equilibrium, calculate the mass of the two liquid phases (l_1, l_2);

(2) If 100g C_6H_5OH is added into the above system, calculate the mass of the two liquid phases (l_1, l_2) when the system reaches liquid-liquid equilibrium again.

Answer: (1)m (l_1) = 179.6g, m (l_2) = 120.4g; (2) m (l_1) = 130.2g, m (l_2) = 269.8g

5. Using the following data to draw roughly a phase diagram of Mg − Cu two-component condensed system and point out the stable phase of every region.

Given: the melting point of Mg and Cu are 648℃ and 1085℃ respectively. Mg and Cu can form two stable compounds Mg_2Cu and $MgCu_2$ whose melting point are 580℃, and 800℃ respectively. Two metal and two compounds can form three kinds of eutectic mixtures, the corresponding composition w(Cu) and melting point of three eutectic points are: 35%, 380℃; 66%, 560℃; and 90.6%, 680℃.

6. As shown in the diagram, A and B are full miscible in liquid phase and full immiscible in solid phase at p = 101.325kPa, the composition of the lowest eutectic mixture is w_B = 0.60. Now there is 180g solution with w_B = 0.40. Try to calculate:

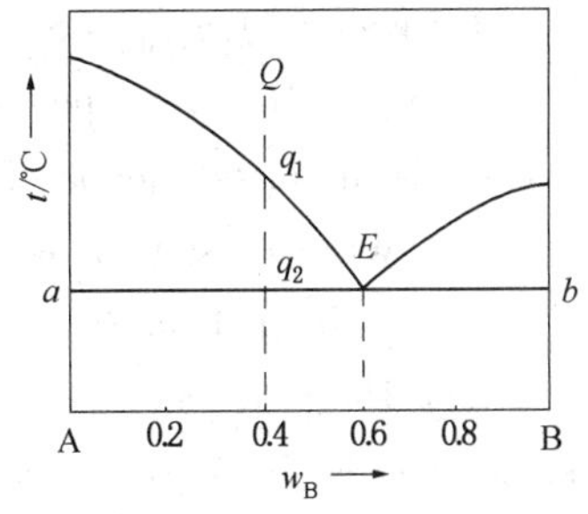

(1) The most mass of pure solid A(s) which we can get when the solution is cooled.

(2) The mass of solid A and solid B when the lowest eutectic solution is 60g and the system reaches triple phase equilibrium.

Answer: (1) 60g; (2) 84g, 36g

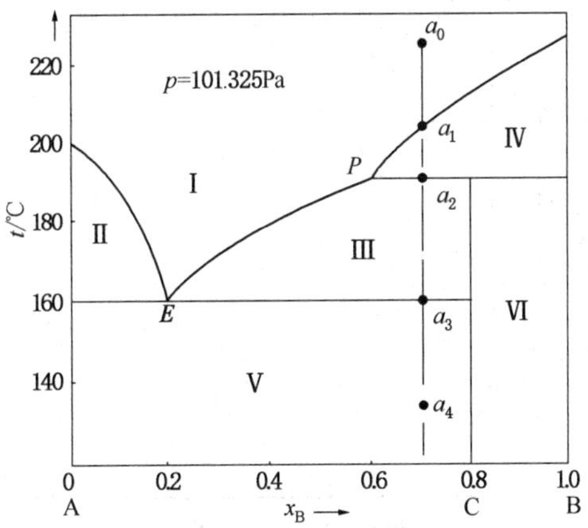

7. As shown in the diagram, A and B are two-component condensed system. t_A^*, and t_B^* are the melting point of A and B respectively.

(1) According to the phase diagram, list table and write P, state of aggregation (g, l or s) and ingredient (A, B or A+B) of every phase and f' of regions from I to VI.

(2) The system point a_0 pass through a_1, a_2, a_3, and a_4 when cooling, write out the changing of phase state at point a_1, a_2, a_3, and a_4 and draw the cooling curve.

Answer:

(1)

Phase region	P	Phase state and ingredient	f'	Phase region	P	Phase state and ingredient	f'
I	1	l(A+B)	2	IV	2	l(A+B) + s(B)	1
II	2	l(A+B) + s(A)	1	V	2	s(A) + s(C)	1
III	2	l(A+B) + s(C)	1	VI	2	s(B) + s(C)	1

(2) point a_1: l $\rightleftharpoons$ s(B); point a_2: l + s(B) $\rightleftharpoons$ s(C); point a_3: l $\rightleftharpoons$ s(A)+s(C); point a_4: s(A) $\rightleftharpoons$ s(C).

8. A and B can form a compound AB of congruent melting point. A, B and AB are full immiscible in solid phase. The melting point of A, AB and B are 200℃, 300℃, and 400℃ respectively. A and AB, B and AB can form two lowest eutectic points (E_1 and E_2): 150℃, x_{B,E_1} = 0.2 and 250℃, x_{B,E_2} = 0.8.

(1) Draw the melting point-composition diagram (t-x_B) of above system;

(2) Draw the cooling curves of the following two systems: system with x_B = 0.1 cooling from

200℃ to 100℃ and system with $x_B = 0.5$ cooling from 400℃ to 200℃;

(3) Mixture formed by 8 mol B and 12 mol A is cooled nearly to 150℃, how many phases are in equilibrium? what is the composition of every phase? how much is the amount of substance of every phase?

Answer: (3) $n_{s(AB)} = 6.67$ mol, $n_{l(A+B)} = 6.67$ mol

9. A and B two-component gas-liquid equilibrium phase diagram is shown in the figure, the abscissa is the mass fraction of B.

(1) Write P, state of aggregation (g, l or s) and ingredient (A, B or A+B) of every phase and f' of regions from ① to ⑥ in the diagram;

(2) The mixture formed by 10.8kg pure liquid A and 7.2kg pure liquid B is heated, when the temperature is closed nearly to t_1 ($t = t_1 + dt$), how many phases are in equilibrium? how much is the mass of every phase?

(3) When the temperature leave t_1 ($t = t_1 + dt$) just now, how many phases are in equilibrium? how much is the mass of every phase?

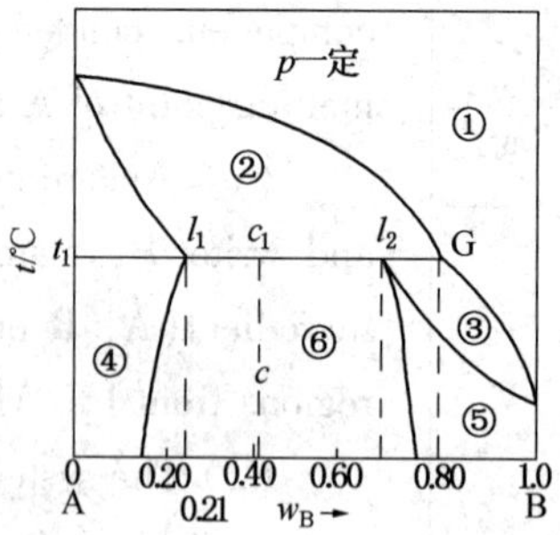

Answer: (1)

Phase region	P	Phase state and ingredient	f'	Phase region	P	Phase state and ingredient	f'
①	1	g(A+B)	2	④	1	l_1(A+B)	2
②	2	g(A+B) + l_1(A+B)	1	⑤	1	l_2(A+B)	2
③	2	g(A+B) + l_2(A+B)	1	⑥	2	l_1(A+B) + l_2(A+B)	1

(2) $m(l_2) = 6.98$kg, $m(l_1) = 11.02$kg;

(3) $m(l_1) = 12.2$kg, $m(g) = 5.8$kg

Chapter 7 Electrolyte solution
(电解质溶液)

Leaning objectives:

(1) Define the *activity*, the *activity factor*, and the *mean activity factor of ions* in solution.

(2) Define *ionic strength*.

(3) Derive and use *the Debye-Hückel Limiting Law* for the *mean activity factor* and indicate how it may be extended to more concentrated solutions.

An electrolytic solution refers to a solution formed by completely dissociated or partially dissociated solute into ions after it is dissolved in solvent. Electrolyte solutions are ubiquitous in nature and organisms. Electrolyte solutions are often involved in chemical experiments and chemical production, especially in the galvanic cell, chemical power supply and electrolytic process.

The research of electrolyte solution mainly includes two aspects: the thermodynamic properties of electrolyte solution and the conductive properties of electrolyte solution.

(1) Thermodynamic properties of electrolyte solution

Electrolyte solution with different molecular solution, the electrostatic force between the positive and negative ions belongs to long-range force, so even very dilute electrolyte solution still deviate from the ideal dilute solution to comply with the laws of thermodynamics, so discuss the balance of electrolyte solution must be the introduction of ionic mean activity and concepts of ionic mean activity factor.

(2) The conductivity of an electrolyte solution is different from that of an electronic conductive body, which conducts electricity not by electrons, but by ions, so it is called ionic conductor. Under the action of an applied electric field, the positive ions in electrolyte solution move toward the negative electrode, and the negative ions move toward the positive electrode to realize the current transmission. In a device that converts chemical and electrical energy into each other (galvanic cell, electrolytic cell), that is, in a multiphase system such as an electrochemical system, there must be at least one phase through which electrons can not pass, the charge of the phase through which the electrons can not pass must be carried by ions, such ionic conductors being electrolyte solutions or molten salts. Therefore, the study of the conductivity of electrolyte solution is of great significance to understand the mechanism of the mutual conversion of chemical energy and electric energy in the electrochemical system.

7.1 Electrolyte and types of electrolyte
(电解质及其分类)

7.1.1 Definition of electrolyte(电解质的定义)

Electrolyte—an electrolyte is a substance that can partly or totally dissociate to positive and negative ions in solution or melting, as evidenced by the solution showing electrical conductivity. (电解质是指在溶液中或融化时能全部或部分解离为正负离子，从而使所形成的液相具有导电能力的物质。)

7.1.2 Conducting mechanism of electrolyte solution(电解质溶液的导电机理)

(1) Conducting mechanism of electronic conductor

The current is carried by the motion of electrons. Such as metals, graphite or some metal oxides.

(2) Conducting mechanism of electrolyte solution

The current is carried by the motion of ions in solution under electric field. The ions are set in motion by an applied electric potential difference between the two electrodes.

7.1.3 Types of electrolyte(电解质的类型)

(1) According to the degree of ionization(根据解离度来划分)

According to the degree of ionization, electrolyte can be divided into strong electrolyte and weak electrolyte.

① **Strong Electrolyte**—complete dissociation when dissolved in solvent. (在溶剂中发生完全解离。)

For instance: strong acid(HCl), strong base(NaOH) and a large amount of salts(NaCl) etc.

② **Weak Electrolyte**—partial dissociation when dissolved in solvent. (在溶剂中仅发生部分解离。)

For instance: acetic acid(CH_3COOH).

The classification of strong and weak electrolytes depends not only on the nature of electrolytes, but also on the nature of solvents. For example, CH_3COOH is a weak electrolyte in water and fully ionized in liquid NH_3 and is a strong electrolyte. KI is a strong electrolyte in water and a weak electrolyte in acetone.

(2) According to the valence type (根据电解质的价型来划分)

Let electrolyte S yields the ions X^{z+} and Y^{z-}:

$$S \rightarrow \upsilon_+ X^{z+} + \upsilon_- Y^{z-}$$

The z's are the charges of ions, the υ's are the numbers of ions in the chemical formula.

For example

$NaNO_3$	$z_+=1$	$\lvert z_-\rvert=1$	is a 1 : 1 electrolyte
$BaSO_4$	$z_+=2$	$\lvert z_-\rvert=2$	is a 2 : 2 electrolyte
Na_2SO_4	$z_+=1$	$\lvert z_-\rvert=2$	is a 1 : 2 electrolyte
$Ba(NO_3)_2$	$z_+=2$	$\lvert z_-\rvert=1$	is a 2 : 1 electrolyte

7.1.4 Faraday's Law(法拉第定律)

Faraday's Law: Extent of electrode reaction $\Delta\xi$ is proportional to the quantity of electricity passed through the solution Q, and inversely proportional to charge of the reaction z when electricity is passing through electrolyte solution.

Michael Faraday

Mathematical formula:

$$\Delta\xi = \frac{Q}{zF} \tag{7.1.1}$$

where Q is the quantity of electricity passed through the solution, z is the magnitude of the charge of the ions in terms of the electronic charge as a unit, and F, called Faraday's constant, is the charge of an Avogadro's number of electrons, $F = Ne = 96485\text{C} \cdot \text{mol}^{-1}$, $\Delta\xi$ is the extent of reaction.

7.2 Transference numbers of ions (离子的迁移数)

7.2.1 Electromigration(电迁移)

According to the Faraday's law, for each electrode, the amount of charge flowing out equals the amount of charge flowing in equals the total amount of charge flowing across any cross section of the circuit for a given period of time. In a metal wire, the electric current is carried entirely by electrons, but in solution it is carried by both cations and anions.

Electromigration—migration of ion under the action of the outer electric field. (离子在电场作用下的运动称为电迁移。)

Electric mobility—the speed of ion in a field of unit strength. (通常将在单位电场强度 E = 1V · m^{-1}下离子在溶液中的运动速度称为离子的电迁移率，以 u_B表示。)

$$u_B = v_B/E \qquad \text{Unit: m}^2 \cdot \text{V}^{-1} \cdot \text{s}^{-1} \tag{7.2.1}$$

7.2.2 Transference Numbers of ions(离子迁移数)

In order to determine the fraction of the current carried by each ion in a given electrolyte, the transference number t_+ and t_- are introduced according to the definitions.

Transference number—the fraction of the current carried by an ion. (每种离子所运载的电流的分数。)

When an electric current is applied to the electrolyte solution, the anions and cations in the solution, which are responsible for conducting electricity, move toward the positive and negative electrodes respectively, and oxidation or reduction takes place at the corresponding interface between the two electrodes, the concentration of the solution adjacent to the electrodes also changes.

t_+ = fraction of current carried by cation. (正离子所运载的电流的分数。) $t_+ = I_+/I$

t_- = fraction of current carried by anion. (负离子所运载的电流的分数。) $t_- = I_-/I$

In order to find the relationship between the migration of ions and the transference number, let us consider the following sample cell(As shown in figure 7.1):

① The electrolytic cell is divided into three parts: the cathode region, the anode region, and the middle region.

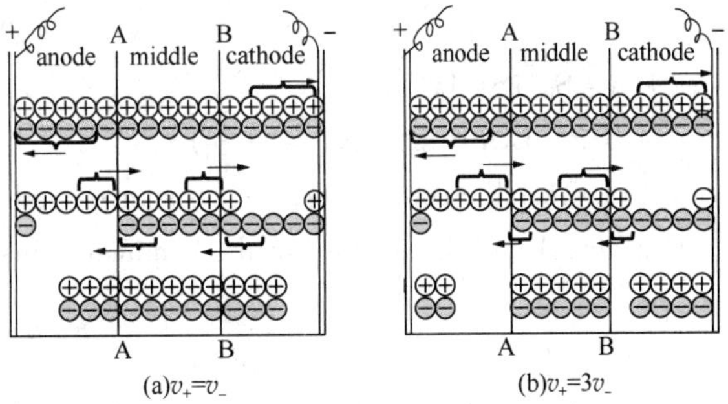

Fig 7.1 The phenomenon of electromigration

② The solution contains 15 mol negative ions and 15 mol positive ions.

③ The total charge passed through the solutions equals $4F$.

The results indicate that after the electrolysis the solution is electrically neutral in all three parts. The concentration changes at the two electrode regions are different and are proportional to the velocities of the ions.

That is

$$\frac{\Delta n_+}{\Delta n_-}=\frac{v_+}{v_-}=\frac{u_+}{u_-} \tag{7.2.2}$$

$$\frac{\Delta n_-}{\Delta n_+}=\frac{v_-}{v_+}=\frac{u_-}{u_+} \tag{7.2.3}$$

Where Δn_+ is the change at the anode and Δn_- is the change at the cathode.

So we can get

$$t_+=\frac{I_+}{I}=\frac{Q_+}{Q_++Q_-}=\frac{\Delta n_+}{\Delta n_++\Delta n_-}=\frac{v_+}{v_++v_-}=\frac{u_+}{u_++u_-} \tag{7.2.4}$$

$$t_-=\frac{I_-}{I}=\frac{Q_-}{Q_++Q_-}=\frac{\Delta n_-}{\Delta n_++\Delta n_-}=\frac{v_-}{v_++v_-}=\frac{u_-}{u_++u_-} \tag{7.2.5}$$

As easily seen from the preceding equations

$$t_++t_-=1 \tag{7.2.6}$$

7.3 Conductance, conductivity and molar conductivity (电导、电导率、摩尔电导率)

7.3.1 Conductance(电导)

The fundamental measurement used to study the motion of ions is that of the electrical resistance, R, of the solution. The conductance, G, of a solution is the inverse of its resistance R: $G=1/R$. As resistance is expressed in ohms, Ω, the conductance of a sample is expressed in Ω^{-1}. The reciprocal ohm used to be called the Siemens, S.

Conductance—the conductive ability of the conductor or the electrolyte solution. (描述导体或电解质溶液导电能力的物理量。)

It is calculated from the measured resistance as

$$G=\frac{1}{R} \tag{7.3.1}$$

Unit: S, $1S=1\Omega^{-1}$.

7.3.2 Conductivity(电导率)

As for metallic conductors, the resistance, and therefore the conductance, depends on the cross section area A and the length l of the conductor. Just for a metallic conductor one has

$$R=\rho\frac{l}{A} \tag{7.3.2}$$

Where ρ is the specific resistance, it is the proportionality factor that corresponds to the resistance of a conductor of unit cross-section area and unit length. Thus

$$G=\frac{1}{R}=\kappa\frac{A}{l} \qquad (7.3.3)$$

Where κ, the conductivity is the proportionality factor that corresponds to the conductance of a conductor of unit cross-section area and unit length. Unit: $S \cdot m^{-1}$

In principle, the conductivity may be obtained directly from equation (7.3.3) if the distance l between the electrodes and the areas A of the electrodes are known. However, it seems to be more convenient to standardize the conductance cell - to find the ratio l/A - with a solution of known conductivity, for example, KCl solution.

Definition of cell constant(电导池常数的定义):

$$K_{cell}=\frac{l}{A} \qquad (7.3.4)$$

The cell constant K_{cell} remains fixed as long as the distance between the two electrodes and their effective areas is unchanged.

So
$$G=\frac{1}{R}=\kappa/K_{cell} \qquad (7.3.5)$$

7.3.3　Molar conductivity(摩尔电导率)

Since the number of the charge carriers per unit volume usually increases with increasing electrolyte concentration, the conductivity κ usually increases as the electrolyte's concentration increases at low concentration. To get a measure of the current—carrying ability of a given amount of electrolyte, the molar conductivity Λ_m of an electrolyte in solution is defined.

Molar conductivity—molar conductivity is the quotient of κ and c. (摩尔电导率是该溶液的电导率与其浓度之比。)

$$\Lambda_m \overset{\text{def}}{=\!=} \frac{\kappa}{c} \qquad (7.3.6)$$

Unit: $S \cdot m^2 \cdot mol^{-1}$.

7.3.4　Calculation of conductivity and molar conductivity(电导率和摩尔电导率的计算)

(1) Calculate K_{cell}(计算电导池常数)

We can calculate the K_{cell} from a measurement of G, when the cell is filled with a solution of known conductivity(such as KCl aq).

$$K_{cell}=\kappa/G$$

Table 7.1 shows the conductivities of KCl aqueous solutions at different concentration at different concentration.

Tab. 7.1　Conductivities of KCl aqueous solutions at different concentration

$c/mol \cdot m^{-3}$	$\kappa/S \cdot m^{-1}$		
	273.15K	291.15K	298.15K
1	6.643	9.820	11.173
0.1	0.7154	1.1192	1.2886
0.01	0.7751	0.1227	0.14114

(2) Calculate κ(计算电导率)

Put the test electrolyte solution into the same conductance cell, and then measure its conductance. From $\kappa = GK_{cell}$, we can get κ.

(3) Calculate Λ_m(计算摩尔电导率)

Use $\Lambda_m \overset{def}{=\!=} \dfrac{\kappa}{c}$ to calculate Λ_m.

Example 7.1

At 25℃, a cell contains 0.01 mol · dm^{-3} KCl aqueous solution, which at that concentration has a conductivity of 0.141 S · m^{-1}. The measured resistance was 112.3Ω. When the same cell was filled with electrolyte solution x of 0.01 mol · dm^{-3}, the resistance was 2184Ω(Omit the conductivity of water).

Find: (1) Cell constant K_{cell};

(2) Conductivity κ of solution x;

(3) Molar conductivity Λ_m of solution x.

***Answer*:**

(1) $K_{cell} = \kappa R = 0.141\text{S} \cdot \text{m}^{-1} \times 112.3\Omega = 15.85\text{m}^{-1}$

(2) $\kappa = \dfrac{K_{cell}}{R} = \dfrac{15.85\text{m}^{-1}}{2184\Omega} = 7.257 \times 10^{-3}\text{S} \cdot \text{m}^{-1}$

(3) $\Lambda_m = \dfrac{\kappa}{c} = \dfrac{7.257 \times 10^{-3}\text{S} \cdot \text{m}^{-1}}{0.01 \times 10^{3}\text{mol} \cdot \text{m}^{-3}} = 7.257 \times 10^{-4}\text{S} \cdot \text{m}^{2} \cdot \text{mol}^{-1}$

7.3.5 Relationship between κ and c or Λ_m and c(电导率及摩尔电导率与电解质的物质的量浓度的关系)

(1) Relationship between κ and c(电导率与电解质的物质的量浓度的关系)

Figure 7.2 plots κ vs. c for some electrolytes in aqueous solution. As can be seen in the figure that the conductivity increases as concentration increases at low concentration, until it reaches the peak value, and then it decreases with concentration increases at high concentration.

(2) Relationship between Λ_m and c(摩尔电导率与电解质的物质的量浓度的关系)

Figure 7.3 plots Λ_m vs. $c^{1/2}$ for some electrolytes in aqueous solution. The rapid increase in Λ_m for weak electrolyte CH_3COOH as $c \to 0$ is due to an increase in the degree of dissociation of this weak acid as c decreases. The slow decrease in Λ_m for strong electrolyte HCl and NaOH as c increase is due to attractions between oppositely charge ions, which reduce the conductivity.

Let Λ_m^∞ denote the infinite-dilute value. $\Lambda_m^\infty = \lim\limits_{c \to 0} \Lambda_m$. The relationship of the Λ_m and $c^{1/2}$ of the strong electrolytes at dilute concentration can be expressed by the relation $\Lambda_m = \Lambda_m^\infty - A\sqrt{c}$. Then Λ_m^∞ value of the strong electrolyte is obtained by extrapolation. The extrapolation procedure cannot be applied to weakly dissociated electrolytes. Such solutions have lower conductance at higher concentrations, but the values increases greatly with increasing dilution. Because of this steep increase in Λ_m at high dilution, the extrapolation to zero concentration is uncertain and may result in large errors.

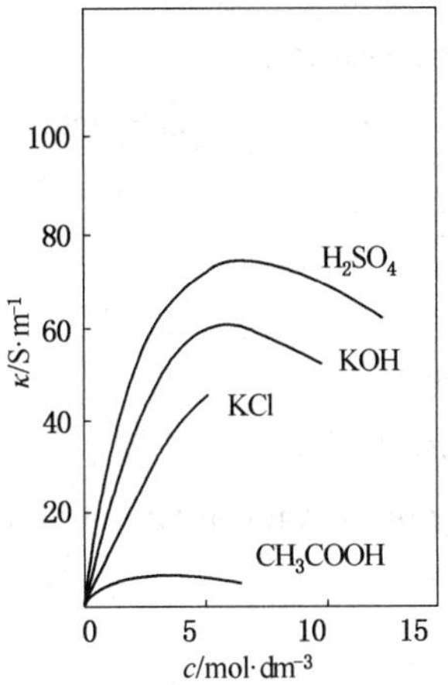

Fig. 7. 2 Relationship between κ and c

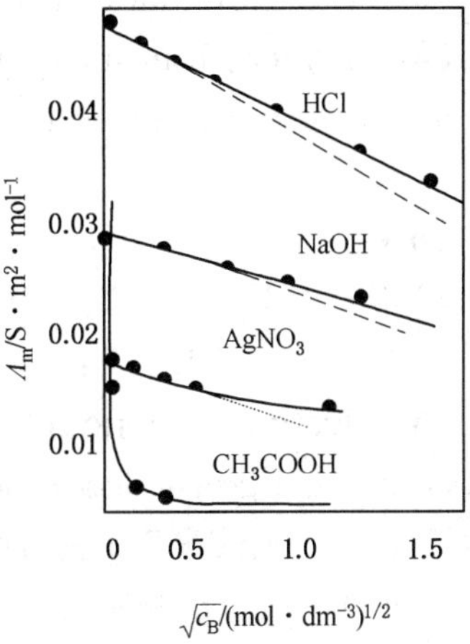

Fig. 7. 3 Relationship between Λ_m and $c^{1/2}$

For this reason, Kohlrausch recommended a different procedure for weakly dissociated solutes.

7. 3. 6 Law of the independent migration of ions—Kohlrausch's law(离子独立运动定律)

Kohlrausch discovered that the difference in Λ_m^∞ for pairs of electrolytes having a common ion is always approximately a constant. Table 7. 2 shows Λ_m^∞ for some strong electrolytes at 298. 15K.

Friedrich Kohlrausch

Tab. 7. 2 Λ_m^∞ for some strong electrolytes at 298. 15K

Electrolyte	Λ_m^∞/ S · m^2 · mol^{-1}	$\Delta\Lambda_m^\infty$/ S · m^2 · mol^{-1}	Electrolyte	Λ_m^∞/ S · m^2 · mol^{-1}	$\Delta\Lambda_m^\infty$/ S · m^2 · mol^{-1}
KCl	0. 014986	0. 00348	HCl	0. 042616	0. 00049
LiCl	0. 011503		HNO_3	0. 04213	
$KClO_4$	0. 014004	0. 00351	KCl	0. 014986	0. 00049
$LiClO_4$	0. 010598		KNO_3	0. 01450	
KNO_3	0. 01450	0. 00349	LiCl	0. 011503	0. 00049
$LiNO_3$	0. 01101		$LiNO_3$	0. 01101	

From Table 7. 2, we can see that

$$\Lambda_m^\infty(KCl)-\Lambda_m^\infty(LiCl)=\Lambda_m^\infty(KClO_4)-\Lambda_m^\infty(LiClO_4)=\Lambda_m^\infty(KNO_3)-\Lambda_m^\infty(LiNO_3)$$

$$\Lambda_m^\infty(HCl)-\Lambda_m^\infty(HNO_3)=\Lambda_m^\infty(KCl)-\Lambda_m^\infty(KNO_3)=\Lambda_m^\infty(LiCl)-\Lambda_m^\infty(LiNO_3)$$

Law of the independent migration of ions—Λ_m of an electrolyte is made up of independent contributions from the cationic and anionic species in limiting dilute solution. (离子独立运动定律——无论是强电解质还是弱电解质，在无限稀薄时电解质的摩尔电导率为正、负离子的摩尔电导率之和。)

That is

$$\Lambda_m^\infty=\upsilon_+\Lambda_{m,+}^\infty+\upsilon_-\Lambda_{m,-}^\infty \tag{7.3.7}$$

where υ_+ and υ_- are the numbers of cations and anions per formula unit of electrolyte (for example, $\upsilon_+ = \upsilon_- = 1$ for HCl, NaCl, and $CuSO_4$, but $\upsilon_+ = 1$, $\upsilon_- = 2$ for $MgCl_2$).

Applicable condition: both strong electrolyte and weak electrolyte.

The immediate practical application of it is that it can be used to calculate Λ_m^∞ of weak electrolyte.

For instance:

$$\Lambda_m^\infty(CH_3COOH) = \Lambda_m^\infty(H^+) + \Lambda_m^\infty(CH_3COO^-)$$
$$= \Lambda_m^\infty(HCl) + \Lambda_m^\infty(CH_3COONa) - \Lambda_m^\infty(NaCl)$$

7.3.7 Application of conductance measurement(电导测定的应用)

(1) Calculate the degree of dissociation and dissociation equilibrium constant of weak electrolyte (计算弱电解质的解离度及解离平衡常数)

① Degree of dissociation(解离度)

$$\alpha = \frac{\Lambda_m}{\Lambda_m^\infty} \tag{7.3.8}$$

where Λ_m can be calculated from $\Lambda_m = \frac{\kappa}{c}$

Λ_m^∞ can be calculated from $\Lambda_m^\infty = \upsilon_+ \Lambda_{m,+}^\infty + \upsilon_- \Lambda_{m,-}^\infty$

② Dissociation equilibrium constant(解离平衡常数)

For example

$$CH_3COOH \rightleftharpoons CH_3COO^- + H^+$$

$t=0$	c	0	0
$t=t$	$c(1-\alpha)$	$c\alpha$	$c\alpha$

$$K^\ominus = \frac{\alpha^2}{1-\alpha}\frac{c}{c^\ominus}$$

Example 7.2

At 25℃, 0.05 mol · dm^{-3} aqueous solution of CH_3COOH has a conductivity of 3.68×10^{-2} s · m^{-1}. Calculate the degree of dissociation α and the dissociation equilibrium constant $K^\ominus$ of CH_3COOH solution.

Example 7.2

Given that $\Lambda_m^\infty(H^+) = 349.82\times10^{-4}$ S · m^2 · mol^{-1};

$\Lambda_m^\infty(CH_3COO^-) = 40.9\times10^{-4}$ S · m^2 · mol^{-1}.

Answer:

$$CH_3COOH \rightleftharpoons CH_3COO^- + H^+$$

$t=0$	c	0	0
$t=\infty$	$c(1-\alpha)$	$c\alpha$	$c\alpha$

The dissociation equilibrium constant: $K^\ominus = \frac{\alpha^2}{1-\alpha}\frac{c}{c^\ominus}$

The degree of dissociation: $\alpha = \frac{\Lambda_m}{\Lambda_m^\infty}$

From the law of the independent migration of ions, we can get

$$\Lambda_m^\infty(CH_3COOH) = \Lambda_m^\infty(H^+) + \Lambda_m^\infty(CH_3COO^-)$$
$$= (349.82+40.9)\times10^{-4} = 390.72\times10^{-4}\ S \cdot m^2 \cdot mol^{-1}$$

$$\Lambda_m = \frac{\kappa_{(CH_3COOH)}}{c_{(CH_3COOH)}} = \frac{0.0368}{0.05\times10^3} = 7.36\times10^{-4} S \cdot m^2 \cdot mol^{-1}$$

$$\alpha = \frac{\Lambda_m}{\Lambda_m^\infty} = \frac{7.36\times10^{-4}}{390.72\times10^{-4}} = 0.0188$$

So

$$K^\ominus = \frac{\alpha^2}{1-\alpha}\frac{c}{c^\ominus} = \frac{(0.0188)^2}{1-0.0188}\cdot\frac{0.05}{1} = 1.81\times10^{-5}$$

(2) Calculate the ionic concentrations produced by insoluble salt(计算难溶盐的溶解度)

Since the solubility of insoluble salt is quite small, the molar conductivity of a saturated solution will be little different from that of infinite dilution.

That is

$$\Lambda_m \approx \Lambda_m^\infty$$

Example 7.3

At 298.15K, the conductivity of the saturated aqueous solution of AgCl is $3.41\times10^{-4} S \cdot m^{-1}$, while the conductivity of pure water is $1.60\times10^{-4} S \cdot m^{-1}$. Given that at 298.15K,

$$\Lambda_m^\infty(Ag^+) = 61.92\times10^{-4} S \cdot m^2 \cdot mol^{-1}$$

$$\Lambda_m^\infty(Cl^-) = 76.34\times10^{-4} S \cdot m^2 \cdot mol^{-1}$$

Calculate the concentration of AgCl saturated solution at 298.15K.

***Answer*:**

$$\kappa(AgCl) = \kappa(\text{solution}) - \kappa(H_2O) = (3.41-1.60)\times10^{-4} = 1.81\times10^{-4} S \cdot m^{-1}$$

$$\Lambda_m(AgCl) \approx \Lambda_m^\infty(AgCl) = \Lambda_m^\infty(Ag^+) + \Lambda_m^\infty(Cl^-)$$

$$= (61.92+76.34)\times10^{-4}$$

$$= 138.26\times10^{-4} S \cdot m^2 \cdot mol^{-1}$$

$$c(AgCl) = \frac{\kappa(AgCl)}{\Lambda_m(AgCl)} = \frac{1.81\times10^{-4}}{138.26\times10^{-4}} = 1.309\times10^{-5} mol \cdot dm^{-3} = 1.309\times10^{-2} mol \cdot m^{-3}$$

7.4　Mean ionic activity of electrolyte (电解质的平均离子活度)

7.4.1　Mean ionic activity and mean ionic activity factor(平均离子活度及平均离子活度因子)

For strong electrolyte S:

$$S \rightarrow v_+ X^{Z^+} + v_- Y^{Z^-}$$

We can get:

$$\mu_B = v_+\mu_+ + v_-\mu_- \quad (7.4.1)$$

μ_B, μ_+, μ_- are the chemical potential of the whole electrolyte, positive ion and negative ion respectively.

They are expressed as follows:

$$\mu_B = \mu_B^\ominus + RT\ln a_B$$
$$\mu_+ = \mu_+^\ominus + RT\ln a_+ \qquad (7.4.2)$$
$$\mu_- = \mu_-^\ominus + RT\ln a_-$$

Where, a_B, a_+, a_- are the activity of the whole electrolyte, positive ion and negative ion respectively.

Let b, b_+ and b_- are the molalities of the whole electrolyte, positive ion and negative ion respectively.

We can get

$$b_+ = v_+ b \qquad b_- = v_- b \qquad b^\ominus = 1\text{mol}\cdot\text{kg}^{-1}$$

Definition of ionic activity factor(正、负离子的活度因子定义):

$$\gamma_+ \overset{\text{def}}{=\!=\!=} \frac{a_+}{b_+/b^\ominus} \qquad \gamma_- \overset{\text{def}}{=\!=\!=} \frac{a_-}{b_-/b^\ominus} \qquad (7.4.3)$$

Substitution of(7.4.3)for a_+ and a_- into(7.4.2)gives μ_+, μ_-

$$\mu_+ = \mu_+^\ominus + RT\ln a_+ = \mu_+^\ominus + RT\ln(\gamma_+ b_+/b^\ominus)$$
$$\mu_- = \mu_-^\ominus + RT\ln a_- = \mu_+^\ominus + RT\ln(\gamma_- b_-/b^\ominus) \qquad (7.4.4)$$

Substitution of(7.4.4)for μ_+, and μ_- into(7.4.1)gives μ

$$\mu_B = (v_+\mu_+^\ominus + v_-\mu_-^\ominus) + RT\ln(a_+^{v_+}a_-^{v_-})$$
$$= \mu_B^\ominus + RT\ln\{\gamma_+^{v_+}\gamma_-^{v_-}(b_+/b^\ominus)^{v_+}(b_-/b^\ominus)^{v_-}\} \qquad (7.4.5)$$

where $\mu_B^\ominus \overset{\text{def}}{=\!=\!=} v_+\mu_+^\ominus + v_-\mu_-^\ominus$; $v = v_+ + v_-$

To simplify the appearance of(7.4.5), we define $a_\pm$, $\gamma_\pm$, $b_\pm$ as

$a_\pm \overset{\text{def}}{=\!=\!=} (a_+^{v_+}a_-^{v_-})^{1/v}$ mean ionic activity(平均离子活度) (7.4.6)

$\gamma_\pm \overset{\text{def}}{=\!=\!=} (\gamma_+^{v_+}\gamma_-^{v_-})^{1/v}$ mean ionic activity coefficient(平均离子活度因子) (7.4.7)

$b_\pm \overset{\text{def}}{=\!=\!=} (b_+^{v_+}b_-^{v_-})^{1/v}$ mean ionic molality(平均离子质量摩尔浓度) (7.4.8)

So

$$\mu_B = \mu_B^\ominus + RT\ln a_\pm^v$$
$$= \mu_B^\ominus + RT\ln\{\gamma_\pm^v(b_\pm/b^\ominus)^v\} \qquad (7.4.9)$$

That is

$$a_\pm = \gamma_\pm(b_\pm/b^\ominus) \qquad (7.4.10)$$

Combine(7.4.9), (7.4.6)and(7.4.2), we can get

$$a_\pm = a_B^{1/v} = \gamma_\pm(v_+^{v_+}\cdot v_-^{v_-})^{1/v} b/b^\ominus \qquad (7.4.11)$$

For example:

1 : 1 electrolyte and 2 : 2 electrolyte $\quad a_\pm = a_B^{1/2} = \gamma_\pm b/b^\ominus$

1 : 2 electrolyte and 2 : 1 electrolyte $\quad a_\pm = a_B^{1/3} = 4^{1/3}\gamma_\pm b/b^\ominus$

1 : 3 electrolyte and 3 : 1 electrolyte $\quad a_\pm = a_B^{1/4} = 27^{1/4}\gamma_\pm b/b^\ominus$

The mean ionic activity factors of some electrolytes in water at 25℃ is shown in Table 7.3.

Tab. 7.3 Mean ionic activity factors of electrolytes in water at 25℃

b/mol · kg^{-1}	HCl	KCl	$CaCl_2$	$LaCl_3$	H_2SO_4	$In_2(SO_4)_3$
0.001	0.966	0.966	0.888	0.853	—	—
0.005	0.930	0.927	0.798	0.715	0.643	0.16
0.01	0.906	0.902	0.732	0.637	0.545	0.11
0.05	0.833	0.816	0.584	0.417	0.341	0.035
0.10	0.798	0.770	0.524	0.356	0.266	0.025
0.50	0.769	0.652	0.510	0.303	0.155	0.014
1.00	0.811	0.607	0.725	0.583	0.131	—
2.00	1.011	0.577	—	0.954	0.125	—

7.4.2 Ionic strength of electrolyte solution(电解质溶液的离子强度)

Electrolytes containing ions with multiple charges have greater effects on the activity coefficients of ions than electrolytes containing only singly charged ions. To express electrolyte concentrations in a way that takes this into account, G. N. Lewis introduced the ionic strength.

Definition of ionic strength(离子强度的定义)

$$I \overset{\text{def}}{=\!=} \frac{1}{2}\sum b_B z_B^2 \tag{7.4.12}$$

I—ionic strength, unit: mol · kg^{-1};

b_B—molality of ion B;

z_B—charge of ion B.

For example:

$$S \rightarrow v_+ X^{z+} + v_- X^{z-}$$

$$I = \frac{1}{2}(b_+ z_+^2 + b_- z_-^2) = \frac{1}{2}(v_+ z_+^2 + v_- z_-^2)b$$

Example 7.4

Please calculate the ionic strength of the following solutions: (1) 0.025mol · kg^{-1} NaCl; (2) 0.025mol · kg^{-1} $CuSO_4$; (3) 0.5mol · kg^{-1} K_2SO_4.

Answer:

(1) For NaCl, $b_+ = b_- = 0.025$mol · kg^{-1}, $z_+ = 1$, $z_- = -1$

$$I = \frac{1}{2}\sum b_B z_B^2 = \frac{1}{2}[0.025\times1^2 + 0.025\times(-1)^2] = 0.025\text{mol} \cdot \text{kg}^{-1}$$

(2) For $CuSO_4$, $b_+ = b_- = 0.025$mol · kg^{-1}, $z_+ = 2$, $z_- = -2$

$$I = \frac{1}{2}\sum b_B z_B^2 = \frac{1}{2}[0.025\times2^2 + 0.025\times(-2)^2] = 0.1\text{mol} \cdot \text{kg}^{-1}$$

(3) For K_2SO_4, $v_+ = 2$, $v_- = 1$; $z_+ = 1$, $z_- = -2$

$$I = \frac{1}{2}\sum b_B z_B^2 = \frac{1}{2}[2b \cdot 1^2 + b \cdot (-2)^2] = 3b = 3\times0.5 = 1.5\text{mol} \cdot \text{kg}^{-1}$$

7.5 Ionic mutual attraction theory of strong electrolyte and Debye-Hückel limiting law
(强电解质离子互吸理论及德拜-休克尔极限定律)

Interactions between ions are so strong that the approximation of replacing activities by molalities is valid only in very dilute solutions(less than 1 mmol · kg^{-1} in total ion concentration) and in precise work activities themselves must be used. We need, therefore, to pay special attention to the activities of ions in solution, especially in preparation for the discussion of electrochemical phenomena.

7.5.1 Ionic atmosphere model(离子氛模型)

The first successful theory of strong electrolytes was suggested by Debye-Hückel. The Debye-Hückel theory is based on a few assumptions, of which the following seem to be of great importance:

① The dissociation of the solute molecules is complete. Consequently, it applies to dilute solutions only.

② The dominant forces acting between ions are Coulomb force.

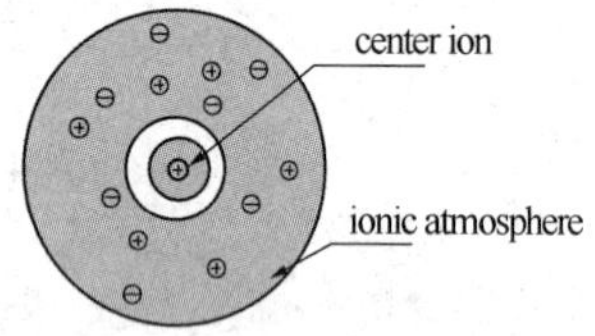

Fig. 7.4 Schematic of ionic atmosphere

③ Since the solution as a whole is always electrically neutral—the total positive charge and the total negative charge have equal magnitude—it might be incorrectly assumed that, on the average, the mutual action between the ions results in no net accumulation of positive and negative charges in the solution. This is not so; as the figure 7.4 will show.

There are two opposing (competing) factors operating in ionic solution: the Coulomb force factor, which arranges the ions into a certain organized structure, and the factor arising from the thermal motion of the solvent molecules and ions, which tends to prevent any kind of organized structure in the solution. As a result of thermal collisions, no perfect organization can ever be reached in the solution. The higher the temperature the less organized the structure will be. Because of these opposing factors, ions tend toward an arrangement on which negative ions will predominate as the nearest neighbors of any chosen central(positive) ion and vice versa. Around a certain cation there will be more anions than cations, and around a certain anion there will be more cations than anions. A central ion is surrounded by a group of ions called the ionic atmosphere.

④ Since ions and solvent molecules are in steady motion, the central ion-ionic atmosphere concept refers to the time average of the ionic configurations. Solutions of strongly dissociated electrolytes can therefore be treated as systems composed of central ions and ionic atmospheres in solvents.

7.5.2 Debye-Hückel limiting law(德拜-休克尔极限定律)

Debye-Hückel limiting law: Debye and Hückel were able to show that in dilute solutions the activity coefficient of an ion species with a charge number of z_i is given by

$$\lg\gamma_i = -Cz_i^2 I^{1/2} \tag{7.5.1}$$

Equation 7. 5. 1 gives the activity coefficient of a single ion, but the quantity that is accessible to experimental determination is the mean ionic activity coefficient, which for the electrolyte $A^{z+}B^{z-}$ is given by equation

$$\lg\gamma_{\pm} = -C\,|z_+ z_-|\,I^{1/2} \qquad (7.5.2)$$

where I is ionic strength, unit: $mol \cdot kg^{-1}$; C is constant. For aqueous solution at 25℃, $C = 0.509 mol^{-1/2} \cdot kg^{1/2}$. (7. 5. 1) and (7. 5. 2) are called the **Debye-Hückel limiting law**, since they are valid only in the limit of infinite dilution ($b < 0.01 \sim 0.001 mol \cdot kg^{-1}$).

Example 7. 5

Use Debye-Hückel limiting law to calculate the $\gamma_{\pm}$ of $0.005 mol \cdot kg^{-1}$ $ZnCl_2$ solution at 25℃.

Answer: For $ZnCl_2$:

$$b_+ = b = 0.005 mol \cdot kg^{-1};\ b_- = 2b = 0.01 mol \cdot kg^{-1};\ z_+ = 2,\ z_- = -1$$

$$I = \frac{1}{2}\sum b_B z_B^2 = \frac{1}{2}[0.005\times 2^2 + 0.01\times(-1)^2] = 0.015 mol \cdot kg^{-1}$$

For $C = 0.509 mol^{-1/2} kg^{1/2}$, so we can get

$$\lg\gamma_{\pm} = -Cz_+|z_-|\sqrt{I} = -0.509\times 2\times|-1|\sqrt{0.015} = -0.1246$$

$$\gamma_{\pm} = 0.7505$$

Supplementary Examples of Chapter 7

EXERCISES

(习题)

1. At 25℃, a cell contains $0.02 mol \cdot dm^{-3}$ KCl aqueous solution, which at that concentration has a conductivity of $0.277 S \cdot m^{-1}$. The measured resistance was 82. 4Ω. When the same cell was filled with $0.005 mol \cdot dm^{-3}$ K_2SO_4 aqueous solution, the resistance was 326. 0Ω (Omit the conductivity of water).

Find: (1) Cell constant K_{cell};

(2) Conductivity κ of K_2SO_4 solution;

(3) Molar conductivity Λ_m of K_2SO_4 solution.

Answer: $22.8 m^{-1}$; $0.07 S \cdot m^{-1}$; $0.014 S \cdot m^2 \cdot mol^{-1}$

2. At 25℃, $0.02 mol \cdot dm^{-3}$ aqueous solution of KCl has a conductivity of $0.277 S \cdot m^{-1}$. This solution was filled in a cell, the measured resistance was 453Ω. The same cell was filled with $0.555 g \cdot dm^{-3}$ aqueous solution of $CaCl_2$ with the same volume, the resistance was 1050Ω.

Find: (1) Cell constant K_{cell};

(2) Conductivity κ of $CaCl_2$ solution;

(3) Molar conductivity Λ_m of $CaCl_2$ solution.

Answer: $125.4 m^{-1}$; $0.1194 S \cdot m^{-1}$; $0.02388 S \cdot m^2 \cdot mol^{-1}$

3. At 298. 15K, KCl aqueous solution whose conductivity is $0.141 s \cdot m^{-1}$ was filled in a cell, the resistance was 525Ω. The same cell was filled with $0.1 mol \cdot dm^{-3}$ NH_4OH aqueous solution, the resistance was 2030Ω. Given that at 298. 15K.

$\Lambda_m^\infty(NH_4^+) = 73.4\times10^{-4} S\cdot m^2\cdot mol^{-1}$, $\Lambda_m^\infty(OH^-) = 198\times10^{-4} S\cdot m^2\cdot mol^{-1}$.

Calculate the degree of dissociation α and the dissociation equilibrium constant K° of NH_4OH solution.

Answer: 0.01344; 1.83×10^{-5}

4. Use Debye-Hückel limiting law to calculate $\gamma(Ca^{2+})$, $\gamma(Cl^-)$ and $\gamma_\pm$ of $0.002 mol\cdot kg^{-1}$ $CaCl_2$ solution at 25℃.

Answer: $\gamma(Ca^{2+}) = 0.6955$; $\gamma(Cl^-) = 0.9132$; $\gamma_\pm = 0.8340$

5. Please calculate the ionic strength of the following solutions: (1) $0.025 mol\cdot kg^{-1}$ KCl; (2) $0.025 mol\cdot kg^{-1}$ $BaSO_4$; (3) $0.025 mol\cdot kg^{-1}$ $LaCl_3$.

Answer: $0.025 mol\cdot kg^{-1}$; $0.1 mol\cdot kg^{-1}$; $0.15 mol\cdot kg^{-1}$

Chapter 8 Electrochemical system
(电化学系统)

Leaning objectives:

(1) Define the terms *anode* and *cathode* when used in *galvanic cells* and in *electrolytic cells.*

(2) Define the *electromotive force(e. m. f.) of a cell.*

(3) Define *electrode potential* and describe the sign convention.

(4) Derive and use the *Nernst equation* for the concentration dependence of the e. m. f. of a cell, and define the term *standard e. m. f.*.

(5) Relate the *standard e. m. f.* to the *equilibrium constant* of the cell reaction.

(6) Use *e. m. f.* data to deduce *Gibbs function of reactions.*

(7) Describe how to use *e. m. f.* measurements to determine *activity factors.*

(8) Relate the *e. m. f. of a cell* to the spontaneous direction of change of the cell reaction.

(9) Define *solubility product* and deduce its value from *e. m. f.* measurements.

(10) Describe how to use *e. m. f.* measurements to determine *pH.*

When two different electrically conducting phases come in contact, a difference in electric potential is usually established between them as a result of transfer of charges between phases and of nonuniform distribution of ions, orientation of molecules with dipole moments, and distortion of charge distributions in molecules near the interface.

Electrochemistry is a science that studies the relationship between electricity and chemical reaction. It mainly involves the study of generating electric energy through chemical reaction and generating chemical reaction through input electric energy.

Electrochemistry in physical chemistry focuses on the basic theory of electrochemistry, which studies the law of mutual conversion between chemical energy and electric energy by means of thermodynamics. It mainly includes two aspects: on the one hand, the use of chemical reactions to generate electricity—will be able to spontaneously carry out chemical reactions through galvanic device to convert chemical energy into electrical energy; The other is to use electricity to drive chemical reactions-by feeding a current into an electrolytic cell to allow reactions that cannot occur spontaneously.

8.1 Cell
(电池)

Electrochemical system—the system in which there is a difference of electric potential between two or more phases.(在两相或数相间存在电势差的系统，称作电化学系统。)

An interphase potential difference $\Delta\varphi$ between two phases α and β can be expressed as follows: $\Delta\varphi=\varphi^{\beta}-\varphi^{\alpha}$.(若$\alpha$、$\beta$两相相接触，则两相间的电势差$\Delta\varphi=\varphi^{\beta}-\varphi^{\alpha}$。)

8.1.1 Cell(电池)

Electrochemical cells can be grouped into two categories: galvanic cell(原电池) and electrolytic cell(电解池).

The two kinds of cells differ in their function. An electrolytic cell is a device which an external supply of electric energy is necessary in order to bring about a physical or chemical change in the cell. In a galvanic cell, the free energy released by a physical or chemical change is transformed into electric energy. Thus, a galvanic cell can serve as a source of electricity, whereas an electrolytic cell cannot.

Figure 8.1 and 8.2 shows the schematics of Cu-Zn galvanic cell and electrolytic cell.

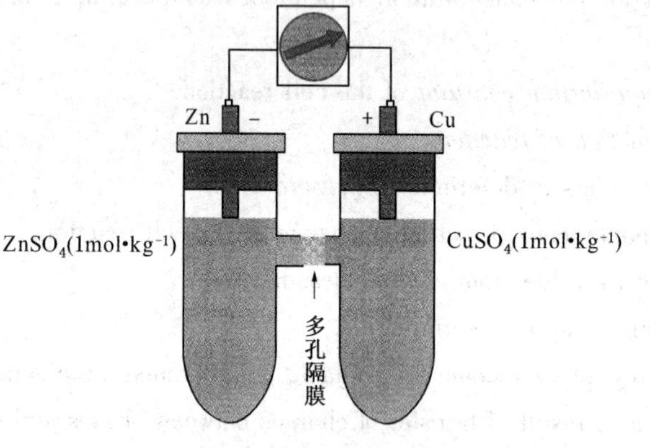

Fig. 8.1 Cu-Zn cell

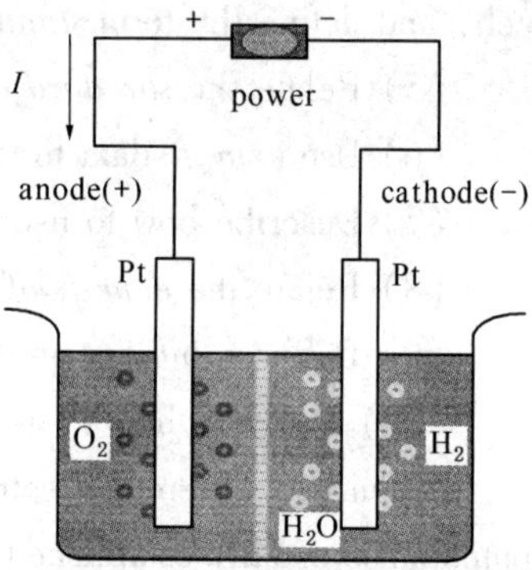

Fig. 8.2 Electrolytic cell

8.1.2 Electrode(电极)

Anode—the electrode at which oxidation reaction occurs.(阳极是发生氧化反应的电极。)

Cathode—the electrode at which reduction reaction occurs.(阴极是发生还原反应的电极。)

For example, in Daniell cell, Zn is the anode, Cu is the cathode.

Positive electrode—the electrode having high potential.(正极是电势高的电极。)

Negative electrode—the electrode having low potential.(负极是电势低的电极。)

For example, in Daniell cell, Zn is negative electrode, Cu is positive electrode.

Summary

Cell	Galvanic cell	Electrolytic cell
energy conversion	chemical energy→electric energy	electric energy→chemical energy
electrode	anode(−); cathode(+)	anode(+); cathode(−)

8.1.3 Cell diagram, electrode reaction and cell reaction of galvanic cell(电池图式、原电池中的电极反应与电池反应及电池图式)

When writing electrode reactions and cell reactions, the quantity and charge balance of the substances must be satisfied. At the same time, the activity of ionic or electrolyte solutions should be indicated, the pressure of gases should be indicated, and the phase state of pure liquid or solid should be indicated.(书写电极反应和电池反应时,必须满足物质的量及电量平衡,同时,离子

或电解质溶液应标明活度，气体应标明压力，纯液体或纯固体应标明相态。)

For Cu-Zn cell:

$(-): Zn(s) - 2e^- \rightarrow Zn^{2+}(a)$ (oxidation)

$(+): Cu^{2+}(a) + 2e^- \rightarrow Cu(s)$ (reduction)

$Zn(s) + Cu^{2+}(a) \rightarrow Zn^{2+}(a) + Cu(s)$

Cell diagram—a galvanic cell can be represented by a diagram.

(一个实际的电池装置可用一简单的符号来表示，称为电池图式。)

For Cu-Zn cell can be represented as follows:

$Zn(s) | ZnSO_4(1mol \cdot kg^{-1}) | CuSO_4(1mol \cdot kg^{-1}) | Cu(s)$

Cell diagram of galvanic cell:

① Left is anode and right is cathode;

② From left to right, phases(s, l, g) and compositions(a, p) of different phases should be written with chemical formula according to real sequence;

③ A phase boundary is denoted by a vertical line "|" or a comma ",", adding salt bridge is indicated by "||".

8.1.4 Types of electrodes(电极的类型)

(1) First-class electrodes(第一类电极)

The characteristic of this type of electrode is that the electrode is in direct contact with its ionic solution, the substances involved in the reaction exist in the two phases, and the electrode has a phase interface. The first type of electrode can be divided into metal electrode and non-metal electrode.

① Metal-metal ion electrodes(金属-金属离子电极)

Metal-metal ion electrode consists of a piece of metal immersed in a solution containing the metal ions. Here, a metal M is in electrochemical equilibrium with a solution containing the metal ions.

Examples include $Cu^{2+}(a) | Cu(s)$; $Zn^{2+}(a) | Zn(s)$

Electrode reaction is $Cu^{2+}(a) + 2e^- \rightarrow Cu(s)$; $Zn(s) \rightarrow Zn^{2+}(a) + 2e^-$

② Gas-ion electrodes(非金属电极)

A gas-ion electrode consists of a gas bubbling about an inert metal bathed in a solution containing ions to which the gas is reversible. Here, a gas is in equilibrium with ions in solution. The function of the inert metal, which usually is platinized platinum or graphite, is to facilitate the establishment of equilibrium between the gas and its ions in the solution and also to serve as the electron carrier for the electrode.

Examples: $Pt | H_2(p) | H^+(a)$; $Pt | Cl_2(p) | Cl^-(a)$

Electrode reaction is $2H^+(a) + 2e^- \rightarrow H_2(p)$; $Cl_2(p) + 2e^- \rightarrow 2Cl^-(a)$

Figure 8.3 shows the schematic of hydrogen electrode.

(2) Second-class electrodes(第二类电极)

① Metal-insoluble salt electrodes(金属-难溶盐电极)

The common characteristic of electrodes of this group is that they all consist of a metal in contact with one of its metal insoluble salts and a solution containing ions present in the salt other than the metal. Electrodes of this type are extremely important in electrochemistry, and they are usually used as standard electrode.

Examples:

$Hg(l)|Hg_2Cl_2(s)|Cl^-(a)$; $Ag(s)|AgCl(s)|Cl^-(a)$

Electrode reaction is $Hg_2Cl_2(s)+2e^- \rightarrow 2Hg(l)+2Cl^-(a)$

$$AgCl(s)+e^- \rightarrow Ag(s)+Cl^-(a)$$

Figure 8.4 shows the schematic of calomel electrode.

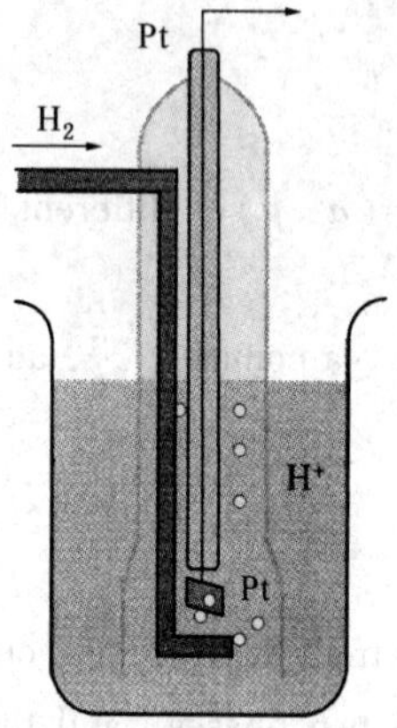

Fig. 8.3 Hydrogen electrode

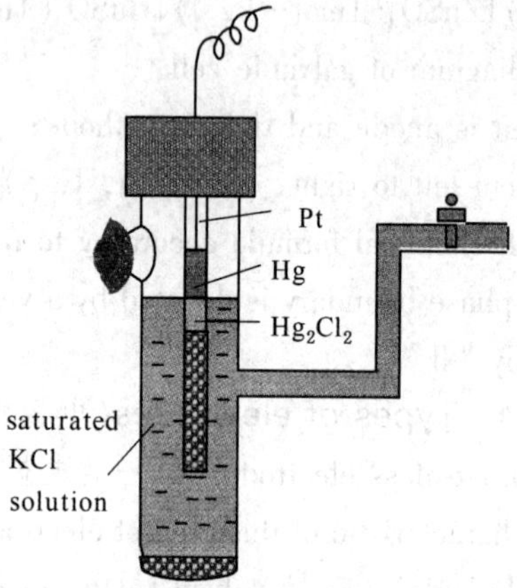

Fig. 8.4 Saturated calomel electrode

② Metal-insoluble oxide electrodes(金属-难溶氧化物电极)

The common characteristic of electrodes of this group is that they all consist of a metal in contact with one of its insoluble oxide is bathed in a solution containing H^+ or OH^- ions.

For instance $H^+(a)$, $H_2O(l)|Sb_2O_3(s)|Sb(s)$

Electrode reaction is $Sb_2O_3(s)+6H^+(a)+6e^- = 2Sb(s)+3H_2O(l)$

$OH^-(a)$, $H_2O(l)|Sb_2O_3(s)|Sb(s)$

Electrode reaction is $Sb_2O_3(s)+3H_2O(l)+6e^- = 2Sb(s)+6OH^-(a)$

(3) Third-class electrodes—Oxidation-reduction electrodes(第三类电极—氧化还原电极)

Although each electrode reaction involves oxidation or reduction, the term "Oxidation-reduction electrodes" refers only to an electrode whose oxidation-reduction reaction occurs between ions in two different states of oxidation in the same solution, the inert metal that dips into this solution serves only to supply or accept electrons.

For instance $Fe^{3+}(a)$, $Fe^{2+}(a)|Pt$

Electrode reaction is $Fe^{3+}(a)+e^- \rightarrow Fe^{2+}(a)$

or quinhydrone electrode $H^+(a)$, $Q \cdot QH_2(a)|Pt$

Q represents quinone($C_6H_4O_2$) and QH_2 represents hydroquinone[$C_6H_4(OH)_2$]

Electrode reaction is $Q(a)+2H^+(a)+2e^- \rightarrow QH_2(a)$

8.1.5 Types of galvanic cells(原电池的分类)

Any combinations of two suitable electrodes result in a galvanic cell. In general, there are two different types of galvanic cells.

(1) Chemical cell(化学电池)

If the electrochemical reactions in the two electrodes differ, the overall cell reaction is a chemical reaction and the cell is a chemical cell.

$$Zn(s)|Zn^{2+}(a)|Cu^{2+}(a')|Cu(s)$$

$$Zn(s)+Cu^{2+}(a')\rightarrow Cu(s)+Zn^{2+}(a)$$

$$Pt|H_2(p)|HCl(a)|AgCl(s)|Ag(s)$$

$$\frac{1}{2}H_2(p)+AgCl(s)\rightarrow Ag(s)+HCl(a)$$

(2) Concentration cell(浓差电池)

If the electrochemical reactions in the two half-cells are the same, but one species B is at a different concentration in each electrode, the overall cell reaction is a physical reaction that amounts to the transfer of B from one concentration to another, the cell is a concentration cell.

① Electrode concentration cell(电极浓差电池)

Electrode concentration cell—the cell in which electrolyte solution is same, but concentrations of electrodes are different.(电解质溶液是相同的，电极的浓度不同。)

For instance: $Pt|H_2(p)|H^+(a)|H_2(p')|Pt$

② Electrolyte concentration cell(电解质浓差电池)

Electrolyte concentration cell—the cell in which electrode is same, but concentrations of electrolyte solutions are different.(电极是相同的,电解质溶液的浓度不同。)

(i) $Pt|H_2(p)|HCl(a)|HCl(a')|H_2(p)|Pt$

$$H^+(a')\rightarrow H^+(a)$$

(ii) $Ag(s)|AgCl(s)|KCl(a)|K(Hg)|KCl(a')|AgCl(s)|Ag(s)$

$$Cl^-(a)\rightarrow Cl^-(a')$$

(iii) $Pt|H_2(p)|HCl(a)|H_2(p')|Pt$

$$H_2(p)\rightarrow H_2(p')$$

8.2 Definition of electromotive force of cell and reversible cell (原电池电动势的定义及可逆电池)

8.2.1 Definition of electromotive force of cell(原电池电动势的定义)

Electromotive force—the potential difference of a cell, measured between the two terminals with same connecting wires and when there is no flow of current, is called the e. m. f. of the cell. It is denoted by E_{MF} or E. (原电池的电动势定义为在没有电流通过的条件下，原电池两极的金属引线为同种金属时，电池两端的电势差，原电池电动势用符号 E_{MF}表示。)

That is
$$E_{MF}\overset{\text{def}}{=\!=}[\varphi(M_{\text{right}})-\varphi(M_{\text{left}})]_{I\rightarrow 0} \qquad (8.2.1)$$

8.2.2 Reversible cell(可逆电池)

Reversible condition:

(1) Electrode reaction and cell reaction must be reversible.

Examples:

① $Zn(s)|ZnSO_4(a)|CuSO_4(a)|Cu(s)$

If $E_{MF}>E_{ex}$

left: $Zn(s)\rightarrow Zn^{2+}(a)+2e^-$

right: $Cu^{2+}(a)+2e^-\rightarrow Cu(s)$

cell reaction: $Zn(s)+Cu^{2+}(a)\rightarrow Zn^{2+}(a)+Cu(s)$

If $E_{MF}<E_{ex}$

left: $Zn^{2+}(a)+2e^-\rightarrow Zn(s)$

right: $Cu(s)\rightarrow Cu^{2+}(a)+2e^-$

cell reaction: $Zn^{2+}(a)+Cu(s)\rightarrow Cu^{2+}(a)+Zn(s)$

Both electrode reaction and cell reaction are reversible.

② $Zn(s)|HCl(a)|AgCl(s)|Ag(s)$

If $E_{MF}>E_{ex}$

left: $Zn(s)\rightarrow Zn^{2+}(a)+2e^-$

right: $2AgCl(s)+2e^-\rightarrow 2Ag(s)+2Cl^-(a)$

cell reaction: $Zn(s)+2AgCl(s)\rightarrow Zn^{2+}(a)+2Ag(s)+2Cl^-(a)$

If $E_{MF}<E_{ex}$

left: $2H^+(a)+2e^-\rightarrow H_2(p)$

right: $2Ag(s)+2Cl^-(a)\rightarrow 2AgCl(s)+2e^-$

cell reaction: $2Ag(s)+2H^+(a)+2Cl^-(a)\rightarrow H_2(p)+2AgCl(s)$

Electrode reaction of right electrode is reversible. Electrode reaction of left electrode and cell reaction are irreversible.

(2) Working condition of cell must be reversible($I\rightarrow 0$).

(3) Other processes must be reversible too.

For example: elimination of liquid-junction potential.

8.3 Electromotive force (电动势)

8.3.1 Electromotive force(电动势)

The non-volume work which is done on the system during a reversible process at constant temperature and pressure equals the change in the Gibbs free energy. That is,

$$\Delta G_{T,p}=W'_r \tag{8.3.1}$$

$$\Delta_r G_m=W'_r/\Delta\xi \tag{8.3.2}$$

The non-volume work

$$W'_r=-QE_{MF}=-zFE_{MF}\Delta\xi \tag{8.3.3}$$

So we can get

$$\Delta_r G_m=-zFE_{MF} \tag{8.3.4}$$

E_{MF} can be measured by a potentiometer.

Notes:

$\Delta_r G_m$ is related to the reaction stoichiometric equation, however, E_{MF} is independent of the reaction stoichiometric equation.

If all the species in the cell reaction are in their standard states, we can get

$$\Delta_r G_m^\ominus = -zFE_{MF}^\ominus \tag{8.3.5}$$

$E_{MF}^\ominus$—standard potential of the cell.

According to the van Hoff's equation: $\Delta_r G_m = \Delta_r G_m^\ominus + RT \ln \prod_B (a_B)^{\nu_B}$

We can get

$$E_{MF} = E_{MF}^\ominus - \frac{RT}{zF} \ln \prod_B (a_B)^{\nu_B} \tag{8.3.6}$$

The Nernst equation(8.3.6) relates the cell's E_{MF} to the activities a_B of the substances in the cell's chemical reaction and to the standard potential of the reaction.

Walther Hermann Nernst

8.3.2 Standard electrode potential(标准电极电势)

(1) Standard electrode potential

The potential of a single electrode can be discussed but cannot be measured, except by introducing a second—that is, a reference—electrode as a point of zero or known absolute potential. The measured potential difference between the two electrodes, which is the electromotive force of the cell when no current is drawn from it, is

$$E_{MF} = E - E_{ref}$$

The absolute potential of a reference electrode E_{ref} is, however, also inaccessible, so an arbitrary scale of electrode potentials has been established by assigning a zero potential to the standard hydrogen electrode, SHE. The reaction taking place on this electrode is

$$H_2(p = 100\text{kPa}) \rightleftharpoons 2H^+(a=1) + 2e^-$$

So it is clearly that the standard potential of the hydrogen electrode is 0V.

$$E^\ominus(\text{Pt}|H_2|H^+) = 0\text{V}$$

Thus, the potential of a single electrode on the hydrogen scale is the electromotive force of a cell, E, consisting of the electrode in question and the standard hydrogen electrode. Table 8.1 shows some standard electrode potentials at 25℃ and $p^\ominus = 100$kPa.

(2) Nernst equation for electrode reaction(电极反应的能斯特方程)

The reaction on the hydrogen electrode represents oxidation and consequently the reaction on the other electrode must be reduction, the potential of this electrode is the reduction potential. Thus a general reaction of the form

$$b_{ox} + ze^- \rightarrow d_{red}$$

$$E = E^\ominus - \frac{RT}{zF} \ln \frac{a_{red}^d}{a_{ox}^b} \tag{8.3.7}$$

(8.3.7) is the **Nernst equation for electrode reaction**, relating the potential of an electrode to the activities of the reactants and products of the electrode reaction.

For example:

① $Cl^-(a)|AgCl(s)|Ag(s)$

Electrode reaction $AgCl(s)+e^- \rightarrow Ag(s)+Cl^-(a)$

The Nernst equation becomes

$$E=E^{\ominus}-\frac{RT}{F}\ln a(Cl^-)$$

② $Cl^-(a)|Cl_2(p)|Pt$

Electrode reaction $\frac{1}{2}Cl_2(p)+e^- \rightarrow Cl^-(a)$

The Nernst equation becomes

$$E=E^{\ominus}-\frac{RT}{F}\ln\frac{a(Cl^-)}{[p(Cl_2)/p^{\ominus}]^{\frac{1}{2}}}$$

The electrode electromotive force of any cell is equal to

$$E_{MF}=E_+-E_- \tag{8.3.8}$$

Tab. 8.1 Standard electrode potentials at 25℃ and $p^{\ominus}=100$kPa

Electrodes	Electrode reactions	$E^{\ominus}$/V
$Li^+|Li$	$Li^++e^- \rightleftharpoons Li$	−3.045
$K^+|K$	$K^++e^- \rightleftharpoons K$	−2.924
$Ba^{2+}|Ba$	$Ba^{2+}+2e^- \rightleftharpoons Ba$	−2.90
$Ca^{2+}|Ca$	$Ca^{2+}+2e^- \rightleftharpoons Ca$	−2.76
$Na^+|Na$	$Na^++e^- \rightleftharpoons Na$	−2.7111
$Mg^{2+}|Mg$	$Mg^{2+}+2e^- \rightleftharpoons Mg$	−2.375
$H_2O, OH^-|H_2(g)|Pt$	$2H_2O+2e^- \rightleftharpoons H_2(g)+2OH^-$	−0.8277
$Zn^{2+}|Zn$	$Zn^{2+}+2e^- \rightleftharpoons Zn$	−0.7630
$Cr^{3+}|Cr$	$Cr^{3+}+3e^- \rightleftharpoons Cr$	−0.74
$Cd^{2+}|Cd$	$Cd^{2+}+2e^- \rightleftharpoons Cd$	−0.4028
$Co^{2+}|Co$	$Co^{2+}+2e^- \rightleftharpoons Co$	−0.28
$Ni^{2+}|Ni$	$Ni^{2+}+2e^- \rightleftharpoons Ni$	−0.23
$Sn^{2+}|Sn$	$Sn^{2+}+2e^- \rightleftharpoons Sn$	−0.1366
$Pb^{2+}|Pb$	$Pb^{2+}+2e^- \rightleftharpoons Pb$	−0.1265
$Fe^{3+}|Fe$	$Fe^{3+}+3e^- \rightleftharpoons Fe$	−0.036
$H^+|H_2(g)|Pt$	$2H^++2e^- \rightleftharpoons H_2(g)$	0.0000
$Hg_2^{2+}|Hg$	$Hg_2^{2+}+2e^- \rightleftharpoons 2Hg$	+0.7959
$Ag^+|Ag$	$Ag^++e^- \rightleftharpoons Ag$	+0.7994
$Hg^{2+}|Hg$	$Hg^{2+}+2e^- \rightleftharpoons Hg$	+0.851
$Br^-|Br_2(l)|Pt$	$Br_2(l)+2e^- \rightleftharpoons 2Br^-$	+1.065

Continued

Electrodes	Electrode reactions	$E^{\ominus}$/V
H_2O, H^+ | $O_2(g)$ | Pt	$O_2(g)+4H^++4e^- \rightleftharpoons 2H_2O$	+1. 229
Cl^- | $Cl_2(g)$ | Pt	$Cl_2(l)+2e^- \rightleftharpoons 2Cl^-$	+1. 3580
Au^+ | Au	$Au^++e^- \rightleftharpoons Au$	+1. 68
Cu^{2+} | Cu	$Cu^{2+}+2e^- \rightleftharpoons Cu$	+0. 3400
Cu^+ | Cu	$Cu^++e^- \rightleftharpoons Cu$	+0. 522
F^- | $F_2(g)$ | Pt	$F_2+2e^- \rightleftharpoons 2F^-$	+2. 87
$SO_4{}^{2-}$ | $PbSO_4(s)$ | Pb	$PbSO_4(s)+2e^- \rightleftharpoons Pb+SO_4^{2-}$	−0. 356
I^- | AgI(s) | Ag	$AgI(s)+e^- \rightleftharpoons Ag+I^-$	−0. 1521
Br^- | AgBr(s) | Ag	$AgBr(s)+e^- \rightleftharpoons Ag+Br^-$	+0. 0711
Cl^- | AgCl(s) | Ag	$AgCl(s)+e^- \rightleftharpoons Ag+Cl^-$	+0. 2221
Cr^{3+}, Cr^{2+} | Pt	$Cr^{3+}+e^- \rightleftharpoons Cr^{2+}$	−0. 41
Sn^{4+}, Sn^{2+} | Pt	$Sn^{4+}+2e^- \rightleftharpoons Sn^{2+}$	+0. 15
Cu^{2+}, Cu^+ | Pt	$Cu^{2+}+e^- \rightleftharpoons Cu^+$	+0. 158
H^+, Q, QH_2 | Pt	$Q+2H^++2e^- \rightleftharpoons QH_2$	+0. 6993
Fe^{3+}, Fe^{2+} | Pt	$Fe^{3+}+e^- \rightleftharpoons Fe^{2+}$	+0. 770
Tl^{3+}, Tl^+ | Pt	$Tl^{3+}+2e^- \rightleftharpoons Tl^+$	+1. 247
Ce^{4+}, Ce^{3+} | Pt	$Ce^{4+}+e^- \rightleftharpoons Ce^{3+}$	+1. 61
Co^{3+}, Co^{2+} | Pt	$Co^{3+}+e^- \rightleftharpoons Co^{2+}$	+1. 808

8. 3. 3 Calculation of E_{MF}(原电池电动势的计算)

(1) According to the Nernst equation for cell reaction

$$E_{MF}=E^{\ominus}_{MF}-\frac{RT}{zF}\ln\prod_{B}(a_B)^{\nu_B} \qquad \text{where} \qquad E^{\ominus}_{MF}=E^{\ominus}_{+}-E^{\ominus}_{-}$$

(2) According to the Nernst equation for electrode reaction

$$E_{MF}=E_+-E_-$$

$$E=E^{\ominus}-\frac{RT}{zF}\ln\frac{a^d_{red}}{a^b_{ox}}$$

Example 8. 1

Calculate the E_{MF} of the cell at 25℃: Zn(s) | $Zn^{2+}(a=0.\ 1)$ ‖ $Cu^{2+}(a=0.\ 01)$ | Cu(s)

***Answer*:**

The electrode reactions

(−): $Zn(s)-2e^- \longrightarrow Zn^{2+}(a=0.\ 1)$

(+): $Cu^{2+}(a=0.\ 01)+2e^- \longrightarrow Cu(s)$

According to the Nernst equation for electrode reactions, we can get

$$E_-=E^{\ominus}(Zn^{2+}|Zn)-\frac{RT}{2F}\ln\frac{1}{a(Zn^{2+})}$$

$$E_+=E^{\ominus}(Cu^{2+}|Cu)-\frac{RT}{2F}\ln\frac{1}{a(Cu^{2+})}$$

From Table 8.1, we can get

$E^{\ominus}(Zn^{2+}|Zn) = -0.7630V$, $E^{\ominus}(Cu^{2+}|Cu) = 0.3402V$

So $E_- = -0.792V$; $E_+ = 0.281V$

$E_{MF} = E_+ - E_- = 0.281V - (-0.792V) = 1.07V$

According to Nernst equation for cell reaction, we can get the same result.

Example 8.2

Calculate the E_{MF} of the concentration cell at 25℃:

$Pt|Cl_2(p^{\ominus})|Cl^-(a=0.1)\|Cl^-(a'=0.001)|Cl_2(p^{\ominus})|Pt$

Answer:

$$\text{left}(-):\ Cl^-(a=0.1) - e^- \rightarrow \frac{1}{2}Cl_2(p^{\ominus})$$

$$\text{right}(+):\ \frac{1}{2}Cl_2(p^{\ominus}) + e^- \rightarrow Cl^-(a'=0.001)$$

cell reaction: $Cl^-(a=0.1) \rightarrow Cl^-(a'=0.001)$

According to Nernst equation for cell reaction, we can get

$$E_{MF} = E^{\ominus}_{MF} - \frac{RT}{F}\ln\frac{a'}{a}$$

$$E^{\ominus}_{MF} = E^{\ominus}_+ - E^{\ominus}_- = 0$$

$$E_{MF} = -\frac{RT}{F}\ln\frac{a'}{a} = -\frac{8.314\times298.15}{96485}\times\ln\frac{0.001}{0.1} = 0.1183V$$

According to Nernst equation for electrode reaction, we can get the same result.

8.4 Design of galvanic cell
(原电池的设计)

Separate the given reaction into two parts, write oxidation electrode as negative electrode, reduction electrode as positive electrode, then connect them from left to right, and write the cell diagram, for double fluid cell, add "‖".

Example 8.3

Design cells in which the following reaction could take place

(1) $H^+(a) + OH^-(a) = H_2O(l)$

(2) $Ag^+(a) + I^-(a) \rightarrow AgI(s)$

(3) $H_2(g, p_1) = H_2(g, p_2)$

Design of galvanic cell

Answer:

(1) $H^+(a) + OH^-(a) = H_2O(l)$ can be divided into

anode: $1/2H_2(g, p) + OH^-(a) - e^- = H_2O(l)$

cathode: $H^+(a) + e^- = 1/2H_2(g, p)$

galvanic cell diagram: $Pt|H_2(g, p)|OH^-(a)\|H^+(a)|H_2(g, p)|Pt$

(2) $Ag^+(a) + I^-(a) \rightarrow AgI(s)$ can be divided into

anode: $Ag(s)+I^-(a)-e^- = AgI(s)$

cathode: $Ag^+(a)+e^- = Ag(s)$

galvanic cell diagram: $Ag(s)|AgI(s)|I^-(a)\|Ag^+(a)|Ag(s)$

(3) $H_2(p_1)\rightarrow H_2(p_2)$

anode: $H_2(p_1)-2e^-\rightarrow 2H^+(a)$

cathode: $2H^+(a)+2e^-\rightarrow H_2(p_2)$

galvanic cell diagram: $Pt|H_2(p_1)|H^+(a)|H_2(p_2)|Pt$

8.5 Applications of E_{MF} measurements (原电池电动势测定的应用)

8.5.1 Determination of $\Delta_r G_m$, $\Delta_r H_m$, $\Delta_r S_m$, $Q_{r,m}$ for the cell reaction (测定电池反应的 $\Delta_r G_m$, $\Delta_r H_m$, $\Delta_r S_m$, $Q_{r,m}$)

$\Delta_r G_m$ for the cell reaction can be found from the experimentally determined E_{MF}

$$\Delta_r G_m = -zFE_{MF} \tag{8.5.1}$$

From $\left(\frac{\partial G}{\partial T}\right)_p = -S$, we can get

$$\left(\frac{\partial \Delta_r G_m}{\partial T}\right)_p = -\Delta_r S_m \tag{8.5.2}$$

$$\Delta_r S_m = -\left(\frac{\partial \Delta_r G_m}{\partial T}\right)_p = -\left[\frac{\partial(-zFE_{MF})}{\partial T}\right]_p = zF\left(\frac{\partial E_{MF}}{\partial T}\right)_p \tag{8.5.3}$$

Where $\left(\frac{\partial E_{MF}}{\partial T}\right)_p$ **is called the temperature coefficient of E_{MF} of galvanic cell.** (原电池电动势的温度系数。)

At constant T

$$\Delta_r H_m = \Delta_r G_m + T\Delta_r S_m = -zFE_{MF} + zFT\left(\frac{\partial E_{MF}}{\partial T}\right)_p \tag{8.5.4}$$

$Q_{r,m}$ can then be found from

$$Q_{r,m} = T\Delta_r S_m = zFT\left(\frac{\partial E_{MF}}{\partial T}\right)_p \tag{8.5.5}$$

Since the cell system does electrical work in addition to volume work, $Q_{r,m}\neq\Delta_r H_m$, as is obvious from the equations (8.5.5) and (8.5.4).

Example 8.4

For cell $Cd(s)|CdCl_2\cdot 2.5H_2O$(saturated solution)$|AgCl(s)|Ag(s)$, given that

$E_{MF}=0.6753V$, $\left(\frac{\partial E_{MF}}{\partial T}\right)_p = -6.5\times10^{-4} V\cdot K^{-1}$

Please calculate $\Delta_r G_m$, $\Delta_r H_m$, $\Delta_r S_m$, $Q_{r,m}$ of above cell reaction when $T=298.15K$.

Answer:

left(-): $Cd(s)+2Cl^-(a)+2.5H_2O(l)-2e^- \rightarrow CdCl_2 \cdot 2.5H_2O(s)$

right(+): $2AgCl(s)+2e^- \rightarrow 2Ag(s)+2Cl^-(a)$

cell reaction: $Cd(s)+2.5H_2O(l)+2AgCl(s) \rightarrow CdCl_2 \cdot 2.5H_2O(s)+2Ag(s)$

$z=2$

$$\Delta_r G_m = -zFE_{MF} = -2\times96485\times0.6753 = -130.32\text{kJ}\cdot\text{mol}^{-1}$$

$$\Delta_r S_m = zF\left(\frac{\partial E_{MF}}{\partial T}\right)_p = 2\times96485\times(-6.5\times10^{-4})$$

$$= -125.5\text{J}\cdot\text{K}^{-1}\cdot\text{mol}^{-1}$$

$$\Delta_r H_m = zF\left[T\left(\frac{\partial E_{MF}}{\partial T}\right)_p - E_{MF}\right]$$

$$=2\times96485\times[298.15\times(-6.5\times10^{-4})-0.67533]$$

$$=-167.7\text{kJ}\cdot\text{mol}^{-1}$$

$$Q_r = T\Delta_r S_m = 298.15\times(-125.4) = -37.39\text{kJ}\cdot\text{mol}^{-1}$$

讨论：$\Delta_r H_m=-167.7\text{kJ}\cdot\text{mol}^{-1}$是指反应在一般容器中进行时的反应放出的热量 Q_p，而 $Q_r=T\Delta_r S_m=zFT\left(\frac{\partial E_{MF}}{\partial T}\right)_p=-37.38\text{kJ}\cdot\text{mol}^{-1}$。若反应在电池中可逆地进行，则 Q_p 与 Q_r 之差为电功，$W'_r=-167.7-(-37.38)=-130.32\text{kJ}\cdot\text{mol}^{-1}$

若$\left(\frac{\partial E_{MF}}{\partial T}\right)_p=0$，则化学能($\Delta_r H_m$)将全部转化为电功。

注意：$\Delta_r G_m$，$\Delta_r H_m$，$\Delta_r S_m$，$Q_{r,m}$均与电池反应的化学计量方程写法有关，若上式 $z=1$，于是 $\Delta_r G_m$，$\Delta_r H_m$，$\Delta_r S_m$，$Q_{r,m}$的数值都要减半。

8.5.2 Determination of $K^\ominus$ of reaction(测定反应的标准平衡常数)

Design galvanic cell according to the reaction, and look for $E_+^\ominus$ and $E_-^\ominus$, then calculate $E_{MF}^\ominus$, $E_{MF}^\ominus=E_+^\ominus-E_-^\ominus$. Once $E_{MF}^\ominus$ for the cell has been found, the standard equilibrium constant can be found from $\Delta_r G_m^\ominus=-zFE_{MF}^\ominus$ and $\Delta_r G_m^\ominus(T)=-RT\ln K^\ominus$. Combining these two equations, we have

$$\ln K^\ominus = \frac{zFE_{MF}^\ominus}{RT}$$

Example 8.5

Design a galvanic cell for the following reaction and calculate $K^\ominus$ of the reaction at 25℃.

$$Zn(s)+Cu^{2+}(a)=Zn^{2+}(a)+Cu(s)$$

Given that $E^\ominus(Cu^{2+}|Cu)=0.3402\text{V}$, $E^\ominus(Zn^{2+}|Zn)=-0.7626\text{V}$.

Answer:

$$(-): Zn(s)-2e^- \rightarrow Zn^{2+}(a)$$

$$(+): Cu^{2+}(a)+2e^- \rightarrow Cu(s)$$

Cell diagram: $Zn(s)|Zn^{2+}(a)\|Cu^{2+}(a)|Cu(s)$

$$E_{MF}^\ominus=E^\ominus(Cu^{2+}|Cu)-E^\ominus(Zn^{2+}|Zn)=0.3402-(-0.7626)=1.103\text{V}$$

$$\ln K^{\ominus}(298.15\text{K}) = \frac{zFE^{\ominus}_{\text{MF}}}{RT} = \frac{2\times96485\times1.103}{8.314\times298.15} = 37.3$$

$$K^{\ominus}(298.15\text{K}) = 2\times10^{37}$$

8.5.3 Determination of solubility product $K^{\ominus}_{sp}$ of insoluble salt(测定难溶盐的溶度积)

Write the dissolution reaction, and design galvanic cell according to the dissolution reaction, then look for $E^{\ominus}_{+}$ and $E^{\ominus}_{-}$, calculate $E^{\ominus}_{\text{MF}}$, $E^{\ominus}_{\text{MF}} = E^{\ominus}_{+} - E^{\ominus}_{-}$, at last calculate $K^{\ominus}_{sp}$.

Example 8.6

At 25℃, the standard electrode potentials of the electrodes are as follows:

$E^{\ominus}(Ag^{+}|Ag) = 0.7994\text{V}$; $E^{\ominus}(I^{-}|AgI|Ag) = -0.1521\text{V}$

Try to calculate the solubility product of AgI.

***Answer*:**

Dissolution reaction: $AgI(s) \rightarrow Ag^{+}(a) + I^{-}(a)$,

$(-)$: $Ag(s) - e^{-} \rightarrow Ag^{+}(a)$

$(+)$: $AgI(s) + e^{-} \rightarrow Ag(s) + I^{-}(a)$

Cell diagram:

$Ag(s)|Ag^{+}(a)\|I^{-}(a)|AgI(s)|Ag(s)$

$E^{\ominus}_{\text{MF}} = E^{\ominus}[I^{-}(a)|AgI(s)|Ag(s)] - E^{\ominus}[Ag^{+}(a)|Ag(s)] = -0.1521 - 0.7994 = -0.9515\text{V}$

$$\ln K^{\ominus}_{sp} = \frac{zFE^{\ominus}}{RT} = \frac{1\times96500\times(-0.9515)}{8.314\times298.15} = -37.04$$

$K^{\ominus}_{sp} = 8.232\times10^{-17}$

8.5.4 Determination of mean ionic activity factor $\gamma_{\pm}$(测定离子平均活度因子 $\gamma_{\pm}$)

Since the E_{MF} of a cell depends on the activities of the ions in solution, it is easy to use E_{MF} to calculate activity coefficients.

$$E_{\text{MF}} = E^{\ominus}_{\text{MF}} - \frac{RT}{zF}\ln\prod_{B}(a_{B})^{\nu_{B}}$$

Example 8.7

For cell $Zn(s)|ZnCl_2(0.555\text{mol}\cdot\text{kg}^{-1})|AgCl(s)|Ag(s)$, at 25℃, $E = 1.015\text{V}$. Given that $E^{\ominus}(Zn^{2+}|Zn) = -0.763\text{V}$, $E^{\ominus}(Cl^{-}|AgCl|Ag) = 0.2223\text{V}$.

(1) Write out the cell reaction($z = 2$)

(2) Calculate $K^{\ominus}$ of the above cell reaction.

(3) Calculate $\gamma_{\pm}$ of the $ZnCl_2$ solution.

Example 8.7

***Answer*:** (1) Cell reaction

$Zn(s) + 2AgCl(s) = 2Ag(s) + ZnCl_2(b = 0.555\text{mol}\cdot\text{kg}^{-1})$

(2) $\lg K^{\ominus} = \dfrac{zFE^{\ominus}}{2.303RT}$

$$\lg K^{\ominus} = \frac{2\times96485\times(0.2223+0.763)}{2.303\times8.314\times298.15}$$

$K^{\ominus} = 2.067\times10^{33}$

(3) $E=E^\ominus-\frac{RT}{zF}\ln\prod_B(a_B)^{\nu_B}=E^\ominus-\frac{RT}{zF}\ln\left(\frac{a_{ZnCl_2}\cdot a^2_{Ag}}{a_{Zn}\cdot a^2_{AgCl}}\right)=E^\ominus-\frac{RT}{2F}\ln a(ZnCl_2)$

For $E^\ominus(Zn^{2+}|Zn)=-0.763V$, $E^\ominus(Cl^-|AgCl|Ag)=0.2223V$, $E=1.015V$

We can get

$a_{ZnCl_2}=0.0990$

$a^3_\pm=a_{ZnCl_2}\Rightarrow a_\pm=0.463$, $b_\pm=(b_+^{\nu+}b_-^{\nu-})^{1/\nu}=[0.555\times(2\times0.555)^2]^{1/3}=0.881mol\cdot kg^{-1}$

$a_\pm=\gamma_\pm\cdot(b_\pm/b^\ominus)\Rightarrow\gamma_\pm=0.5255$

8.5.5 Determination of pH(测定溶液的 pH 值)

Quinhydrone electrode (quinhydrone, $Q\cdot QH_2$, is a complex of quinone ($C_6H_4O_2=Q$, 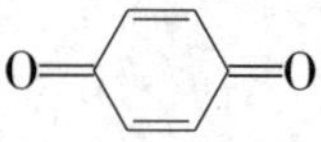 and hydroquinone, $C_6H_4O_2H_2=QH_2$, 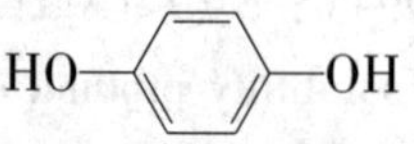).

Electrode reaction: $Q[a(Q)]+2H^+[a(H^+)]+2e^-\rightarrow QH_2[a(QH_2)]$

① Equal concentrations of both Q and QH_2 in the solution.

② Being nonelectrolytes, activity coefficients of dilute Q and QH_2 is unity.

So $$a(Q)\approx a(QH_2)$$

$$E(H^+|Q\cdot QH_2|Pt)=E^\ominus(H^+|Q\cdot QH_2|Pt)-\frac{RT}{2F}\ln\frac{a(QH_2)}{a(Q)a(H^+)^2}$$

$$E(H^+|Q\cdot QH_2|Pt)=E^\ominus(H^+|Q\cdot QH_2|Pt)-\frac{RT}{F}\ln\frac{1}{a(H^+)}$$

$$pH\overset{def}{=\!=}-\lg a(H^+)$$

So

$$E=E^\ominus(H^+|Q\cdot QH_2|Pt)-\frac{RT\ln10}{F}pH$$

$E^\ominus=(H^+|Q\cdot QH_2|Pt)=0.6997V$

So $E(H^+|Q\cdot QH_2|Pt)=(0.6997-0.05916pH)V$ at 25℃

Table 8.2 shows the E of calomel electrode of different concentrations at 25℃.

Tab. 8.2 E of calomel electrode at 25℃

Electrode	E/V	Electrode	E/V
KCl(saturated solution) $\vert Hg_2Cl_2\vert$ Hg	0.2415	KCl(0.1mol · dm^{-3}) $\vert Hg_2Cl_2\vert$ Hg	0.3337
KCl(1mol · dm^{-3}) $\vert Hg_2Cl_2\vert$ Hg	0.2807		

We can obtain the acidity of a solution from a measurement of the electromotive force of a suitable galvanic cell.Such as example 8.8.

Example 8.8

The E_{MF} of the cell:

$Hg(l)|Hg_2Cl_2(s)|KCl(sat)\|Q\cdot QH_2|H^+(pH=?)|Pt$ is 0.025V at 25℃.

Find: the pH value of the solution.

Given that at 25℃, $E^\ominus\{Hg(l)|HgCl_2(s)|KCl(saturated)\}=0.2415V$,

Example 8.8

$E^{\ominus}(Q|H_2Q) = 0.6997V$.

Answer:

$E\{Hg(l)|HgCl_2(s)|KCl(\text{saturated})\} = 0.2415V$

electrode reaction of quinhydrone electrode:

$$Q(s)+2H^+ +2e^- \rightarrow H_2Q(s)$$

According to Nernst equation for electrode reaction, we can get

$$E_+ = E^{\ominus}(Q|H_2Q) - \frac{RT}{2F}\ln\frac{a(H_2Q)}{a(Q)\cdot a(H^+)^2} = 0.6997 - \frac{RT}{F}\ln\frac{1}{a(H^+)}$$

$$= 0.6997 - \frac{RT\ln 10}{F}\text{pH} = 0.6997 - 0.05916\text{pH}$$

$$E = E_+ - E_-$$

$$0.025V = (0.6997 - 0.05916\text{pH} - 0.2415)V$$

$$\text{pH} = 7.3$$

8.5.6 Judgement on the direction of reaction(判断反应方向)

$$aA + bB \Leftrightarrow lL + mM$$

Design galvanic cell according to the reaction, and then calculate E_{MF} according to Nernst equation.

if $E_{MF}>0 \Rightarrow \Delta_r G_m<0 \Rightarrow$forward spontaneously

if $E_{MF}<0 \Rightarrow \Delta_r G_m>0 \Rightarrow$backward spontaneously

Example 8.9

One important reaction in the corrosion of iron in an acidic environment is:

$$Fe(s)+2H^+(a)+(1/2)O_2(p) \rightarrow Fe^{2+}(a)+H_2O(l)$$

which is the direction of this reaction can proceed when $a(H^+)=1$, $a(Fe^{2+})=1$ and $p(O_2)=p^{\ominus}$ at 25℃? Given that $E^{\ominus}(H^+|O_2|Pt)=1.229V$, $E^{\ominus}(Fe^{2+}|Fe)=-0.409V$.

Answer:

$(-)$: $Fe(s) \rightarrow Fe^{2+}(a)+2e^-$

$(+)$: $2H^+(a)+\frac{1}{2}O_2(p^{\ominus})+2e^- \rightarrow H_2O(l)$

Cell diagram: $Fe|Fe^{2+}(a)||H^+(a)|O_2(p^{\ominus})|Pt$

For $a(H^+)=1$, $a(Fe^{2+})=1$, $p(O_2)/p^{\ominus}=1$, $a(Fe)=1$, $a(H_2O)\approx 1$

$E_{MF} = E^{\ominus}_{MF} = E^{\ominus}(H^+|O_2|Pt) - E^{\ominus}(Fe^{2+}|Fe)$

Given that $E^{\ominus}(H^+|O_2|Pt)=1.229V$, $E^{\ominus}(Fe^{2+}|Fe)=-0.409V$

$E_{MF} = 1.229-(-0.409) = 1.676V>0$

$\Delta_r G_m = -zFE_{MF} = -323.4kJ\cdot mol^{-1}<0$

So the reaction can proceed to the right spontaneously.

8.6 Decomposition voltage (分解电压)

Two platinum electrodes are inserted into the KOH solution to form an electrolytic cell for electrolyzing water as shown in figure 8. 5. As the applied voltage is gradually increased, the voltage-current curve is measured as shown in figure 8. 6.

Decomposition voltage—the minimum potential difference which must be applied between electrodes before decomposition occurs and a current flows.（电解时在两电极上显著析出电解产物所需的最低外加电压称为分解电压。）

产生上述现象的原因是由于电极上析出的 H_2 和 O_2 构成的电池的反电动势的存在，此反电动势也称为理论分解电压(theoretical decomposition voltage)。

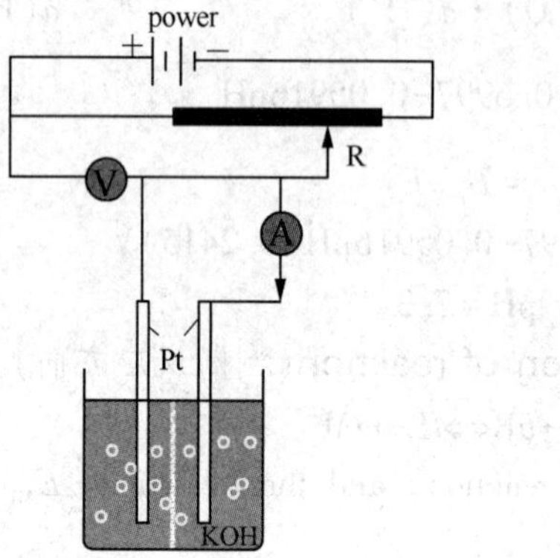

Fig. 8. 5 Measurement of decomposition voltage

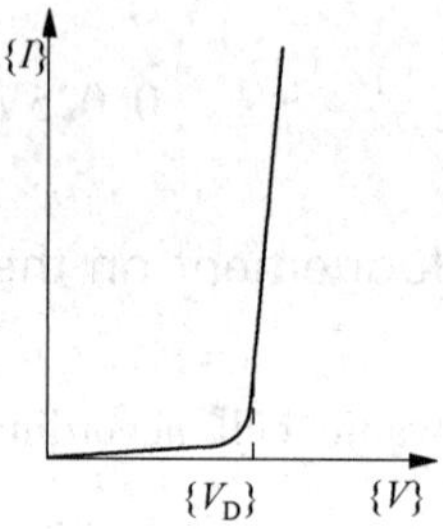

Fig. 8. 6 $I \sim V$ curve

Theoretical decomposition voltage equals to the reversible electromotive force of the cell.

Real decomposition voltage is higher than the theoretical decomposition voltage

$$E(\text{real}) = E(\text{theoretical}) + (\eta_a + |\eta_c|) + IR \qquad (8.6.1)$$

8.7 Polarization
（极化作用）

8. 7. 1 Polarization（电极的极化）

The phenomenon of which the potentials at two electrodes deviate from their equilibrium values while current passing through the electrochemical system is called polarization of the electrode.（当电化学系统中有电流通过时，两个电极上的实际电极势偏离其平衡电极势的现象叫作电极的极化。）

8. 7. 2 Types of polarization（极化的类型）

Polarization is the result of slow reactions or processes such as diffusion in the cell.

(1) Diffusion polarization（浓差极化）

Reason: Ion concentration near the electrode is difference to the ion concentration in the solution.

Example: cathode $Zn(s)|Zn^{2+}(a)$

When the current passes through the electrode, the concentration of Zn^{2+} near the cathode is reduced due to the deposition of Zn^{2+} in the liquid layer near the cathode surface. If it is too late to replenish the Zn^{2+} in the liquid layer near the cathode surface from the bulk solution, the concentration of Zn^{2+} will be lower than that in the bulk solution. It is as if the electrode is immersed in a solution of low concentration, and the equilibrium electrode potential is usually referred to in relation to the bulk solution. Obviously, the electrode potential is below its equilibrium value. This phenomenon is called concentration polarization.

The concentration polarization can be reduced by agitation, but it is impossible to remove it completely because of the diffusion layer on the electrode surface.

(2) Electrochemical polarization(电化学极化)

Reason: Electrochemistry reaction is slow.

Example: cathode $Zn(s)|Zn^{2+}(a)$

When the electric current passes through the electrode, as a result of the electrode reaction rate is slow, the electrons on the electrode surface cannot be consumed by reduction of Zn^{2+} in time. The electrode potential will decrease with the increase of the number of free electrons on the electrode surface. This polarization due to the slowness of the electrochemical reaction itself is called electrochemical polarization.

Elimination: can't be eliminated.

8.7.3 Polarization curve(极化曲线)

The tendency of actual electrode potential E deviate from equilibrium electrode potential E_e can be shown by experimentally measured polarization curves, as shown in figure 8.7.

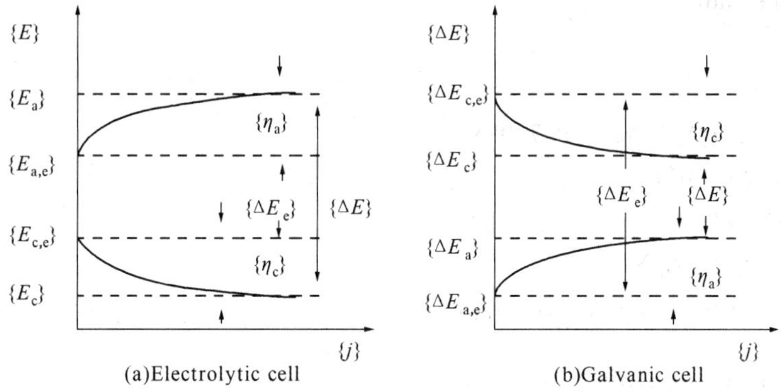

Fig. 8.7 Polarization curve

As the result of polarization, the anode potential increases($E_a > E_{a,e}$) and the cathode potential decreases($E_{c,e} > E_c$). The deviation of actual electrode potential from equilibrium electrode potential increases with the increase of current density.

8.7.4 Overpotential(超电势)

Definition of overpotential:

The discrepancy between reversible potential and irreversible potential is called overpotential(η).

$$\eta_a \overset{\text{def}}{=\!=} E_a - E_{a,e}$$

$$\eta_c \overset{\text{def}}{=\!=} E_{c,e} - E_c \qquad (8.7.1)$$

Where η_a, η_c, are anode overpotential and cathode overpotential respectively.

Notes:

Polarization causes decrease in electromotive force of galvanic cell and increase in decomposition voltage of electrolytic cell.

electrolytic cell: $E_a > E_c$

$$\Delta E = E_a - E_c = (E_{a,e} - E_{c,e}) + (\eta_a + \eta_c) \quad (8.7.2)$$

galvanic cell: $E_c > E_a$

$$\Delta E = E_c - E_a = (E_{c,e} - E_{a,e}) - (\eta_a + \eta_c) \quad (8.7.3)$$

In 1905, Tafel reported the $I \sim V$ curves of hydrogen evolution on different metal surfaces. The relationship between the overpotential and current density is:

$$\eta = a + b\lg J \quad (8.7.4)$$

(8.7.4) is called **Tafel equation**, a and b are called Tafel constants.

8.8 Competition of electrode reaction
(电解时电极反应的竞争)

Example 8.10

Suppose a aqueous solution in an electrolytic cell containing Ag^+, Cu^{2+}, and Zn^{2+} of 1 molarity. If the potential is initially very high and is gradually turned down, in which order will the metals be plated out onto the cathode.

Answer:

$Ag^+(a=1) + e^- \rightarrow Ag(s)$

$E(Ag^+|Ag) = E^{\ominus}(Ag^+|Ag) = 0.7998V$

$Cu^{2+}(a=1) + 2e^- \rightarrow Cu(s)$

$E(Cu^{2+}|Cu) = E^{\ominus}(Cu^{2+}|Cu) = 0.3402V$

$Zn^{2+}(a=1) + 2e^- \rightarrow Zn$

$E(Zn^{2+}/Zn) = E^{\ominus}(Zn^{2+}/Zn) = -0.7630V$

$H^+(a=10^{-7}) + e^- \rightarrow \frac{1}{2}H_2(p^{\ominus})$

$E(H^+|H_2) = -0.05916 \times \ln\frac{1}{10^{-7}} = -0.414V$

For liberation of metal, the overpotential is usually very low, and the reversible potential can be used in stead of irreversible potential.

For evolution of gas, the overpotential is relatively large; therefore, the overpotential should be taken into consideration.

Ag^+, Cu^{2+}, H^+, and Zn^{2+} will liberates at 0.7998V, 0.3402V, −0.414V, −0.7630V respectively without consideration of overpotential; the liberates consequence is $Ag \rightarrow Cu \rightarrow H_2 \rightarrow Zn$.

But the overpotential of hydrogen liberation on Zn is more than 1V, even when the current density is very low; the liberates consequence is $Ag \rightarrow Cu \rightarrow Zn \rightarrow H_2$, when the overpotential of hydrogen can not be omitted.

EXERCISES

(习题)

1. For cell $Zn(s)|ZnCl_2(0.05mol\cdot kg^{-1})|AgCl(s)Ag(s)$, the temperature T dependence of electromotive E_{MF} is as follows:

$$E_{MF}/V=1.015-4.92\times10^{-4}(T/K-298.15)$$

(1) Write out electrode reaction and cell reaction of above cell.

(2) Calculate Δ_rG_m, Δ_rH_m, Δ_rS_m, $Q_{r,m}$ of above cell reaction when $T=298.15K$.

Answer: (1) (−): $Zn(s)-2e^-=Zn^{2+}\{a(Zn^{2+})\}$;

(+): $2AgCl(s)+2e^-=2Ag(s)+2Cl^-\{a(Cl^-)\}$

Cell reaction: $Zn(s)+2AgCl(s)=2Ag(s)+ZnCl_2\{a(ZnCl_2)\}$

(2) $\Delta_rG_m=-195.89kJ\cdot mol^{-1}$; $\Delta_rS_m=-94.956J\cdot K^{-1}\cdot mol^{-1}$;

$\Delta_rH_m=-224.20kJ\cdot mol^{-1}$; $Q_{r,m}=28.3kJ\cdot mol^{-1}$

2. At 298.15K, for cell $Pt|H_2(g, 100kPa)|HI\{a(HI)=1\}|I(s)|Pt$, two reactions take place as follows:

(1) $H_2(g, 100kPa)+I_2(s)=2HI\{a(HI)=1\}$

(2) $1/2H_2(g, 100kPa)+1/2I_2(s)=HI\{a(HI)=1\}$

Given $E^\ominus(I^-|I_2(s)|Pt)=0.535V$, $E^\ominus(H^+|H_2(g)|Pt)=0V$, calculate $E^\ominus$, $\Delta_rG_m^\ominus$ and $K^\ominus$ of the two reactions.

Answer: (1) $E^\ominus=0.535V$; $\Delta_rG_m^\ominus=-103.2kJ\cdot mol^{-1}$; $K^\ominus=1.22\times10^{18}$

(2) $E^\ominus=0.535V$; $\Delta_rG_m^\ominus=-51.6kJ\cdot mol^{-1}$; $K^\ominus=1.10\times10^{9}$

3. For cell $Pt|H_2(g, 101.325kPa)|HCl(0.1mol\cdot kg^{-1})|Hg_2Cl_2(s)|Hg(l)$, the temperature dependence of electromotive force is as follows:

$$E/V=0.0694+1.881\times10^{-3}(T/K)-2.9\times10^{-6}(T/K)^2$$

(1) Write out the cell reaction.

(2) Calculate Δ_rG_m, Δ_rH_m, Δ_rS_m, $Q_{r,m}$ of the reaction at 25℃.

Answer: (1) cell reaction: $H_2(101.325kPa)+Hg_2Cl_2(s)=2Hg(l)+2HCl(b=0.1mol\cdot kg^{-1})$;

(2) $\Delta_rG_m=-71.86kJ\cdot mol^{-1}$; $\Delta_rS_m=-29.23J\cdot K^{-1}\cdot mol^{-1}$; $\Delta_rH_m=-63.14kJ\cdot mol^{-1}$; $Q_{r,m}=8.73kJ\cdot mol^{-1}$

4. For $Sb(s)|Sb_2O_3(s)|$solution x‖saturated $KCl|Hg_2Cl_2(s)|Hg(l)$, at 25℃, When the solution is a buffer solution pH=3.98 whose electromotive force $E_1=0.228V$. While the pH of the solution is unknown whose electromotive force $E_2=0.345V$, try to calculate pH of the solution.

Answer: pH=5.96

5. For the cell $Pt|H_2(g, 100kPa)|HCl(b=0.1mol\cdot kg^{-1})|Cl_2(g, 100kPa)|Pt$, at 25℃ $E=1.4881V$. Please calculate the ionic mean activity coefficient of HCl in HCl solution. Given $E^\ominus(Cl^-|Cl_2|Pt)=1.3580V$.

Answer: $\gamma_\pm=0.796$

6. At 25℃ $K_{sp}(AgBr) = 4.88\times10^{-13}$, $E^\ominus(Ag^+|Ag) = 0.7994V$, $E^\ominus(Br_2|Br^-) = 1.065V$, calculate at 25℃:

(1) $E^\ominus(AgBr|Ag)$

(2) The standard molar Gibbs function of formation of AgBr(s).

Answer: (1) $E^\ominus(AgBr|Ag) = 0.0715V$; (2) $\Delta_f G_m^\ominus(AgBr(s)) = -95.88kJ\cdot mol^{-1}$

7. Design galvanic cells for following reaction and calculate $\Delta_r G_m^\ominus$ and $K^\ominus$ of the reaction at 25℃. Given

$E^\ominus(Ag^+|Ag) = 0.7994V$, $E^\ominus(Cu^{2+}|Cu) = 0.3400V$, $E^\ominus(Cd^{2+}|Cd) = -0.4028V$,

$E^\ominus(Pb^{2+}|Pb) = -0.1265V$, $E^\ominus(Sn^{4+}|Sn^{2+}) = 0.15V$

(1) $2Ag^+ + H_2(g) = 2Ag(s) + 2H^+$

(2) $Cd + Cu^{2+} = Cd^{2+} + Cu$

(3) $Sn^{2+} + Pb^{2+} = Sn^{4+} + Pb$

Answer:

(1) $Pt|H_2(p)|H^+(a_{H^+})\|Ag^+(a_{Ag^+})|Ag$, $\Delta_r G_m^\ominus = -154.284kJ\cdot mol^{-1}$, $K^\ominus = 1.08\times10^{27}$

(2) $Cd|Cd^{2+}(a_{Cd^{2+}})\|Cu^{2+}(a_{Cu^{2+}})|Cu$, $\Delta_r G_m^\ominus = -143.3kJ\cdot mol^{-1}$, $K^\ominus = 1.30\times10^{25}$

(3) $Sn^{2+}(a_{Sn^{2+}})|Sn^{4+}(a_{Sn^{4+}})\|Pb^{2+}(a_{Pb^{2+}})|Pb$, $\Delta_r G_m^\ominus = 53.53kJ\cdot mol^{-1}$, $K^\ominus = 4.56\times10^{-10}$

8. Write out the cell reactions of the following cells. Calculate the electromotive force and molar Gibbs function changes of the cell reaction at 25℃ and point out whether the cell reaction could proceed spontaneously or not. Given

(1) $Pt|H_2(g, 100kPa)|HCl\{a(HCl) = 1\}|Cl_2(g, 100kPa)|Pt$

(2) $Zn(s)|ZnCl_2\{a(ZnCl_2) = 0.5\}|AgCl(s)|Ag(s)$

$E^\ominus(Cl^-|Cl_2|Pt) = 1.3580V$, $E^\ominus(Zn^{2+}|Zn) = -0.7630V$, $E^\ominus\{Cl^-|AgCl(s)|Ag(s)\} = 0.2221V$

Answer: (1) $H_2(g, 100kPa) + Cl_2(g, 100kPa) = 2HCl\{a(HCl) = 1\}$

$E = 1.3580V$, $z = 2$, $\Delta_r G_m = -262.1kJ\cdot mol^{-1}$, can proceed spontaneously.

(2) $Zn(s) + 2AgCl(s) = 2Ag(s) + ZnCl_2\{a(ZnCl_2 = 0.5)\}$

$E = 0.9940V$, $z = 2$, $\Delta_r G_m = -191.813kJ\cdot mol^{-1}$, can proceed spontaneously.

Chapter 9 Interface phenomena
(界面现象)

Leaning objectives:

(1) Define *surface tension.*

(2) Derive and use *Kelvin equation.*

(3) Distinguish between *physisorption* and *chemisorption.*

(4) Define *surface excess* and derive and use the *Gibbs adsorption isotherm.*

(5) Derive and use the *Langmuir adsorption isotherm.*

The three-dimensional region of contact between phased α and β in which molecules interact with molecules of both phases is called the inter facial layer or surface layer. The physical boundary of any condensed phase object is called a surface, and there doesn't seem to be anything special about them. The top of a desk, the blacktop of a road, are surfaces that we encounter daily, and there does not seem to be any unusual behavior associated with them. Perhaps on a mundane level this is true. But now that we understand that matter is composed of atoms, that these atoms behave according to the laws of thermodynamics and quantum mechanics, and that gas and solid phases themselves behave in some understandable fashion, we should be willing to think that surfaces are worthy of special attention. A surface represents a series of points making a plane where one material or phase ends and another begins. This discontinuity of matter means that the bulk properties of the material will not necessarily be found at the surface. In order to understand how surface properties differ from bulk properties, we need to consider some of the ways surfaces are defined and how they are different from the bulk material. Therefore, we conclude our presentation of physical chemistry by considering surfaces.

9.1 Surface tension
(表面张力)

9.1.1 Surface tension, Surface work and Surface Gibbs function(表面张力、表面功及表面吉布斯函数)

(1) Interface layer(界面层的定义)

In a system composed of the phases α and β, moleculars at or very near the region of contact of phases α and β clearly have a different molecular environment than molecules in the interior of either phase. The three-dimensional region of contact between phases α and β in which molecules interact with molecules of both phases α and β is called the interfacial layer or surface layer. The surface layer is a few molecules thick(The term interface refers to the apparent two-dimensional geometrical boundary surface separating the two phases.)

(2) Surface tension(表面张力)(l-g)

The surface layer of atoms/molecules of a liquid can also be considered such a film, shown diagrammatically in figure 9.1. Further, we might suggest that this surface layer would have different properties than the bulk material. This is because the surface layer isn't really "bulk". Bulk atoms or molecules are surrounded on all sides by other molecules of the same material. At the surface, atoms or molecules are surrounded by the same molecules on one side but different molecules (or nothing) on the other. Forces between different materials (or between one material and nothing) are different, implying that the forces on the single surface layer of molecules are different from those in the bulk. Therefore, surface molecules aren't really bulk species and their behavior might not be the same as the bulk material. Suppose we want to increase or decrease the amount of surface available, perhaps by changing the shape of the liquid so that more surface area is exposed. Because of the differing forces acting at the surface, it will require work to change the surface area.

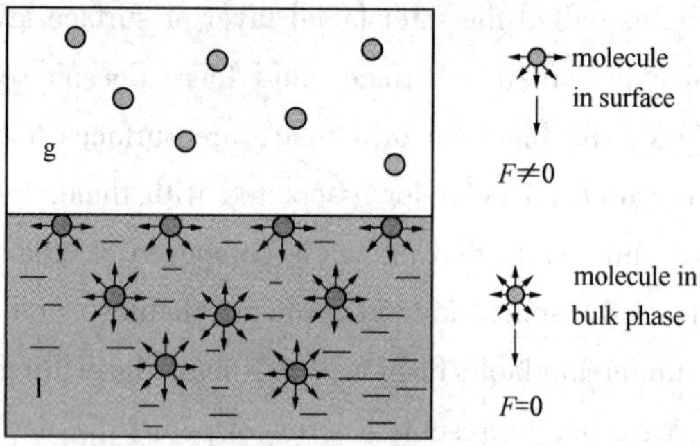

Fig. 9.1 Molecules in surface and molecules in bulk phase

$$F = 2\gamma l \qquad (9.1.1)$$

$$\gamma = \frac{F}{2l} \qquad (9.1.2)$$

γ is the surface tension which can be taken as the force exerted on unit length.

Unit: $N \cdot m^{-1}$.

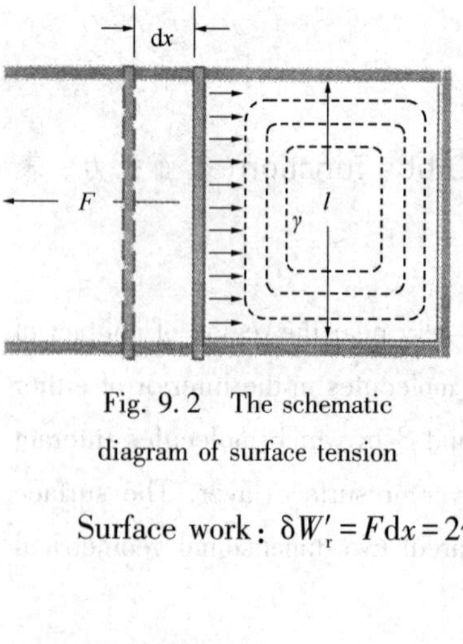

Fig. 9.2 The schematic diagram of surface tension

Figure 9.2 shows a diagram of what we are trying to accomplish for an idealized system. If we want to increase the size of the rectangular surface area, then we have to do work on the liquid and against the unbalanced forces that exist at the surface. (Again, increasing surface area of a liquid requires that work be done on the liquid. Conversely, if the surface area is decreased, work is done by the liquid on the surroundings.) If the magnitude of the unbalanced force is represented by F, then the infinitesimal amount of work needed to increase the rectangular area by moving one boundary out by an infinitesimal amount dx is

Surface work: $\delta W'_r = F\mathrm{d}x = 2\gamma l \mathrm{d}x = \gamma \mathrm{d}A_s$

$$\gamma = \frac{\delta W'_{\mathrm{r}}}{\mathrm{d}A_{\mathrm{s}}} \tag{9.1.3}$$

γ is the work which is needed to reversibly change per m^2 of the surface. where the factor of 2 in the denominator is there because we want to consider surface tension as a force per length per surface. In this case, F—the unbalanced force experienced by the surfaces of the film—is twice the force for a single surface.

Unit: $J \cdot m^{-2} = N \cdot m^{-1}$.

Work for a reversible process at constant T and p in a system is equals to the Gibbs function change in this process. That is

$\delta W'_{\mathrm{r}} = \mathrm{d}G_{T,p} = \gamma \mathrm{d}A_{\mathrm{s}}$ rev. proc. at const. T, p

so
$$\gamma = \left(\frac{\partial G}{\partial A_{\mathrm{s}}}\right)_{T,p} \tag{9.1.4}$$

γ is the Gibbs function change when reversibly change per m^2 of the surface occurs.

Unit: $J \cdot m^{-2}$.

So we can see

$$\gamma = \left(\frac{\partial G}{\partial A_{\mathrm{s}}}\right)_{T,p} = \frac{\delta W'_{r}}{\mathrm{d}A_{\mathrm{s}}} = \frac{F}{2l} \tag{9.1.5}$$

表面张力、单位面积的表面功、单位面积的表面吉布斯函数三者的数值、量纲等同，但它们有不同的物理意义，是从不同角度说明同一问题。

(3) Direction and the effect of the surface tension(表面张力的作用方向与效果)

由图9.3可见，表面张力是垂直作用于表面上单位长度的收缩力，其作用的结果使液体表面缩小，其方向对于平液面是沿着液面并与液面平行，对于弯曲液面则与液面相切。

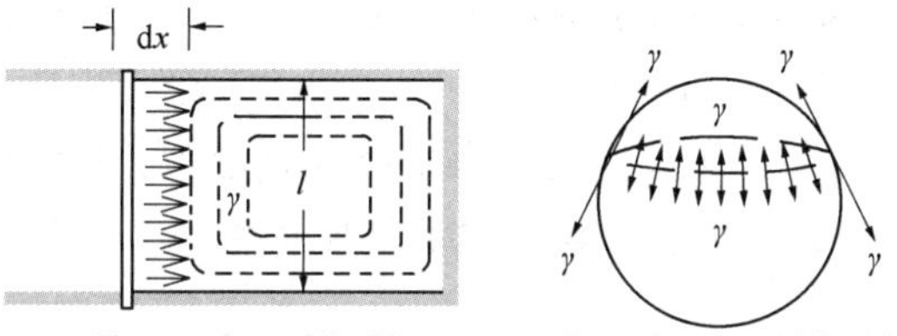

Fig.9.3 Surface tension of the plane surface and the curved surface

It has long been known that a sphere is the most compact solid object: It has the minimum surface area for any given volume. Therefore, the effects of surface tension require that liquids assume a spherical shape if no additional forces are acting on them. In the absence of gravity, this is indeed what happens, and it is ultimately caused by the surface tension of the liquid. In many instances, liquid amounts are large enough that effects due to gravity distort the ideal spherical shape of liquids. However, for small amounts—like small drops of water on a plastic surface—the tendency toward a spherical shape can be obvious.

(4) Effecting factors(影响表面张力的因素)

① Intermolecular force(分子间力的影响)

Surface tension increases as the intermolecular force increases(or chemical bond force). That is

γ(metallic bond) > γ(ionic bond) > γ(polarized covalent bond) > γ(non-polarized covalent bond)

Table 9.1 shows the surface tensions of some substances.

Tab. 9.1 Surface tensions of the substances

Substance	$\gamma/(10^{-3}\mathrm{N\cdot m^{-1}})$	T/K	Substance	$\gamma/(10^{-3}\mathrm{N\cdot m^{-1}})$	T/K
water(l)	72.75	293	W(s)	2900	2000
ethanol(l)	22.75	293	Fe(s)	2150	1673
benzene(l)	29.99	293	Fe(l)	1990	1909
acetone(l)	23.7	293	Hg(l)	495	293
n-octanol(l/water)	9.5	293	NaCl(s)	227	299
n-octanone(l)	27.5	293	KCl(s)	110	299
n-hexane(l/water)	51.1	293	MgO(s)	1200	299
n-hexane(l)	19.4	293	CaF_2(s)	450	79
n-octane(l/water)	50.9	293	He(l)	0.309	2.5
n-octane(l)	21.9	293	Xe(l)	19.6	163

② Temperature(温度)

Surface tension decreases as the temperature increases. As the temperature is raised, the two phases become more and more alike until at the critical temperature T_c, the liquid-vapor interface disappears and only one phase is present. At T_c, the value of surface tension must therefore become 0.

γ(low temperature) > γ(high temperature)

③ Pressure(压力)

Surface tension decreases as the pressure increases.

γ (low pressure) > γ(high pressure)

表面张力一般随压力增加而下降。这是由于随压力增加，气相体积质量增大，同时气体分子更多地被液面吸附，并且气体在液体中的溶解度也增加，以上三种效果均使表面张力随压力增加而下降。

9.1.2 The fundamental equation of thermodynamics of high disperse system (高度分散系统的热力学基本方程)

For high disperse system, the phenomenon of surface is apparent, so we can't omit it, that is

$$G=f(T,\ p,\ n_B,\ \cdots\cdots A_s)$$

For

$$\mathrm{d}G_{T,p}=\gamma\mathrm{d}A_s$$

So we can get

$$\mathrm{d}G=-S\mathrm{d}T+V\mathrm{d}p+\sum_{\alpha}\sum_{B}\mu_{B(\alpha)}\mathrm{d}n_{B(\alpha)}+\gamma\mathrm{d}A_s \tag{9.1.6}$$

$$\mathrm{d}U=T\mathrm{d}S-p\mathrm{d}V+\sum_{\alpha}\sum_{B}\mu_{B(\alpha)}\mathrm{d}n_{B(\alpha)}+\gamma\mathrm{d}A_s \tag{9.1.7}$$

$$\mathrm{d}H=T\mathrm{d}S+V\mathrm{d}p+\sum_{\alpha}\sum_{B}\mu_{B(\alpha)}\mathrm{d}n_{B(\alpha)}+\gamma\mathrm{d}A_s \tag{9.1.8}$$

$$dA = -SdT - pdV + \sum_{\alpha}\sum_{B}\mu_{B(\alpha)}dn_{B(\alpha)} + \gamma dA_s \tag{9.1.9}$$

$$\gamma = \left(\frac{\partial G}{\partial A_s}\right)_{T,\,p,\,n_{B(\alpha)}} = \left(\frac{\partial U}{\partial A_s}\right)_{S,\,V,\,n_{B(\alpha)}} = \left(\frac{\partial H}{\partial A_s}\right)_{S,\,p,\,n_{B(\alpha)}} = \left(\frac{\partial A}{\partial A_s}\right)_{T,\,V,\,n_{B(\alpha)}} \tag{9.1.10}$$

γ 等于在定温、定容、定组成(或定温、定压、定组成)下，增加单位表面时系统亥姆霍茨自由能(或吉布斯自由能)的增加，因此 γ 又称为单位表面亥姆霍茨自由能或单位表面吉布斯自由能，简称为单位表面自由能。

在定温、定压、定组成下

$$dG_{T,\,p,\,n(B)} = \gamma dA_s \tag{9.1.11}$$

$dG_{T,\,p}<0$ 的过程是自发过程，所以定温、定压下凡是使 A_s 变小(表面收缩)或使 γ 下降的过程都会自发进行，这是产生表面现象的热力学原因。

9.2 Excess pressure of curved liquid surface (弯曲液面的附加压力)

9.2.1 Excess pressure of curved liquid surface—Laplace equation(弯曲液面的附加压力——拉普拉斯方程)

Liquid surface can be divided into **plane liquid surface** and **curved liquid surface**.

Curved liquid surface can be divided into **convex liquid surface** and **concave liquid surface**.

When the liquid surface between phases α and β is curved, the surface tension causes the equilibrium pressures in the bulk phases α and β to differ. This can be seen from figure 9.4.

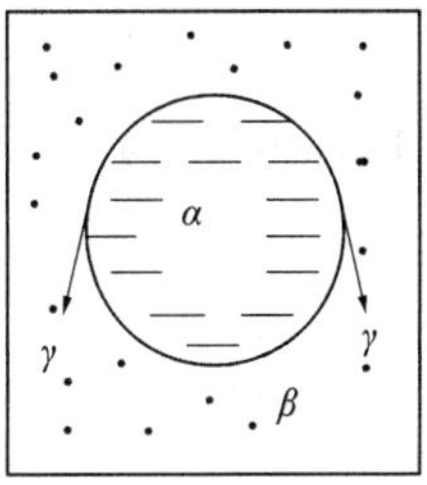

Fig. 9.4 Excess pressure of curved liquid surface

Definition of excess pressure Δp

$$\Delta p = p_\alpha - p_\beta > 0$$

For small drop in gas phase, we can get

$$\Delta p = p_l - p_g > 0 \Rightarrow p_l > p_g$$

For small bubble in liquid phase, we can get

$$\Delta p = p_g - p_l > 0 \Rightarrow p_g > p_l$$

For plane liquid surface, we can get $\Delta p = 0$

Laplace equation(拉普拉斯方程)

If the system under consideration was a droplet

$$\Delta p = \frac{2\gamma}{r} \tag{9.2.1}$$

Equation(9.2.1) was derived independently by Young and by Laplace. As $R\rightarrow\infty$ in(9.2.1), the pressure difference goes zero, as it should for a planar interface.

If the system under consideration was a bubble (that is, a film with an inner and an outer surface) instead of a droplet, then both surfaces would contribute a surface energy (that is, surface tension) and equation 9.21 would be

$$\Delta p = \frac{4\gamma}{r}$$

9.2.2 Capillary phenomenon(毛细管现象)

Equation(9.2.1) is the basis for the capillary-rise method of measuring the surface tension of liquid-vapor and liquid-liquid interfaces. When a capillary tube is inserted in the liquid, the liquid will rise or fall in the tube. This phenomenon is called capillary phenomenon. From measurement of the height to which the liquid rise or fall in the tube, the surface tension can be calculated.

Suppose that $0°<\theta<90°$. Figure 9.5(a) shows the situation immediately after a capillary tube has been inserted into a wide dish of liquid. Point 1 and 2 are at the same height in liquid, so $p_1=p_2$.

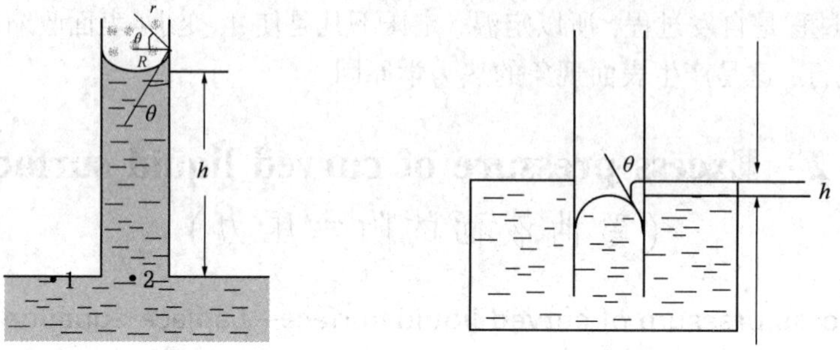

(a) Wetting: $\theta<90° \Rightarrow$ concave liquid (b) Not wetting: $\theta>90° \Rightarrow$ convex liquid surface

Fig. 9.5 Capillary phenomenon

$$\rho gh=\Delta p=\frac{2\gamma}{r} \tag{9.2.2}$$

From figure 9.5(a), we can get

$$\cos\theta=\frac{R}{r} \tag{9.2.3}$$

Substitution(9.2.3) in(9.2.2) gives

$$h=\frac{2\gamma\cos\theta}{\rho gR} \tag{9.2.4}$$

γ—surface tension;

ρ—volume mass of liquid;

R—the radius of capillaries;

Conclusion:

Wetting: concave liquid surface $\theta<90° \Rightarrow h>0$, liquid rise in capillary.

Not wetting: convex liquid surface $\theta>90° \Rightarrow h<0$, liquid fall in capillary.

9.2.3 Saturated vapor pressure of curved liquid surface—Kelvin Equation(弯曲液面的饱和蒸气压)

由热力学推导，可以得出曲率半径为 r 的液滴，其饱和蒸气压与曲率半径 r 的关系为

$$\ln\frac{p_r^*}{p^*}=\pm\frac{2\gamma}{r}\frac{M}{\rho RT} \tag{9.2.5}$$

(+: convex liquid surface; −: concave liquid surface in capillary)

p_r^*—the saturated vapor pressure of curved liquid surface;

p^*—the saturated vapor pressure of plane liquid surface;

ρ—volume mass of liquid, that is density of liquid;

M—molar mass of liquid;

γ—surface tension;

r—the radius of curvature.

Equation(9.2.5) is called Kelvin equation.

From(9.2.5) we can get that

for convex liquid surface: $\ln(p_r^*/p^*)>0$, $p_r^*>p^*$

for concave liquid surface in capillary: $\ln(p_r^*/p^*)<0$, $p_r^*<p^*$

p_r^*(convex liquid surface) $>p_r^*$(plane liquid surface) $>p_r^*$(concave liquid surface)

Example 9.1

A drop equilibriums with it's vapor at 25℃, the radius of the drop is 1μm, calculate Δp and the saturated vapor pressure of the drop. (γ of water is $71.97\times10^{-3}\text{N}\cdot\text{m}^{-1}$, volume mass of water is $0.9971\text{g}\cdot\text{cm}^{-3}$, the vapor pressure of water is 3.168kPa and molar mass is $18.02\text{g}\cdot\text{mol}^{-1}$ at 25℃.)

Answer:

$$(1)\ \Delta p=\frac{2\gamma}{r}=\frac{2\times71.97\times10^{-3}\text{N}\cdot\text{m}^{-1}}{1\times10^{-6}\text{m}}=143.9\times10^{3}\text{Pa}=143.9\text{kPa}$$

$$(2)\ \ln\left(\frac{p_r^*}{p^*}\right)=\frac{2\gamma M}{RT\rho r}$$

$$=\frac{2\times71.97\times10^{-3}\text{N}\cdot\text{m}^{-1}\times18.02\times10^{-3}\text{kg}\cdot\text{mol}^{-1}}{8.314\text{J}\cdot\text{K}^{-1}\cdot\text{mol}^{-1}\times298.15\text{K}\times0.9971\times10^{3}\text{kg}\cdot\text{m}^{-3}\times1\times10^{-6}\text{m}}$$

$$=1.049\times10^{-3}$$

$$\frac{p_r^*}{p^*}=1.001$$

$$p_r^*=1.001\times p^*=1.001\times3.168\text{kPa}=3.1712\text{kPa}$$

9.2.4 Metastable state and new phase formation(亚稳状态和新相的生成)

Metastable state—state of supersaturated vapor, supersaturated solution, superheated liquid and supercooled liquid.

(1) Supersaturated vapor(过饱和蒸气)

Supersaturated vapor—a vapor which is heated or compressed to a pressure higher than the vapor pressure of the bulk liquid, condensation does not occur.(蒸气的过饱和现象——在一定温度下，当蒸气分压超过该温度下的饱和蒸气压，而蒸气仍不凝结的现象。此时的蒸气称为过饱和蒸气。)

Elimination: add the formation center of small drop.

e.g. manual rainfall: add AgI small grains.

(2) Supersaturated solution(过饱和溶液)

Supersaturated solution—a solution in which the concentration is higher than the concentration in the saturated solution but the solute does not separate out. (溶液的过饱和现象——在一定温度、压力下，当溶液中溶质的浓度已超过该温度、压力下的溶质的溶解度，而溶质仍不析出

的现象。此时的溶液称为过饱和溶液。)

Elimination: add the growth center of small crystal.

e. g. small grains.

(3) Superheated liquid(过热液体)

Superheated liquid—the liquid which does not boil when temperature is higher than the boiling point at the certain pressure. (液体的过热现象——在一定的压力下，当液体的温度高于该压力下的沸点，而液体仍不沸腾的现象。此时的液体称为过热液体。)

Elimination: add zeolite or capillaries.

(4) Supercooled liquid(过冷液体)

Supercooled liquid—the liquid which does not freeze when temperature is lower than the freezing point at the certain pressure.(液体的过冷现象——在一定压力下，当液体的温度已低于该压力下液体的凝固点，而液体仍不凝固的现象。此时的液体称为过冷液体。)

Elimination: add small grains or stirring.

9.3 Soild Surface
(固体表面)

9.3.1 Adsorption(吸附)

In this section, the interphase region of an essentially involatile solid in contact with a gas is considered. Commonly used gases in adsorption studies include He, H_2, N_2, CO, CO_2, CH_4, C_2H_6, NH_3, and SO_2; commonly used solids include metals, metal oxides, and carbon in the form of charcoal. The solid on whose surface adsorption occurs is called adsorbent. The adsorbed gas is the adsorbate.

9.3.2 Difference between physical adsorption and chemisorption(物理吸附与化学吸附的区别)

Adsorption on solids is classified into physical adsorption and chemical adsorption (or chemisorption) according to the adsorption force; the dividing line between the two is not always sharp. The differences between physical adsorption and chemisorption are shown in Table 9.2.

Tab. 9.2 Differences between physical adsorption and chemisorption

Characteristics	Physical adsorption	Chemisorption
adsorption force	intermolecular force	chemical bond force
adsorption molecule layer	monolayer or multilayer	monolayer
adsorption temperature	low	high
adsorption heat	small	large
adsorption rate	quick	slow
adsorption selectivity	have not	have
adsorption stability	bad	good
activation energy	small	large
electron transfer	have not	have

9.3.3 Adsorption quantity(吸附量)

Adsorption equilibrium—the dynamic equilibrium with $v_a = v_d$.

吸附平衡是动态平衡，即达吸附平衡时，吸附和脱附过程同时进行，只不过此时吸附速率和脱附速率相同。

Adsorption quantity—the amount of gas adsorbed on a given weight of adsorbent. It is usually expressed as the volume V(at standard state) adsorbed per gram of adsorbent. (吸附量：在一定量的吸附剂上所吸附的气体的量。吸附量的大小，一般用单位质量吸附剂所吸附的气体在标准状况下所占有的体积 V 来表示。)

$$V^a \stackrel{\text{def}}{=\!=} \frac{V}{m} \tag{9.3.1}$$

m—mass of solid;

V—volume of gas adsorbed at adsorption temperature and pressure.

9.3.4 Adsorption curve(吸附曲线)

Adsorption isotherm: A plot of V^a vs. p at constant T, $V^a \sim p$.

Adsorption isobar: A plot of V^a vs. T at constant p, $V^a \sim T$.

Adsorption isostere: A plot of T vs. p at constant V^a, $T \sim p$.

9.3.5 Types of adsorption isotherm(吸附等温线的类型)

Figure 9.6 shows five typical types of adsorption isotherm.

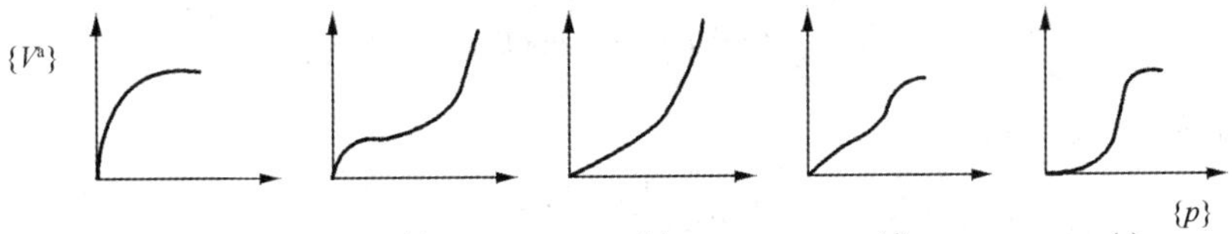

Fig. 9.6 Different types of adsorption isotherm

9.3.6 Theory of Langmuir adsorption of unimolecular layer(兰谬尔单分子层吸附理论)

Irving Langmuir

Theoretical basis(basic hypothesis):

① Adsorption is unimolecular layer;

② Surface of solid is uniform;

③ There is on interaction between one absorbed molecule and another;

④ Adsorption equilibrium is dynamic equilibrium.

Adsorption process

$$A(g) + M \underset{k_d}{\overset{k_a}{\rightleftharpoons}} AM$$

k_a—rate coefficients for adsorption;

k_d—rate coefficients for desorption;

A—gas molecule;

M—solid surface.

Extent of surface coverage(θ)—the fraction of surface that is covered. (θ 为固体表面被覆盖

的分数，称为表面覆盖度。)

That is
$$\theta=\frac{\text{surface that is covered}}{\text{total surface}}=\frac{V^a}{V_m^a} \quad (9.3.2)$$

Where V^a is the volume of gas that adsorbed at p and V_m^a is the volume of gas that adsorbed in the high-pressure limit when a monolayer covers the entire surface.

Extent of vacancy $(1-\theta)$—fractional vacancy of solid surface.[$(1-\theta)$代表固体空白表面的分数。]

The rate of adsorption v_a is proportional to pressure p and the number of vacant sites on the surface $(1-\theta)$; the rate of desorption v_d is proportional to the number of adsorbed species on the surface θ.

So
$$v_a=k_a(1-\theta)p$$
$$v_d=k_d\theta$$

When the adsorption reaches to equilibrium, the two rates are equal
$$v_a=v_d$$

So we can get
$$k_a(1-\theta)p=k_d\theta$$

Solving for θ gives the Langmuir isotherm
$$\theta=\frac{k_a p}{k_d+k_a p} \quad (9.3.3)$$

Let
$$b=\frac{k_a}{k_d}\text{(adsorption coefficient)}$$

Equation(9.3.3) becomes
$$\theta=\frac{bp}{1+bp}=\frac{V^a}{V_m^a} \quad (9.3.4)$$

(9.3.3) and (9.3.4) is Langmuir adsorption isotherm.

Comment:

(1) When p is low or adsorption is weak, $bp\ll 1$, then
$$\theta=bp \text{ or } V^a=V_m^a bp$$

(2) When p is high or adsorption is strong, $bp\gg 1$, then
$$\theta=1 \text{ or } V^a=V_m^a$$

(9.3.4) can be write in another form
$$V^a=\frac{V_m^a bp}{1+bp} \quad (9.3.5)$$
$$\frac{1}{V^a}=\frac{1}{V_m^a bp}+\frac{1}{V_m^a} \quad (9.3.6)$$

A plot of $\frac{1}{V^a}$ vs. $\frac{1}{p}$ is a straight line. b and V_m^a can be obtained from the slope and the intercept of the plot.

Applicable condition: gas adsorbed by solid, at low-, middle-and high-pressure.

Two gases A and B mixed adsorbed by solid

$$\theta_A = \frac{b_A p_A}{1+b_A p_A+b_B p_B} \tag{9.3.7}$$

$$\theta_B = \frac{b_B p_B}{1+b_A p_A+b_B p_B} \tag{9.3.8}$$

For $A_2+2* = 2(A-*)$

$$\theta = \frac{\sqrt{bp}}{1+\sqrt{bp}} \tag{9.3.9}$$

9.3.7 Empirical formula of adsorption—Freundlich formula(吸附经验式——弗罗因德利希公式)

$$V^a = kp^n \tag{9.3.10}$$

k, n—empirical constant.

$\lg V^a = \lg k + n\lg p$; the constants k and n evaluated from the intercept and solpe of a plot of $\lg V^a$ vs. $\lg p$.

The Freundlich formula is not valid at very high pressures but is frequently more accurate than the Langmuir isotherm for intermediate pressures.

The Freundlich formula is often applied to adsorption of solutes from liquid solutions onto solids. Here, the solute's concentration c replaces p, and the mass adsorbed per unit mass of adsorbent replaces V^a.

9.3.8 BET adsorption isotherm of polymolecular layer(多分子层吸附的BET吸附等温线)

吸附等温线有以上五种类型，朗缪尔等温式仅能较好地说明其中的第一种类型。对于其他四种类型等温线却无法解释。布鲁诺尔(Brunauer)、埃米特(Emmett)和特勒(Teller)三人在1938年提出的多分子层吸附理论(BET理论)较成功地解释了其他类型的吸附等温线。

Theoretical basis(basic hypothesis)

① Adsorption is multiple molecular layers;

② Surface of solid is uniform;

③ There aren't interactions among the gas molecules absorbed on the surface of solid;

④ Adsorption equilibrium is dynamic equilibrium.

BET equation:

$$\frac{V^a}{V_m^a} = \frac{c(p/p^*)}{(1-p/p^*)\{1+(c-1)p/p^*\}}$$

Where V^a—the volume adsorbed at p;

V_m^a—the volume adsorbed when a monolayer covers the entire surface;

p^*—the saturated pressure of the absorbate at the temperature of the experiment;

c—constant relative with the absorption heat.

BET equation can be write below:

$$\frac{p}{V^a(p^*-p)} = \frac{1}{V_m^a} + \frac{c-1}{cV_m^a} \cdot \frac{p}{p^*} \tag{9.3.11}$$

The constants c and V_m^a can be obtained from the slope and intercept of a plot of $\frac{p}{V^a(p^*-p)}$ vs. $\frac{p}{p^*}$

由 V_m^a 及公式 $a_s=\frac{V_m^a}{V_0}La_m$，可求吸附剂比表面积。其中 V_0 为在吸附温度、压力下的摩尔体积，L 是阿伏加德罗常数，a_m 是分子的截面积。

Example 9.2

The largest adsorption quantity is 93.8dm^3 · kg^{-1} when $CHCl_3$ was adsorbed by active carbon when $T=273.15$K, known the absorption quantity is 82.5dm^3 · kg^{-1} when the partial pressure of $CHCl_3$ is 1.34×10^4Pa.

Example 9.2

Calculate:

(1) Equilibrium constant of adsorption b in Langmuir isotherm;

(2) The absorption quantity when the partial pressure of $CHCl_3$ is 6.67×10^3Pa.

Answer:

(1) According to Langmuir isotherm we can get

$$\theta=\frac{bp}{1+bp}=\frac{V^a}{V_m^a}$$

$$b=\frac{V^a}{(V_m^a-V^a)p}=\frac{82.5}{(93.8-82.5)\times1.34\times10^4}=5.4\times10^{-4}\text{Pa}^{-1}$$

(2) $\theta=\frac{bp}{1+bp}=\frac{V^a}{V_m^a}$

$$V^a=\frac{V_m^a bp}{1+bp}=\frac{93.8\times5.45\times10^{-4}\times6.67\times10^3}{1+5.45\times10^{-4}\times6.67\times10^3}=73.5\text{dm}^3\cdot\text{kg}^{-1}$$

9.4 Liquid-solid interface
(液固表面)

9.4.1 Types of wetting(润湿及其类型)

$$\gamma=\left(\frac{\partial G}{\partial A_s}\right)_{T,\,p,\,n_B} \qquad (9.4.1)$$

(1) Wetting(润湿)

Wetting—the phenomenon which the gas(liquid) on the solid surface was substituted by liquid (another liquid). [润湿是指:固体表面上的气体(或液体)被液体(或另一种液体)取代的现象。]

(2) Types of wetting(润湿的类型)

Wetting include adhesion wetting, dipping wetting and spreading wetting(As shown in figure 9.7~9.9).

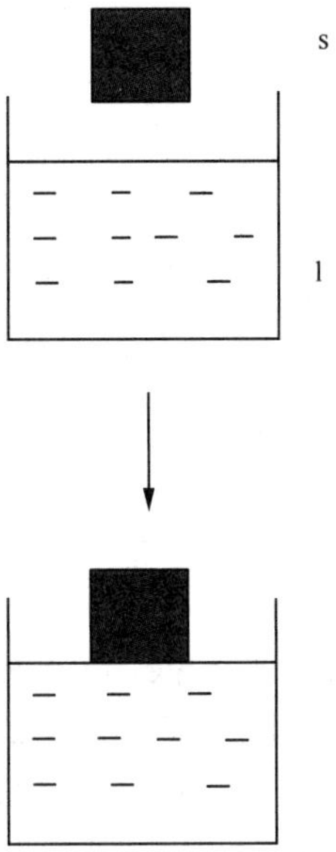

Fig. 9.7 Adhesion wetting

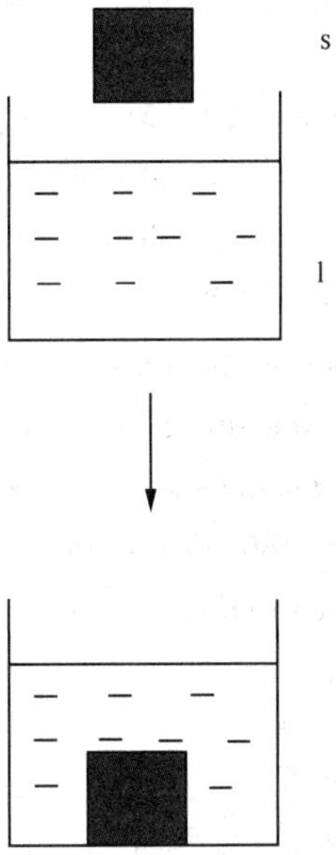

Fig. 9.8 Dipping wetting

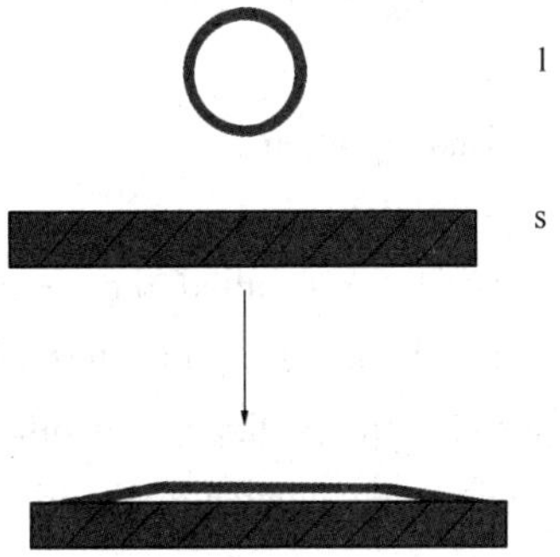

Fig. 9.9 Spreading wetting

设被取代的界面为单位面积，单位界面自由能分别为γ(s/g)、γ(l/g)及γ(s/l)，则三种润湿过程系统在定温、定压下吉布斯自由能的变化分别为：

① Adhesion wetting(沾湿)

$$\Delta G_{a,w}=\gamma(s\text{-}l)-[\gamma(s\text{-}g)+\gamma(l\text{-}g)]<0 \quad (9.4.2)$$

② Dipping wetting(浸湿)

$$\Delta G_{d,w}=\gamma(s\text{-}l)-\gamma(s\text{-}g)<0 \quad (9.4.3)$$

③ Spreading wetting(铺展)

$$\Delta G_{s,w}=[\gamma(s\text{-}l)+\gamma(l\text{-}g)]-\gamma(s\text{-}g)<0 \quad (9.4.4)$$

Spreading coefficient(铺展系数)

Definition:

$$s \stackrel{\text{def}}{=\!=} -\Delta G_{s,w}=\gamma(s\text{-}g)-[\gamma(s\text{-}l)+\gamma(l\text{-}g)] \quad (9.4.5)$$

So, the prerequisite of spreading wetting is $s>0$

From(9.4.2) to (9.4.4) we can see that for a given system:

$-\Delta G_{s,w}<-\Delta G_{d,w}<-\Delta G_{a,w}$

$\Delta G_{a,w}<\Delta G_{d,w}<\Delta G_{s,w}$

Conclusion:

If spreading wetting can produce spontaneously, then adhesion wetting and dipping wetting can produce spontaneously.

Example 9.3

At 20℃, the surface tension of water and mercury is 72.8mN · m^{-1} and 486.5mN · m^{-1}, while the interfacial tension of water and mercury is 375mN · m^{-1}. Please judge: (1) Can water spread on the surface of mercury?

(2) Can mercury spread on the surface of water?

Answer:

Spreading coefficient $S=-\Delta G_s=\gamma^s-\gamma^{ls}-\gamma^l$

(1) water on the surface of mercury(mercury-s, water-l)

$S=\gamma(\text{mercury})-\gamma(\text{water-mercury})-\gamma(\text{water})$

$=(486.5-375-72.8)\,\text{mN}\cdot\text{m}^{-1}=38.7\,\text{mN}\cdot\text{m}^{-1}>0$

Water can spread on the surface of mercury spontaneously.

(2) mercury on the surface of water(mercury-l, water-s)

$S=\gamma(\text{water})-\gamma(\text{water-mercury})-\gamma(\text{mercury})$

$=(72.8-375-486.5)\,\text{mN}\cdot\text{m}^{-1}=-788.3\,\text{mN}\cdot\text{m}^{-1}<0$

Mercury can spread on the surface of water spontaneously.

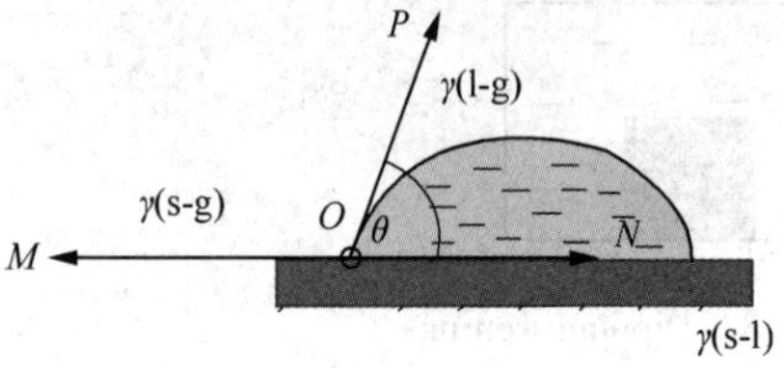

Fig. 9.10 Contact angle

9.4.2 Contact angle(接触角)

A liquid on a solid surface has a behavior dictated by three interfaces: the one between liquid and solid, the one between liquid and vapor, and the one between vapor and solid. The tangential angle that the liquid's edge makes with the surface is defined as the contact angle θ(As shown in figure 9.10).

This equation is

$$\gamma(\text{s-g})=\gamma(\text{s-l})+\gamma(\text{l-g})\cos\theta \tag{9.4.6}$$

(9.4.6) is called the Young's equation. It can actually be considered as a balance of three vectors: the solid-liquid interfacial tension pulling in one direction, and the liquid-solid and liquid-vapor tensions pulling in the other direction. But in the case of the liquid-vapor surface tension, only its component along the liquid-solid interface contributes to the balance of forces. The term $\cos\theta$ accounts for that component.

$\theta<90°$—wetting; $\theta>90°$—not wetting.

Example 9.4

At 293.15K, the surface tensions of ether-water, ether-Hg and water-Hg are $0.0107\,\text{N}\cdot\text{m}^{-1}$, $0.379\,\text{N}\cdot\text{m}^{-1}$, $0.375\,\text{N}\cdot\text{m}^{-1}$. If a drop of water was placed on the ether-Hg interface, please calculate the contact angle.

Answer:

Supposing γ_1, γ_2, γ_3 are the surface tensions of ether-Hg, ether-water and water-Hg respectively, then we can get

$$\gamma_1=\gamma_2\cos\theta+\gamma_3$$

$$\cos\theta=\frac{\gamma_1-\gamma_3}{\gamma_2}=\frac{0.379-0.375}{0.0107}=0.3738$$

$$\theta=68.05°$$

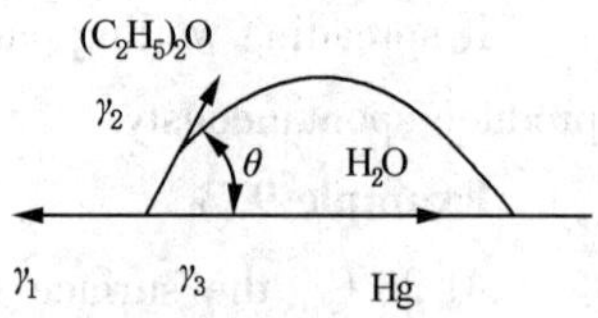

9.5 Adsorption phenomenon of solution surface (溶液界面上的吸附)

9.5.1 Adsorption phenomenon of solution surface(溶液表面上的吸附)

溶质在溶液的表面层(或表面相)中的浓度与它在溶液本体中的浓度不同的现象，称为溶液表面的吸附。对于纯液体，无所谓吸附，恒温、恒压下，表面张力为定值。对于溶液来讲，由于表面效应，溶质在表面层的浓度与在本体中的浓度不同，从而能改变溶液的表面张力，所以溶液的表面张力是温度、压力与浓度的函数。在一定温度的纯水中，加入不同种类的溶质时，溶质浓度对表面张力影响，大致有如图 9.11 所示三种类型。

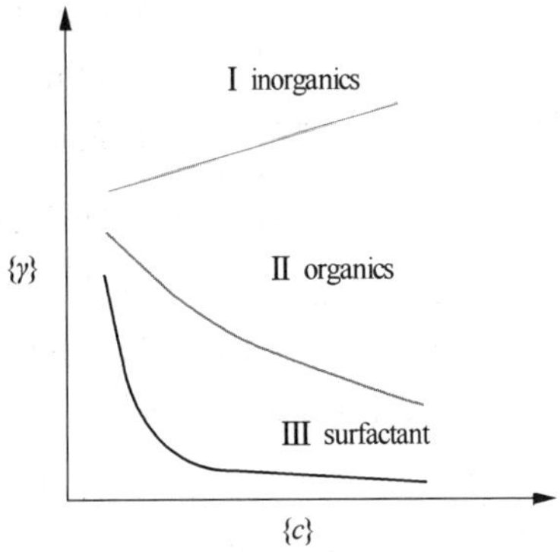

Fig. 9.11 γ vs. c

Three types of surface adsorption(figure 9.11):

Type Ⅰ: As the concentration increases, the surface tension of the solution increases slightly.

For Example: salts, non-volatile acids and bases, sucrose etc.

Type Ⅱ: As the concentration increases, the surface tension of the solution decreases slowly.

For Example: nonionic solvable organic molecule with low molecular weight/short chains and containing polar groups such as hydroxyl, amine groups, etc.

Type Ⅲ: As the concentration increases, the surface tension of the solution decreases sharply at low concentration. But the surface tension keeps steady when the concentration reaches to a certain degree.

Ionic/nonionic solvable organic molecule with high molecular weight/long chains and containing polar ionic groups such as $—COO^-$, $—SO^{3-}$, $—NR^{4+}$, etc.

For example: the sodium salts of long-chain fatty acids($n>10$) and sodium dodecyl sulfate.

产生溶液表面吸附现象的原因是:恒温恒压下，溶液表面吉布斯函数有自动减小的趋势。因为，在恒温恒压下，$dG=d(\gamma A)=Ad\gamma+\gamma dA$，当面积不变，$dA=0$ 时，要使 $dG=Ad\gamma<0$，只能由表面张力的减小来达到。

Positive adsorption—solute concentration of interface is higher than that of the bulk phase. (溶质在表面层浓度大于本体浓度，形成所谓的正吸附。)

Negative adsorption—solute concentration of interface is lower than that of the bulk phase. (溶质在表面层的浓度低于本体浓度。这种现象即为负吸附。)

Surface active agent—the substances that can drastically lower the surface tension of water even at low concentrations are called surface-active compounds/agent or surfactant. (把能显著降低液体表面张力的物质称为该液体的表面活性剂)。

There are two different models of systems in which surface effects are significant. Guggenheim in 1940 treated the interfacial layer as a three-dimensional phase having certain volume. Gibbs in

1878 replaced the actual system by hypothetical one in which the presence of the interface region is allowed for by a two-dimensional properties. Compared with the Gibbs model, the Guggenheim method is easier to visualize and more closely corresponds to the actual physical situation. However, the Gibbs method is more widely used and is the one we shall adopt.

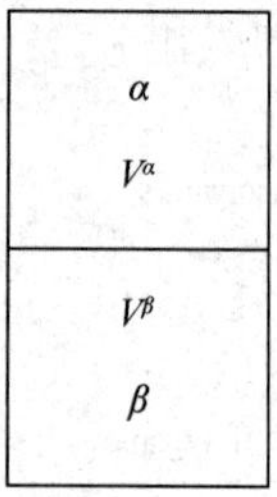

Fig. 9. 12 The Gibbs model system

Figure 9. 12 shows the Gibbs model system. In the model system, phases α and β are separated by a surface of zero thickness, the Gibbs dividing surface; phases α and β on either side of this dividing surface are defined to have the same concentration. c_B^α and c_B^β are the concentrations of component B in phases α and β; V^α and V^β are the volumes of phases α and β; n_B^α and n_B^β are the numbers of moles of component B in phases α and β, n_B is the total numbers of moles of component B in the system. The quantity n_B^σ, called the surface excess amount of component B, can be positive, negative, or zero. The definition states that the surface excess amount n_B^σ is difference between the total amount of B in the real system and the amount of B that would be in the system if the homogeneity of the bulk phases α and β persisted right up to the dividing surface.

$$n_B^\sigma \overset{\text{def}}{=\!=} n_B - (c_B^\alpha V^\alpha + c_B^\beta V^\beta) \tag{9.5.1}$$

The surface excess Γ_B of component B is defined as

$$\Gamma_B \overset{\text{def}}{=\!=} \frac{n_B^\sigma}{A_s} \tag{9.5.2}$$

Unit: mol · m^{-2}.

9. 5. 2 Gibbs adsorption isotherm(吉布斯吸附等温式)

$$\Gamma = -\frac{a}{RT}\frac{d\gamma}{da} \tag{9.5.3}$$

Γ—surface excess concentration of substance, unit: mol · m^{-2};

γ—surface tension;

a—activity of solute in solution.

For dilute solution $c \approx a$, (9. 5. 3) becomes

$$\Gamma = -\frac{c}{RT}\frac{d\gamma}{dc} \tag{9.5.4}$$

Equation (9. 5. 3) and (9. 5. 4) states that Γ is positive if the surface tension decreases with increasing solute concentration and is negative if the surface tension increases with increasing solute concentration

Conclusion:

$\frac{d\gamma}{dc} > 0$ $\Gamma < 0$, negative adsorption

$\frac{d\gamma}{dc} < 0$ $\Gamma > 0$, positive adsorption

Example 9. 5

The concentration c dependence of surface tension γ of alcohol aqueous solution at 25℃ is:

$$\gamma/10^{-3}\text{N}\cdot\text{m}^{-1}=72-0.5(c/c^{\ominus})+0.2(c/c^{\ominus})^2(c^{\ominus}=1.0\text{mol}\cdot\text{dm}^{-3})$$

Calculate the surface adsorption quantity when $c=0.1\text{mol}\cdot\text{dm}^{-3}$ and $c=0.5\text{mol}\cdot\text{dm}^{-3}$.

***Answer*:**

According to Gibbs adsorption isotherm $\Gamma=-\dfrac{c}{RT}\dfrac{\text{d}\gamma}{\text{d}c}$

$$\frac{\text{d}\gamma}{\text{d}c}=\{-0.5+[0.4(c/c^{\ominus})]\}\times10^{-3}$$

when $c=0.1\text{mol}\cdot\text{dm}^{-3}$

$$\Gamma=-\frac{0.1\text{mol}\cdot\text{dm}^{-3}\times(-0.5+0.4\times0.1)\times10^{-3}\text{N}\cdot\text{m}^{-1}}{8.314\text{J}\cdot\text{mol}^{-1}\cdot\text{K}^{-1}\times298.15\text{K}}=18.6\times10^{-9}\text{mol}\cdot\text{m}^{-2}$$

When $c=0.5\text{mol}\cdot\text{dm}^{-3}$,

$$\Gamma=-\frac{0.5\text{mol}\cdot\text{dm}^{-3}\times(-0.5+0.4\times0.5)\times10^{-3}\text{N}\cdot\text{m}^{-1}}{8.314\text{J}\cdot\text{mol}^{-1}\cdot\text{K}^{-1}\times298.15\text{K}}=60.9\times10^{-3}\text{mol}\cdot\text{m}^{-2}$$

9. 5. 3 Surface active agent(表面活性剂)

(1) Structure features of surface active agent(表面活性剂的结构特征)

Surface active agent(surfactant) is composed of polar group(hydrophilic group) and non-polar group(hydrophobic group). As shown in figure 9. 13.

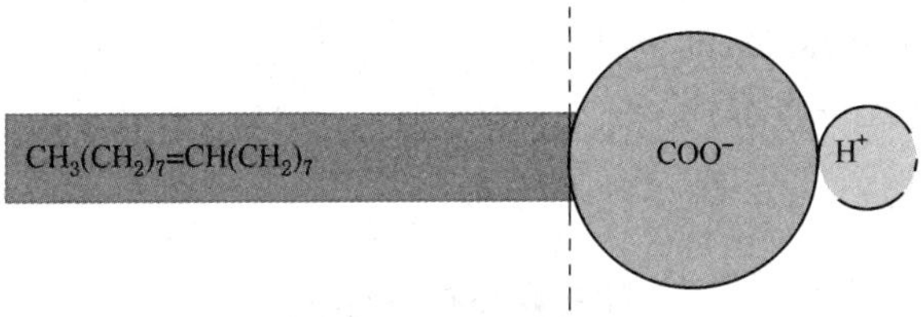

Fig. 9. 13 Structure features of surface active agent

Therefore, when the surfactant molecules are added to the water, in order to escape the water, the hydrophobic groups form two arrangements: a. The hydrophobic group is pushed out of the water into the air, the hydrophilic group remains in the water, that is, the surfactant molecules were oriented at the interface and formed monolayer surface films. b. The surfactant molecules dispersed in the water self-bind with their non-polar parts, known as associated colloids or micelles with hydrophobic group to the inside, hydrophilic group outward.

Micelles: spherical, lamellar and rod-like. As shown in figure 9. 14 and 9. 15.

(2) Types of surface active agent(表面活性剂的分类)

① Ionic surface active agent(离子型表面活性剂)

(ⅰ) Anionic surfactant (阴离子表面活性剂)

e.g. RCOONa

(ⅱ) Cationic surfactant(阳离子表面活性剂)

e.g. $C_{19}H_{37}NH_3{}^+Cl^-$

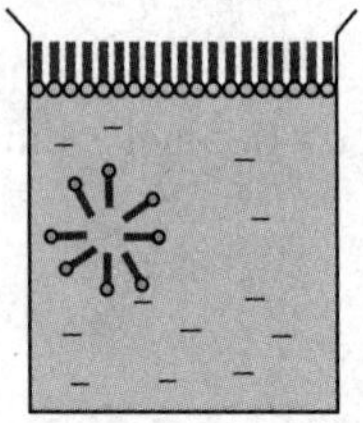

Fig. 9.14 Two arrangements of the surfactant molecules in water

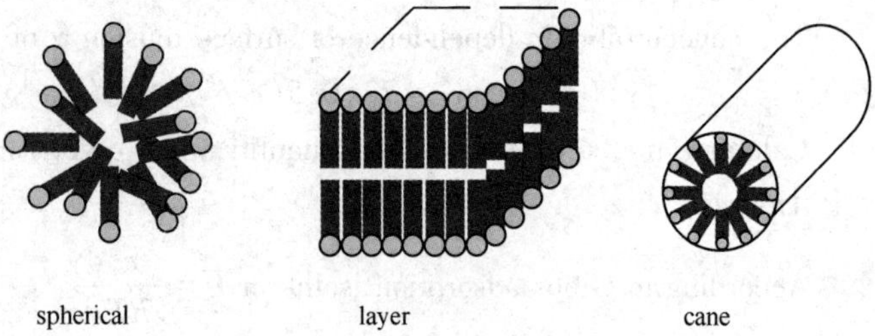

Fig. 9.15 Types of micelles

(ⅲ) Amphoteric ionic surfactant(两性表面活性剂)

e.g. R—NH—CH_2COOH(氨基酸型)

② Nonionic surface active agent(非离子型表面活性剂)

e.g. $HOCOH_2[CH_2OCH_2]_nCH_2OH$(聚乙二醇类)

Action of surface active agent(表面活性剂的应用):

① Wetting action(润湿作用)

Wetting agent, penetrating agent.

② Emulsification, dispersed action, solubilization(乳化、分散、增溶作用)

Emulsifier, dispersant, solvent increasing.

③ Foaming action, do away with foam(发泡作用、消泡作用)

Foaming agent, defoaming agent.

④ Washing action(洗涤作用)

Detergent.

Supplementary Examples of Chapter 9

EXERCISES
(习题)

1. At 293.15K and 101.325kPa, calculate the surface Gibbs function change when the radius of Hg was decreased from 1×10^{-3} m to 1×10^{-9} m. The surface tension of Hg is 0.47N · m^{-1} at 293.15K.

Answer: 5.906J

2. At 293.15K, the saturated vapor pressure of water is 2.337kPa, the density is 998.3kg · m^{-3}, the surface tension is 72.75×10^{-3} N · m^{-1}, Calculate the saturated vapor pressure of water drop whose radius is 10^{-9} m.

Answer: 6.865kPa

3. At 20℃, the benzene vapor condenses to spherical fog drops, the radius of the drop is 10^{-6}m. For benzene, $\gamma = 28.9\times10^{-3}$ N · m^{-1} and $\rho = 879$kg · m^{-3} at 20℃, the normal boiling point of benzene is 80.1℃, $\Delta_{vap}H_m^* = 33.9$kJ · mol^{-1}, calculate the saturated vapor pressure of fog drop of benzene.

Answer: 9526Pa

4. At 298. 15K, a little surface-active substance was dissolved in water. When the surface adsorption of the solution became in equilibrium state, it is measured that the concentration was 0. 2mol · m^{-3}. Quickly shaving off the slice of surface, measured that the adsorption quantity of surface-active substance in the surface slice was 3×10^{-6}mol · m^{-2}. Known that the surface tension of water is 72mN · m^{-1} at 298. 15K. Assuming that in a dilute concentration the surface tension and the concentration of the solution is linear relationship. Calculate the surface tension of the solution.

Exercise 9. 4

Answer: $\gamma=0.06456$N · m^{-1}

5. At 292. 15K, the surface tension of butyric acid aqueous solution can be expressed as follows:

$\gamma=\gamma_0-a\ln(1+bc)$, γ_0 is the surface tension of pure water, a and b are constants.

(1) Try to find the relationship between surface adsorption quantity and concentration of butyric acid in this solution.

(2) If $a=13.1$mN · m^{-1}, $b=19.62$$dm^3$ · mol^{-1}, calculate Γ when $c=0.200$mol · dm^{-3}.

(3) If the concentration of butyric acid is very high, that is $bc\gg1$, what is the saturated adsorption quantity Γ_m? Assuming that butyric acid is unimolecular layer adsorbed on the surface, calculate the transaction areas of every butyric acid molecular.

Answer: (1) $\Gamma=\dfrac{abc}{RT(1+bc)}$

(2) $\Gamma=4.298\times10^{-6}$mol · m^{-2}

(3) $\Gamma_m=5.393\times10^{-6}$mol · m^{-2}, $a_m=0.308$$nm^2$

Chapter 10 Chemical kinetics
(化学动力学)

Leaning objectives:

(1) Define the *rate of reaction* in terms of concentrations.

(2) Define the *rate coefficient* of a reaction.

(3) Define the *order of a reaction*.

(4) Integrate first-order and second-order *rate laws* for concentration as a function of time.

(5) Define the *half-life* of a reaction and relate it to *reaction order*.

(6) Distinguish between *order and molecularity of a reaction*.

(7) Solve the *rate laws for reactions* involving equilibrium and relate the *equilibrium constant* to *the rate*.

This chapter introduces the principles of chemical kinetics. It begins with a discussion of the definition of reaction rate and outlines the techniques for its measurement. The results of such measurements show that reaction rates depend on the concentration of reactants (and products) in characteristic ways that can be expressed in terms of differential equations known as rate laws. The construction of a rate law from a proposed mechanism and its comparison with experiment is main part of this chapter. Simple mathematical expressions tell us how fast a particular chemical reaction will proceed. Rate laws that have similar mathematical forms imply that their reactions behave a certain way as the reaction proceeds in time; we will consider some of those behaviors. Using the tools of calculus, we will be able to derive some simple expressions that will help us predict amounts of reactants and products of reactions that have particular rate laws.

A central part of a rate law for simple elementary steps have simple rate laws, and these rate laws can be combined together by invoking-one or more approximations. These approximations include the concept of the rate determining step of a reaction, the steady-state approximation method and equilibrium-state approximation method.

The rate of a chemical reaction might depend on variables under our control, such as the pressure, the temperature, and the presence of a catalyst, and we may be able to optimize the rate by the appropriate choice of conditions. The study of reaction rates also leads to an understanding of the mechanisms of reactions, their analysis into a sequence of elementary steps.

10.1 Reaction rates and rate equations of chemical reactions
(化学反应的反应速率及速率方程)

Chemical kinetics, also called reaction kinetics, is the study of the rates and mechanism of chemical reactions.

A homogeneous reaction is one that occurs entirely in one phase. A hetergeneous reaction is one that involves species present in two or more phases.

10.1.1 Definition of reaction rate(反应速率的定义)

Consider the reaction

$$aA+bB+\cdots = yY+zZ+\cdots \tag{10.1.1}$$

Where a, b, y, $z\cdots$ are the coefficients in the balanced chemical equation and A, B, Y, Z $\cdots$ are the chemical species.

(10.1.1) can be written as
$$0=\sum_B \nu_B B \tag{10.1.2}$$

The rate of conversion $\dot{\xi}$ for the reaction is defined as

$$\dot{\xi} \overset{\text{def}}{=\!=} \frac{d\xi}{dt} \tag{10.1.3}$$

unit: $mol \cdot s^{-1}$.

where, $\dot{\xi}$— rate of conversion; ξ— reaction extent; t — reaction time

For
$$d\xi = \frac{dn_B}{\nu_B}$$

$$\dot{\xi} \overset{\text{def}}{=\!=} \frac{d\xi}{dt} = \frac{1}{\nu_B}\frac{dn_B}{dt} \tag{10.1.4}$$

The conversion rate $\dot{\xi}$ is an extensive quantity and depends on the system's size. The conversion rate per unit volume $\dot{\xi}/V$ is called the rate of the reaction v.

$$v \overset{\text{def}}{=\!=} \frac{\dot{\xi}}{V} = \frac{1}{\nu_B V}\frac{dn_B}{dt} \tag{10.1.5}$$

Unit: $mol \cdot m^{-3} \cdot s^{-1}$.

v is an intensive quantity and depends on T, p, and the concentrations in the system. In most system studied, the volume is either constant or changes by a negligible amount. When V is a constant, we have $\frac{dn_B}{V} = dc_B$, thus(10.1.5)can be written as

$$v = \frac{1}{\nu_B}\frac{dc_B}{dt} \text{const. } V \tag{10.1.6}$$

Dissipate rate of reacant A is defined as

$$v_A = \frac{-dc_A}{dt} \tag{10.1.7}$$

Increase rate of product Z is defined as

$$v_Z = \frac{dc_Z}{dt} \tag{10.1.8}$$

(The rate should be positive, but the reactants decrease with time, that is, $\frac{dc_A}{dt}<0$. $v_A = \frac{-dc_A}{dt}>0$)

Rate of the reaction v can be expressed by any reactants or products. The reaction rate v is independent of the selection of substance B, so v does not need subscripts. The rate of consumption of the reactants or the rate of formation of the products varies with the selection of substance B. To avoid confusion, the selected substance A or Z is indicated with subscripts, e. g. v_A or v_Z.

The relationship between the reaction rate v, dissipate rate of A (v_A) and increase rate of Z (v_Z) is as follows

$$v=\frac{1}{\nu_A}\frac{dc_A}{dt}=\frac{1}{\nu_B}\frac{dc_B}{dt}=\frac{1}{\nu_Y}\frac{dc_Y}{dt}=\frac{1}{\nu_Z}\frac{dc_Z}{dt} \tag{10.1.9}$$

$$v=\frac{v_A}{-\nu_A}=\frac{v_B}{-\nu_B}=\cdots=\frac{v_Y}{\nu_Y}=\frac{v_Z}{\nu_Z} \tag{10.1.10}$$

$$=\frac{v_A}{|\nu_A|}=\frac{v_B}{|\nu_B|}=\cdots=\frac{v_Y}{|\nu_Y|}=\frac{v_Z}{|\nu_Z|} \tag{10.1.11}$$

If the gas reaction should be carried out under constant temperature and constant volume, the reaction rate, dissipate rate, increase rate, etc., can also be defined by partial pressure. In order to distinguish, p is often added to the subindex.

Reaction rate
$$v_p=\left(\frac{1}{\nu_B}\right)\frac{dp_B}{dt} \tag{10.1.12}$$

Dissipate rate of reacant A
$$v_{p,A}=-\frac{dp_A}{dt} \tag{10.1.13}$$

Increase rate of product Z
$$v_{p,Z}=\frac{dp_Z}{dt} \tag{10.1.14}$$

$$v_p=\frac{1}{\nu_A}\frac{dp_A}{dt}=\frac{1}{\nu_B}\frac{dp_B}{dt}=\cdots=\frac{1}{\nu_Y}\frac{dp_Y}{dt}=\frac{1}{\nu_Z}\frac{dp_Z}{dt} \tag{10.1.15}$$

From
$$p_B=\frac{n_BRT}{V}=c_BRT \quad dp_B=RTdc_B$$

We can get
$$v_p=vRT \tag{10.1.16}$$

10.1.2 Elementary reaction and overall reaction(基元反应和总包反应)

Equation $H_2+I_2\rightarrow 2HI$ gives the overall stoichiometry of the reaction but does not tell us the process or mechanism by which the reaction actually occurs. The reaction is believed to occur by the following three steps:

(1) $I_2+M^0\rightarrow I\cdot+I\cdot+M_0$

(2) $H_2+I\cdot+I\cdot\rightarrow HI+HI$

(3) $I\cdot+I\cdot+M_0\rightarrow I_2+M^0$

The overall stoichiometry of the three steps is $H_2+I_2\rightarrow 2HI$ and does not contain the intermediate specie $I\cdot$. It is formed in the first step of the mechanism and consumed in the subsequent steps.

Elementary reaction—each step in the mechanism of a reaction is called an elementary reaction. (一个宏观的反应过程中的每一个简单的反应步骤，就是一个基元反应。)

A simple reaction consists of a single elementary step. A complex reaction consists of two or more elementary steps.

10.1.3 Rate equation of elementary reaction—law of mass action(基元反应的速率方程——质量作用定律)

The expression for v as a function of concentrations at fixed temperature is called the rate equation of the reaction.

The number of molecules that react in an elementary step is the molecularity of the elementary reaction. Molecularity is defined only for elementary reactions and should not be used to describe overall reactions that consist of more than one elementary step. The elementary reaction A→products is unimolecular. The elementary reaction A+B→products and 2A→products are bimolecular. The elementary reaction A+B+C→products, 2A+B→products, and A+2B→products are trimolecular. No elementary reactions involving more than three molecules are known, because of the very low probability of near-simultaneous collision of more than three molecules. Most elementary reactions are unimolecular or bimolecular reactions, three molecular reactions being uncommon because of the low probability of three-body collisions.

In summary, the rate equation for the elementary aA+bB→products reaction is $v=-\frac{dc_A}{dt}=k\cdot c_A^a\cdot c_B^b$, where $a+b$ is 1, 2 or 3. For an elementary reaction, the orders in the rate law are determined by the reaction's stoichiometry.

For elementary reaction aA+bB ⟶ yY+zZ

$$v=\frac{dc_B}{\nu_B dt}=k\cdot c_A^a\cdot c_B^b \quad (10.1.17)$$

$$v_A=-\frac{dc_A}{dt}=k_A\cdot c_A^a\cdot c_B^b \quad (10.1.18)$$

$$v_B=-\frac{dc_B}{dt}=k_B\cdot c_A^a\cdot c_B^b \quad (10.1.19)$$

$$v_Y=\frac{dc_Y}{dt}=k_Y\cdot c_A^a\cdot c_B^b \quad (10.1.20)$$

$$v_Z=\frac{dc_Z}{dt}=k_Z\cdot c_A^a\cdot c_B^b \quad (10.1.21)$$

k_A, k_B, k_Y, k_Z are independent of the concentrations, but usually dependent on the temperature.

$$\nu_A:\nu_B:\nu_Y:\nu_Z=a:b:y:z$$

$$\Rightarrow k_A:k_B:k_Y:k_Z=a:b:y:z$$

10.1.4 General form of rate equation, order of reaction(化学反应速率方程的一般形式，反应级数)

For many overall reactions aA+bB+⋯⟶yY+zZ+⋯, v is found to have the form

$$v_A=-\frac{dc_A}{dt}=k_A c_A^{n_A} c_B^{n_B}\cdots \quad (10.1.22)$$

Where the exponents n_A, $n_B\cdots$ are called the order with respect to A, order with respect to B. The sum $n_A+n_B+\cdots=n$ is called the overall order of the reaction. The proportionality constant

k, called the rate constant or rate coefficient, is a function of temperature. k has the units $(\text{mol}\cdot\text{m}^{-3})^{1-n}\cdot\text{s}^{-1}$.

From
$$v=\frac{v_A}{-\nu_A}=\frac{v_B}{-\nu_B}=\cdots=\frac{v_Y}{\nu_Y}=\frac{v_Z}{\nu_Z} \tag{10.1.23}$$

We can get
$$\frac{k_A}{-\nu_A}=\frac{k_B}{-\nu_B}=\cdots=\frac{k_Y}{\nu_Y}=\frac{k_Z}{\nu_Z} \tag{10.1.24}$$

For the overall reaction the order of the reaction differs from the coefficient in the chemical reaction. The rate equation must be determined from measurement of reaction rates and cannot be deduced from the reaction stoichiometry.

Some experimentally observed rate equations for the reactions are shown as follows

$$H_2+I_2 \longrightarrow 2HI \qquad \frac{d[HI]}{dt}=k[H_2][I_2]$$

$$H_2+Cl_2 \longrightarrow 2HCl \qquad \frac{d[HCl]}{dt}=k[H_2][Cl_2]^{1/2}$$

$$H_2+Br_2 \longrightarrow 2HBr \qquad \frac{d[HBr]}{dt}=\frac{k[H_2][Br_2]^{1/2}}{1+k'[HBr]/[Br_2]}$$

对于非基元反应，反应级数情况比较复杂，它可以是正数(一级、二级、三级等)，还可以是零级、分数级(如1/2级、3/2级等)，甚至速率方程中还可能出现反应产物的浓度项。

10.1.5 Gas phase reaction(气相反应)

For gas phase reaction, the reaction rate v can also be expressed as a function of pressure.

For example, the rate equation of nth order reaction $aA \longrightarrow p$ can be expressed as

$$\frac{-dc_A}{dt}=kc_A^n \tag{10.1.25}$$

or

$$\frac{-dp_A}{dt}=k_p p_A^n \tag{10.1.26}$$

If A is perfect gas

$$p_A=\frac{n_A}{V}RT=c_ART \quad \text{const. } V,\ T \tag{10.1.27}$$

Substitute(10.1.27) into(10.1.26), we can get

$$-\frac{dp_A}{dt}=-\frac{dc_A}{dt}\cdot(RT)=k_p\cdot p_A^n=k_p\cdot(c_ART)^n$$

$$-\frac{dc_A}{dt}=k_p\cdot c_A^n(RT)^{n-1}=k\cdot c_A^n$$

$$k=k_p(RT)^{n-1} \tag{10.1.28}$$

Where k and k_p are the rate coefficients of the reaction expressed by concentration and pressure respectively.

10.2 Integral form of rate equation (速率方程的积分形式)

10.2.1 Zeroth-order reaction[零级反应(n =0)]

Definition: A zeroth-order reaction is one for which, at a given temperature, the rate of the reaction depends only on the zero power of the concentration of a single reacting species. (定义:若反应 A→产物的反应速率与反应物 A 浓度的零次方成正比,则该反应为零级反应。)

The zeroth-order rate equation can be written as

$$v_A = \frac{-dc_A}{dt} = k_A c_A^0 = k_A \tag{10.2.1}$$

If the initial concentration, at time $t=0$, is $c_{A,0}$ and if at some later time t, the concentration has fallen to c_A, the integration gives

$$-\int_{c_{A,0}}^{c_A} dc_A = k_A \int_0^t dt$$

and

$$c_{A,0} - c_A = k_A t \tag{10.2.2}$$

$$c_A \cdot x_A = k_A t \tag{10.2.3}$$

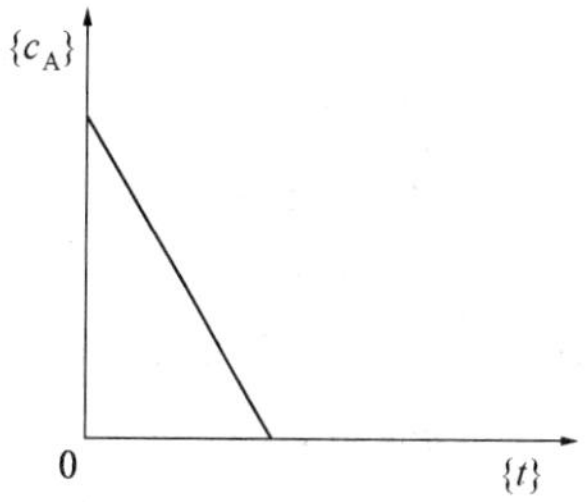

Fig. 10.1 The c_A versus t

Where, x_A is the degree of dissociation of reactant A

$$x_A \stackrel{\text{def}}{=\!=} \frac{n_{A,0} - n_A}{n_{A,0}} = \frac{c_{A,0} - c_A}{c_{A,0}} \tag{10.2.4}$$

Characteristics of zeroth-order reactions:

① Unit of k_A is$[c] \cdot [t]^{-1}$.

② A plot of the time dependent of $\{c_A\}$ is a straight line, the slop = $-k_A$; the intercept = $\{c_{A,0}\}$ (see figure 10.1).

③ Half-life.

Half-life—the time a chemical reaction takes for the concentration of a reactant to fall to half its initial value, that is, the time of $c_A = 0.5\ c_{A,0}$ or $x_A = 0.5$. (定义:c_A变为 $c_{A,0}$一半所需的时间 t 为 A 的半衰期,用 $t_{1/2}$表示。)

For zeroth-order reaction, $t_{1/2} = c_{A,0}/2k$, is proportion to $c_{A,0}$.

10.2.2 First-order reaction[一级反应(n=1)]

Definition: A first-order reaction is one for which, at a given temperature, the rate of the reaction depends only on the first power of the concentration of a single reacting species. (定义:若反应 A→产物的反应速率与反应物 A 浓度的一次方成正比,则该反应为一级反应。)

The first-order rate equation can be written as

$$v_A = \frac{-dc_A}{dt} = k_A c_A \tag{10.2.5}$$

If the initial concentration, at time t =0, is $c_{A,0}$ and at some later time t the concentration has fallen to c_A, the integration gives

$$-\int_{c_{A,0}}^{c_A}\frac{dc_A}{c_A}=k_A\int_0^t dt \tag{10.2.6}$$

and

$$\ln\frac{c_{A,0}}{c_A}=k_A t \tag{10.2.7}$$

or

$$c_A=c_{A,0}e^{-k_A t} \tag{10.2.8}$$

$$\ln\frac{1}{1-x_A}=k_A t \tag{10.2.9}$$

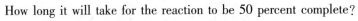

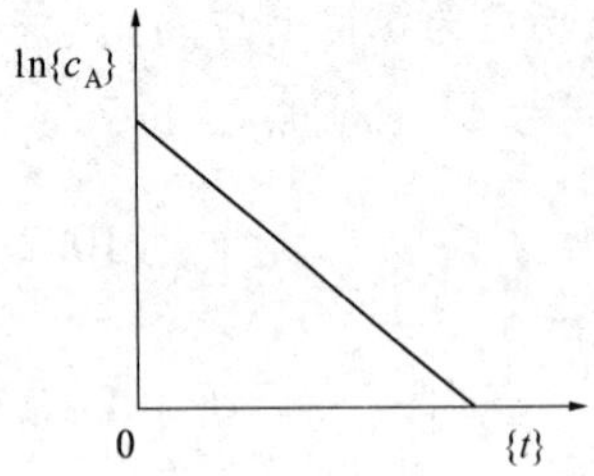

Fig. 10. 2 The In $\{c_A\}$ versus t

Characteristics of first-order reactions:

① Unit of k_A is$[t]^{-1}$. e. g. h^{-1}, min^{-1}, s^{-1}.

② A plot of the time dependent of $\ln\{c_A\}$ is a straight line, the slop $=-k_A$; the intercept $=\ln\{c_{A,0}\}$ (see figure 10. 2).

③ $t_{1/2}=\frac{\ln 2}{k_A}=\frac{0.693}{k_A}$, is independent of $c_{A,0}$.

Example 10. 1

The first-order reaction is 30 percent complete after 10 min. How long it will take for the reaction to be 50 percent complete?

***Answer*:**

Integral rate equation of first-order reaction

$$\ln\frac{1}{1-x}=kt$$

Substitution of $t=10\text{min}$, $x=0.3$ in the former equation, we obtain

$$\ln\frac{1}{1-0.3}=k\cdot 10$$

$$k=0.03567\text{min}^{-1}$$

When $x=0.5$

$$t=\frac{\ln 2}{k}=\frac{\ln 2}{0.03567}=19.4\text{min}$$

10. 2. 3 Second-order reaction[二级反应($n=2$)]

(1) The rate of the reaction is proportional to square of the concentrations of a single reacting species.

$$a\text{A}\rightarrow\text{products}$$

The second-order rate equation can be written as

$$\frac{-dc_A}{dt}=k_A c_A^2 \tag{10.2.10}$$

If the initial concentration, at time $t=0$, is $c_{A,0}$ and if at some later time t the concentration has fallen to c_A, the integration gives

$$-\int_{c_{A,0}}^{c_A}\frac{dc_A}{c_A^2}=k_A\int_0^t dt \tag{10.2.11}$$

and $$\frac{1}{c_A}-\frac{1}{c_{A,0}}=k_A t \qquad (10.2.12)$$

or $$\frac{x_A}{c_{A,0}(1-x_A)}=k_A t \qquad (10.2.13)$$

Characteristics of second-order reactions:

① Unit of k_A is $[c]^{-1}\cdot[t]^{-1}$, e. g. $m^3\cdot mol^{-1}\cdot s^{-1}$.

② A plot of the time dependent of $1/c_A$ is a straight line, the slop$=k_A$, the intercept$=1/c_{A,0}$(see figure 10.3).

③ $t_{1/2}=1/(kc_{A,0})$, $t_{1/2}$ is inversely proportional to $c_{A,0}$.

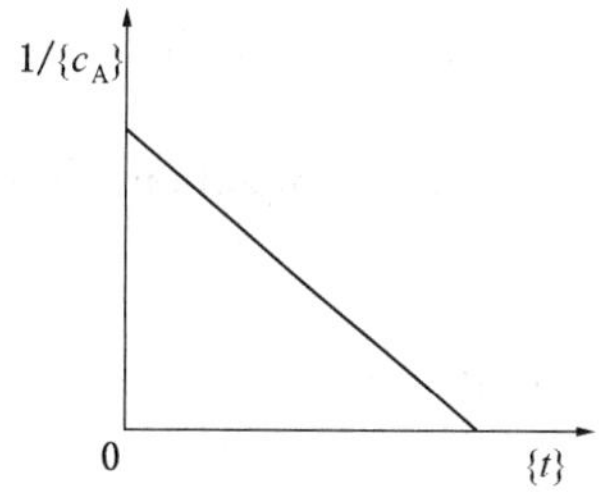

Fig. 10.3 The $1/\{c_A\}$ versus t

(2) Two reactants

Reaction $aA+bB\rightarrow$product

The rate of reaction is proportional to the product of the concentrations of two species of the reagents.

The differential reaction equation

$$\frac{-dc_A}{dt}=k_A c_A c_B \qquad (10.2.14)$$

① A and B are initially present in stoichiometric proportion, $c_{A,0}:c_{B,0}=a:b$

A and B are initially present in stoichiometric proportion, so that A and B will remain in stoichiometric proportion throughout the whole reaction, that is

$$\frac{c_A}{c_B}=\frac{a}{b} \quad \text{or} \quad c_B=\frac{bc_A}{a} \qquad (10.2.15)$$

Substitute(10.2.15)into(10.2.14)gives

$$\frac{-dc_A}{dt}=k_A c_A c_B=\frac{b}{a}k_A c_A^2=k'_A c_A^2\left(k'_A=\frac{b}{a}k_A\right) \qquad (10.2.16)$$

or $$\frac{-dc_B}{dt}=k_B c_A c_B=\frac{a}{b}k_B c_B^2=k'_B c_B^2\left(k'_B=\frac{a}{b}k_B\right) \qquad (10.2.17)$$

(10.2.16) integrates to

$$\frac{1}{c_A}-\frac{1}{c_{A,0}}=k'_A t \qquad (10.2.18)$$

(10.2.17)integrates to

$$\frac{1}{c_B}-\frac{1}{c_{B,0}}=k'_B t \qquad (10.2.19)$$

② One of reactant is present in large excess

The concentration of B is very much larger than that of A, $c_{B,0}\gg c_{A,0}$, so that in the whole reaction $c_B\approx c_{B,0}$

$$-\frac{dc_A}{dt}=k_A\cdot c_A\cdot c_B=k_A\cdot c_A\cdot c_{B,0}=k'_A\cdot c_A(k'_A=k_A\cdot c_{B,0})$$

$$\ln\frac{c_{A,0}}{c_A}=k'_A t \qquad (10.2.20)$$

$$\ln\frac{1}{1-x_A}=k'_A t \tag{10.2.21}$$

③ A and B are not initially present in stoichiometric proportion $\frac{c_{A,0}}{c_{B,0}}\neq\frac{a}{b}$

A and B are not initially present in stoichiometric proportion, so that A and B will not remain in stoichiometric proportions throughout the reaction: $\frac{c_A}{c_B}\neq\frac{a}{b}$

	aA	+	bB	$\longrightarrow$	yY + zZ
$t=0$	$c_A=c_{A,0}$		$c_B=c_{B,0}$		
$t=t$	$c_A=(c_{A,0}-c_{A,x})$		$c_B=\left(c_{B,0}-\frac{b}{a}c_{A,x}\right)$		
or $t=t$	$c_A=c_{A,0}(1-x_A)$		$c_B=\left(c_{B,0}-\frac{b}{a}c_{A,0}x_A\right)$		

$$-\frac{dc_A}{dt}=k_A(c_{A,0}-c_{A,x})\left(c_{B,0}-\frac{b}{a}c_{A,x}\right)$$

or

$$\frac{dx_A}{dt}=k_A(1-x_A)\left(c_{B,0}-\frac{b}{a}c_{A,0}x_A\right)$$

Integral form

$$t=\frac{1}{k_A\left(\frac{b}{a}c_{A,0}-c_{B,0}\right)}\ln\frac{(c_{A,0}-c_{A,x})c_{B,0}}{\left(c_{B,0}-\frac{b}{a}c_{A,x}\right)c_{A,0}} \tag{10.2.22}$$

$$t=\frac{1}{k_A\left(\frac{b}{a}c_{A,0}-c_{B,0}\right)}\ln\frac{c_{B,0}(1-x_A)}{\left(c_{B,0}-\frac{b}{a}c_{A,0}x_A\right)} \tag{10.2.23}$$

Example 10.2

A gas phase reaction $A(g)+2B(g)\rightarrow Y(g)$ generated in a closed container at a constant temperature of 400K, the rate equation was determined to be

$$\frac{-dp_A}{dt}=k_{p,A}p_A^{0.5}p_B^{1.5}$$

Example 10.2

Assuming there were only reactants A(g) and B(g) at the beginning of the reaction(the ratio of initial partial pressure was 1 : 2), the initial total pressure was 3.36kPa, the total pressure was 2.12kPa when the reaction time is 1000s. Please calculate the rate coefficient $k_{p,A}$, k_A and half-life $t_{1/2}$.

***Answer*:**

$$p_{A,0}:p_{B,0}=1:2=a:b\Rightarrow p_B=2p_A$$

Substitution of $p_B=2p_A$ in $\frac{-dp_A}{dt}=k_{p,A}p_A^{0.5}p_B^{1.5}$, we can get

$$\frac{-dp_A}{dt}=k_{p,A}p_A^{0.5}(2p_A)^{1.5}=2^{1.5}k_{p,A}p_A^2=k'_{p,A}p_A^2$$

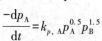

The integration gives $\frac{1}{p_A}-\frac{1}{p_{A,0}}=k'_{p,A}t$

$$A(g)+2B(g)\longrightarrow Z(g)$$

$t=0$	$p_{A,0}$	$2p_{A,0}$	0	$p_0=3p_{A,0}$
$t=t$	p_A	$2p_A$	$p_{A,0}-p_A$	$p_t=p_{A,0}+2p_A$

$p_0=3.36\text{kPa}$, so $p_{A,0}=1.12\text{kPa}$

$t=1000\text{s}$, $p_t=2.12\text{kPa}$

$p_A=(p_t-p_{A,0})/2=(2.12-1.12)/2=0.5\text{kPa}$

$k'_{p,A}=\frac{1}{t}\left(\frac{1}{p_A}-\frac{1}{p_{A,0}}\right)=\frac{1}{1000}\left(\frac{1}{0.5}-\frac{1}{1.12}\right)=1.107\times10^{-3}\text{kPa}^{-1}\cdot\text{s}^{-1}$

$k_{p,A}=2^{-1.5}k'_{p,A}=1.107\times10^{-3}/2^{1.5}=3.914\times10^{-4}\text{kPa}^{-1}\cdot\text{s}^{-1}$

From $k=k_p(RT)^{n-1}$

We can get $k_A=k_{p,A}(RT)^{n-1}=3.914\times10^{-4}\times400\times8.314=1.302\text{dm}^3\cdot\text{mol}^{-1}\cdot\text{s}^{-1}$

Half-life

$$t_{1/2}=\frac{1}{k'_{p,A}p_{A,0}}=\frac{1}{1.107\times10^{-3}\times1.12}=806.6\text{s}$$

10.2.4 *n*th-order reaction(*n* 级反应)

① aA→product;

② aA+bB+cC+…→product, A, B and C are initially present in stoichiometric proportion, so that A, B and C will remain in stoichiometric proportion throughout the whole reaction, that is

$$\frac{c_{A,0}}{a}=\frac{c_{B,0}}{b}=\frac{c_{C,0}}{c}=\cdots,\ \frac{c_A}{a}=\frac{c_B}{b}=\frac{c_C}{c}=\cdots$$

③ aA+bB+cC+…→product, the concentrations of B, C, …are very much larger than that of A, $c_{B,0}\gg c_{A,0}$, $c_{C,0}\gg c_{A,0}\cdots$ so that in the whole reaction $c_B\approx c_{B,0}$, $c_C\approx c_{C,0}\cdots$

The rate equation of the *n*th-order reaction can be expressed

$$\frac{-\mathrm{d}c_A}{\mathrm{d}t}=k'c_A^n \qquad (10.2.24)$$

n can be integers(0, 1, 2, 3…,) or half-integers(1/2, 3/2, …), if $n\neq1$, integration gives

$$-\int_{c_{A,0}}^{c_A}\frac{\mathrm{d}c_A}{c_A^n}=k\int_0^t\mathrm{d}t$$

$$\frac{1}{(n-1)}\left(\frac{1}{c_A^{n-1}}-\frac{1}{c_{A,0}^{n-1}}\right)=kt\quad(n\neq1) \qquad (10.2.25)$$

Characteristics of nth-order reactions

① Unit of k_A is$(\text{mol}\cdot\text{m}^{-3})^{1-n}\cdot\text{s}^{-1}$.

② A plot of the time dependent of $\frac{1}{c_A^{n-1}}$ is a straight line, the slop$=k_A$, the intercept$=\frac{1}{c_{A,0}^{n-1}}$(see figure 10.4).

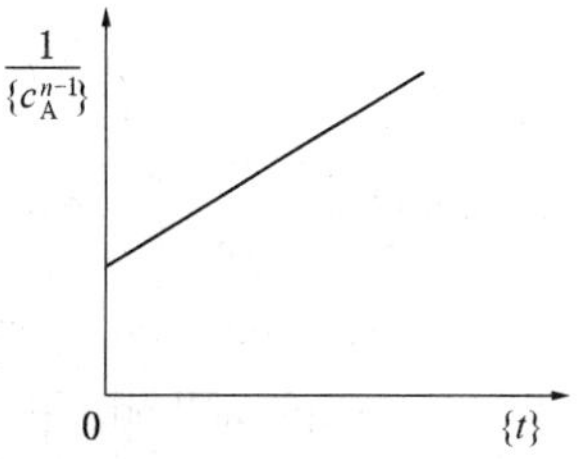

Fig. 10.4 The $1/c_A$ versus t plot

③ $t_{1/2}=\dfrac{2^{n-1}-1}{(n-1)kc_{A,0}^{n-1}}$, $t_{1/2}$ is inverse proportional to$(c_{A,0})^{n-1}$.

10.3 Determination of rate equation
(化学反应速率方程的建立)

10.3.1 c~t curve determination(物质的量浓度–时间曲线的实验测定)

$c_A \sim t$ curve or $x_A \sim t$ curve($c_A \sim t$ 曲线或 $x_A \sim t$ 曲线)

kinetic curve—$c_A \sim t$ curve is usually called kinetic curve.

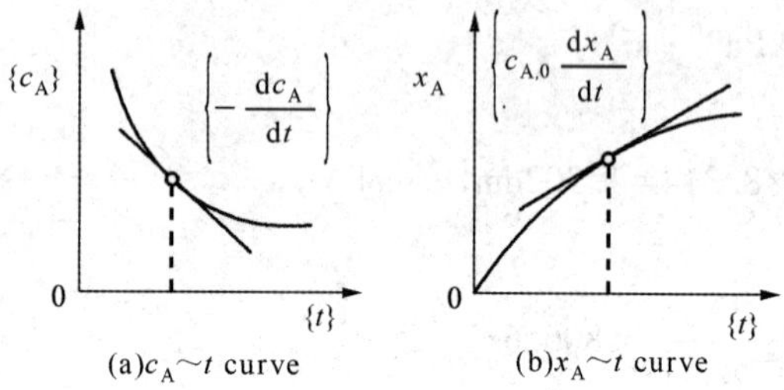

Fig. 10.5 Kinetic curve

The key subject of kinetic study is to measure the concentration of any species after arbitrary time intervals. The concentration of the species can be measured using either chemical or physical methods.

Plots of c_A vs. t or x_A vs. t are shown in figure 10.5.

The rate at a particular time can be obtained by computing the slope of a line tangent to the curve at that time point.

$$v_A=-\frac{dc_A}{dt}$$

10.3.2 Determination of the order of reaction(反应级数的确定)

This section discusses how the rate equation is found from experimental data. We restrict the discussion to cases where the rate law has the form

$$v_A=-\frac{dc_A}{dt}=k_A c_A^{n_A} c_B^{n_B}\cdots$$

It is usually best to find the orders n_A, $n_B\cdots$ first and then find the rate constant k. Three methods for finding the orders are as follows.

(1) Differential method(微分法)

From
$$\frac{-dc_A}{dt}=kc_A^n$$

We can get

$$\lg\left(\frac{-dc_A}{dt}\right)=\lg k+n\lg c_A$$

A plot of $\lg\left(\frac{-dc_A}{dt}\right)$ versus $\lg c_A$ is a straight line, the slop $=n$, the intercept $=\lg k$ (see figure 10.6)

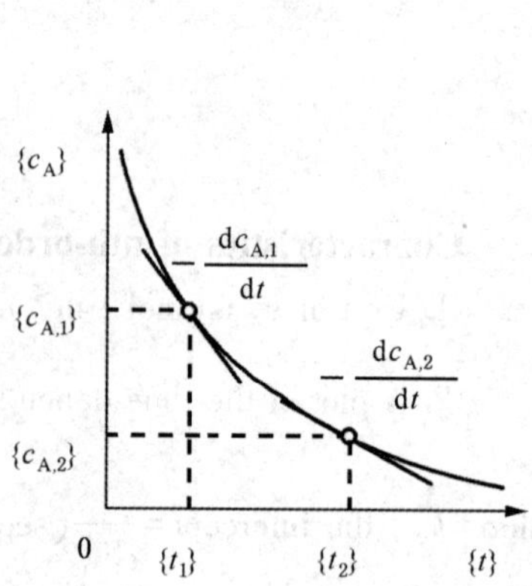

Fig. 10.6 $\{c_A\}\sim t$ curve

(2) Attempt method(尝试法)

尝试法又称试差法，适用于整级数反应。其做法有两种：

① Plot method(作图法)

The linear relationship of reaction with different order is different. One plots c_A vs. t, $\ln c_A$ vs. t and $1/c_A$ vs. t obtains a straight line for one of these plots according to whether $n=0$, 1, or 2.

Order	Linear relationship	Order	Linear relationship
zeroth	$c_A \sim t$	second	$1/c_A \sim t$
first	$\ln c_A \sim t$	nth	$1/c_A^{n-1} \sim t$

This method is dangerous to use, since it is often difficult to decide which of these plots is most nearly linear. Thus, the wrong order can be obtained.

② Substitution method(代入法)

The values of k can be calculated from the selected integrated equation from initial concentration(c_0) and the concentration(c) at various time intervals(t). If the reaction is of the selected order of reaction, the k at different intervals obtained will be the same.

(3) Half-life method(半衰期法)

From

$$t_{1/2}=\frac{2^{n-1}-1}{(n-1)kc_{A,0}^{n-1}}$$

We can get

$$\frac{(t_{1/2})_2}{(t_{1/2})_1}=\left(\frac{c_{A,0,1}}{c_{A,0,2}}\right)^{n-1}$$

$$n=1+\frac{\ln\{t_{1/2}\}_1-\ln\{t_{1/2}\}_2}{\ln\{c_{A,0,2}\}-\ln\{c_{A,0,1}\}}$$

(4) Isolated method

For $a\mathrm{A}+b\mathrm{B}+c\mathrm{C}\cdots\rightarrow$products, the rate equation is

$$-\frac{\mathrm{d}c_A}{\mathrm{d}t}=k_A c_A^{n_A} c_B^{n_B} c_C^{n_C}\cdots$$

Here, one makes the initial concentration of reactant A much less than the concentrations of all other species: $c_{B,0}\gg c_{A,0}$, $c_{C,0}\gg c_{A,0}\cdots$, etc. Then the concentration of all reactants except A will be essentially constant with time. The rate equation becomes

$$-\frac{\mathrm{d}c_A}{\mathrm{d}t}=k'_A c_A^{n_A},\ (k'_A=k_A c_{B,0}^{n_B} c_{C,0}^{n_C}\cdots)$$

where k_A is essentially constant. The reaction has the pseudo order under these conditions. One then analyzed the data from the run using the former method to find n_A. To find n_B we can proceed as we did in finding n_A.

10.4 Temperature dependence of rate equation, activation energy (温度对反应速率常数的影响，活化能)

Rate constants depend strongly on temperature, typically increasing rapidly with increasing temperature(figure 10.7). The effect of temperature on the rate of reaction is shown in figure 10.7.

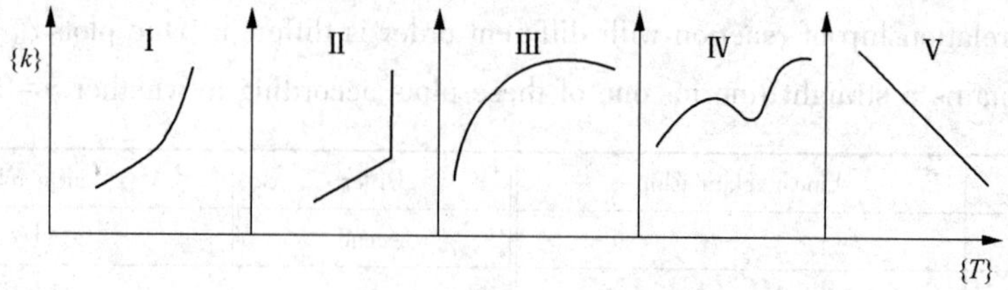

Fig. 10.7 The effect of temperature on the rate of reaction

Type Ⅰ

k increase exponentially with T. This kind of curve can be observed in most of the reactions.

Type Ⅱ

This kind of $k \sim T$ relation was observed in thermal explosions. At ignition temperature, the rate constant makes a sharp increase.

Type Ⅲ

This kind is usually encountered in the catalytic reaction that has an optimum temperature.

Type Ⅳ

This kind can be observed in the oxidation of carbon and gaseous oxidation of hydrocarbons.

Type Ⅴ

The example for this type V is $2NO + O_2 = 2NO_2$

10.4.1 Van't Hoff rule(范特霍夫规则)

Van't Hoff rule

$$k_{T+10K}/k_T \approx 2 \sim 4 \tag{10.4.1}$$

A rough rule, valid for many reactions in solution, is that near room temperature k doubles or four times for each 10℃ increase in temperature.

10.4.2 Arrhenius equation(阿伦尼乌斯方程)

In 1889 Arrhenius pointed out that the k(T) data for many reactions fit the expression

$$k = k_0 \exp\left(-\frac{E_a}{RT}\right) \tag{10.4.2}$$

k_0—pre-exponential parameter(指数前参量); E_a—activation energy(活化能).

Differential equation:

$$\frac{d\ln k}{dT} = \frac{E_a}{RT^2} \tag{10.4.3}$$

If E_a is independent of T, integration of(10.4.3) yields(10.4.4) and(10.4.5) ⋯.

Definite integral form

$$\ln\frac{k_{A,2}}{k_{A,1}} = \frac{E_a}{R}\left(\frac{1}{T_1} - \frac{1}{T_2}\right) \tag{10.4.4}$$

Indefinite integral form

$$\ln\{k_A\} = -\frac{E_a}{RT} + \ln\{k_0\} \tag{10.4.5}$$

If the Arrhenius equation is obeyed, a plot of $\ln\{k_A\}$ vs. $1/\{T\}$ is a straight line, the slop is

$-E_a/R$ and the intercept is $\ln\{k_0\}$. This allows E_a and k_0 to be found.

Note from (10.4.2) that a low activation energy means a fast reaction and a high activation energy means a slow reaction.

Example 10.3

Reaction 1, $E_{a,1} = 100\text{kJ}\cdot\text{mol}^{-1}$; reaction 2, $E_{a,2} = 150\text{kJ}\cdot\text{mol}^{-1}$, when the reaction temperature changed from 300K to 310K. Calculate the $k_{(300+10)k}/k_{300k}$ of the two reactions.

***Answer*:**

Reaction 1: $\dfrac{k_{(300+10)\text{K}}}{k_{300\text{K}}} = e^{-\frac{E_{a,1}\cdot\Delta T}{R(T_1\cdot T_2)}} = 3.64$

Reaction 2: $\dfrac{k_{(300+10)\text{K}}}{k_{300\text{K}}} = e^{-\frac{E_{a,2}\cdot\Delta T}{R(T_1\cdot T_2)}} = 6.96$

Example 10.4

Reaction 1, $E_{a,1} = 100\text{kJ}\cdot\text{mol}^{-1}$; reaction 2, $E_{a,2} = 150\ \text{kJ}\cdot\text{mol}^{-1}$, when the reaction temperature changed from 400K to 410K. Calculate the $k_{(400+10)\text{K}}/k_{400\text{K}}$ of the two reactions.

***Answer*:**

reaction 1: $\dfrac{k_{(400+10)\text{K}}}{k_{400\text{K}}} = e^{-\frac{E_{a,1}\cdot\Delta T}{R(T_1\cdot T_2)}} = 2.08$

reaction 2: $\dfrac{k_{(400+10)\text{K}}}{k_{400\text{K}}} = e^{-\frac{E_{a,2}\cdot\Delta T}{R(T_1\cdot T_2)}} = 3.00$

It can be seen from the results of examples 10.3 and 10.4 that ①when the activation energy is same, the higher the original temperature is, the less the rate constant increases with the same temperature increases. ②when the original temperature is same, the lower the activation energy is, the less the rate constant increases with the same temperature increases.

10.4.3 Definition of activation energy and pre-exponential parameter (活化能 E_a 及指前参量 k_0 的定义)

Equation (10.4.2) is found to hold well for nearly all elementary homogeneous reactions and for most complex reactions. A simple interpretation of (10.4.2) is that two colliding molecules require a certain minimum kinetic energy of relative motion to initiate the breaking of the appropriate bonds and allow new compounds to be formed. The fraction of collisions in which the relative kinetic energy of the molecules along the line of the collision exceeds the value ε_a is proportional to exp $(-\varepsilon_a/kT) = \exp(-E_a/RT)$ where $E_a = L\varepsilon_a$.

The general definition of the active energy E_a of any rate process, applicable whether or not E_a varies with T, is

$$E_a \overset{\text{def}}{=\!=\!=} RT^2\frac{\mathrm{d}\ln\{k_A\}}{\mathrm{d}T} \tag{10.4.6}$$

Whether or not E_a depend on T, the pre-exponential factor k_0 for any rate process is defined as

$$k_0 \overset{\text{def}}{=\!=\!=} k_A\exp(E_a/RT) \tag{10.4.7}$$

E_a 及 k_0 为两个经验参量，可由实验测得的数据计算，而把 E_a 及 k_0 均视为与温度无关，但实质上与温度有关。

当考虑 E_a 及 k_0 与温度关系时，可采用如下的三参量方程：

$$k = k'_0 T^m \exp\left(-\frac{E'}{RT}\right)$$

$$E_a = E' + mRT \qquad (10.4.8)$$

式(10.4.8)表明了活化能 E_a 与温度 T 的关系。由于一般反应 m 较小，加之在温度不太高时 mRT 一项的数量级与 E' 相比可略而不计，此时即可看作 E_a 与温度无关。

Example 10.5

For reaction 1 and reaction 2, $E_{a,1}$, $E_{a,2}$ are different, $k_{0,1}$, $k_{0,2}$ are same, when T=300K:

(1) If $E_{a,1}-E_{a,2}=5\text{kJ}\cdot\text{mol}^{-1}$, please calculate k_2/k_1;

(2) If $E_{a,1}-E_{a,2}=10\text{kJ}\cdot\text{mol}^{-1}$, please calculate k_2/k_1.

***Answer*:**

According to the Arrhenius equation, we get

$k_1 = k_{0,1}\mathrm{e}^{-E_{a,1}/RT}$ and $k_2 = k_{0,2}\mathrm{e}^{-E_{a,2}/RT}$

For $k_{0,1} = k_{0,2}$, $k_2/k_1 = \mathrm{e}^{(E_{a,1}-E_{a,2})/RT}$

(1) If $E_{a,1}-E_{a,2}=5\text{kJ}\cdot\text{mol}^{-1}$

$$k_2/k_1 = \mathrm{e}^{5\times10^3/(8.315\times300)} = 7.42$$

(2) If $E_{a,1}-E_{a,2}=10\text{kJ}\cdot\text{mol}^{-1}$

$$k_2/k_1 = \mathrm{e}^{10\times10^3/(8.315\times300)} = 5.51$$

Conclusion: If the two reactions have the same pre-exponential factors and at the same temperature, the reaction with lower activation energy has a higher rate constant. Moreover, the greater the difference in activation energy, the greater the difference in reaction rate constant.

10.4.4 Physical significance of activation energy(活化能的物理意义)

Only the physical significance of activation energy of the elementary reaction is introduced. In elementary reactions, the reactant molecules are divided into activated molecules (Chemical reaction occurs when reactant molecules collide with each other) and non-activated molecules (Chemical reaction doesn't occur when reactant molecules collide with each other). Non-activated molecules have to absorb a certain amount of energy to become activated molecules. Activation energy is the energy difference between 1 mol activated molecules and 1 mol non-activated molecules.

$$E_a = \langle E^{\neq}\rangle - \langle E\rangle \qquad (10.4.9)$$

$\langle E\rangle$— the average molar energy of the non-activated molecules;

$\langle E^{\neq}\rangle$—the average molar energy of the activated molecules.

10.4.5 Relationship between the activation energy and the reaction heat(活化能 E_a 与反应热的关系)

For reversible reaction $\mathrm{A} \underset{k_{-1}}{\overset{k_1}{\rightleftharpoons}} \mathrm{Y}$, let k_1 and k_{-1} be forward and reverse rate constants, and let $E_{a,1}$ and $E_{a,-1}$ be the corresponding activation energies (see figure 10.8).

$$K_c = k_1/k_{-1}$$

where K_c is the concentration-scale equilibrium constant of the reaction.

Hence $\ln K_c = \ln k_1 - \ln k_{-1}$

Differential with respect to T

$$\frac{\mathrm{d}\ln K_c}{\mathrm{d}T} = \frac{\mathrm{d}\ln k_1}{\mathrm{d}T} - \frac{\mathrm{d}\ln k_{-1}}{\mathrm{d}T}$$

Equation (10.4.6) gives

$$\frac{\mathrm{d}\ln\{k_1\}}{\mathrm{d}T} = \frac{E_{a,1}}{RT^2} \text{ and } \frac{\mathrm{d}\ln\{k_{-1}\}}{\mathrm{d}T} = \frac{E_{a,-1}}{RT^2}$$

For $$\frac{\mathrm{d}\ln K_c}{\mathrm{d}T} = \frac{\Delta U}{RT^2}$$

So we can get

$$E_{a,1} - E_{a,-1} = \Delta U$$

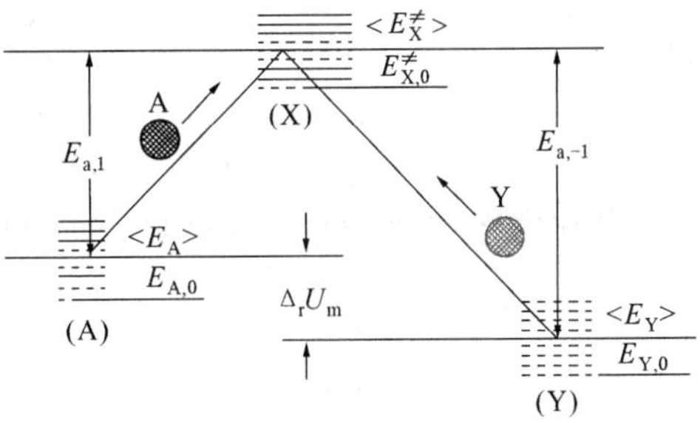

Fig. 10.8 The energy profile for the transformation from reactants to products

Example 10.6

For the first-order reaction, when the conversion reaches to 20% it will take 3.20 min at 340K and 12.6min at 300K. Please calculate the activation energy of the reaction.

***Answer*:**

The rate coefficient of the reaction is denoted by k at 300K, and is denoted by k' at 340K.

Integral rate equation of first-order reaction:

$$k = \frac{1}{t}\ln\frac{1}{1-x}$$

Both the conversion at 300K and 340K are 20%, so

$$\frac{k'}{k} = \frac{t}{t'}$$

$$\ln\frac{k'}{k} = \ln\frac{t}{t'} = \frac{E_a}{R}\left(\frac{1}{T} - \frac{1}{T'}\right)$$

$$E_a = \frac{RTT'}{T'-T}\ln\frac{t}{t'} = \frac{8.314\times300\times340}{340-300}\times\ln\frac{12.6}{3.20} = 29.1\,\mathrm{kJ\cdot mol^{-1}}$$

10.5 Typical complex reactions (典型复合反应)

Complex reaction—reaction which contains more than one elementary reactions.(复合反应通常是指两个或两个以上基元反应的组合。)

Typical complex reaction: parallel reaction, reversible reaction, consecutive reaction.

10.5.1 Parallel first-order reaction(平行一级反应)

Parallel reaction—a species can react in different ways to give a variety of products. (若反应物能同时进行几种不同反应，则这些反应称为平行反应。)

For example:

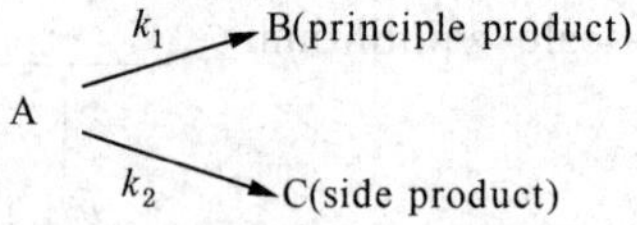

If both reactions are first-order reactions, the rate of product formation as

$$\frac{\mathrm{d}c_\mathrm{B}}{\mathrm{d}t}=k_1c_\mathrm{A} \tag{10.5.1}$$

$$\frac{\mathrm{d}c_\mathrm{C}}{\mathrm{d}t}=k_2c_\mathrm{A} \tag{10.5.2}$$

Since

$$c_\mathrm{A}+c_\mathrm{B}+c_\mathrm{C}=c_{\mathrm{A},0} \tag{10.5.3}$$

Therefore

$$\frac{\mathrm{d}c_\mathrm{A}}{\mathrm{d}t}+\frac{\mathrm{d}c_\mathrm{B}}{\mathrm{d}t}+\frac{\mathrm{d}c_\mathrm{C}}{\mathrm{d}t}=0 \tag{10.5.4}$$

$$-\frac{\mathrm{d}c_\mathrm{A}}{\mathrm{d}t}=\frac{\mathrm{d}c_\mathrm{B}}{\mathrm{d}t}+\frac{\mathrm{d}c_\mathrm{C}}{\mathrm{d}t} \tag{10.5.5}$$

Substitution of (10.5.1) and (10.5.2) in (10.5.5), we have

$$-\frac{\mathrm{d}c_\mathrm{A}}{\mathrm{d}t}=(k_1+k_2)c_\mathrm{A} \tag{10.5.6}$$

$$\ln\frac{c_{\mathrm{A},0}}{c_\mathrm{A}}=(k_1+k_2)t \tag{10.5.7}$$

This equation is the same as (10.2.7) with k_A replaced by (k_1+k_2).

For product, we have

$$\frac{\mathrm{d}c_\mathrm{B}}{\mathrm{d}t}=k_1c_\mathrm{A} \qquad \frac{\mathrm{d}c_\mathrm{c}}{\mathrm{d}t}=k_2c_\mathrm{A}$$

If

$$c_{\mathrm{B},0}=0,\ c_{\mathrm{C},0}=0$$

We can get that at arbitrary time t

$$\frac{c_\mathrm{B}}{c_\mathrm{C}}=\frac{k_1}{k_2} \tag{10.5.8}$$

The amounts of B and C obtained depend on the relative rates of the two competing reactions. Measurement of $c_\mathrm{B}/c_\mathrm{C}$ allows k_1/k_2 to be determined.

Combine (10.5.7) with (10.5.8), we can get k_1 and k_2.

Concentration vs. time for the parallel first-order reaction was shown in figure 10.9

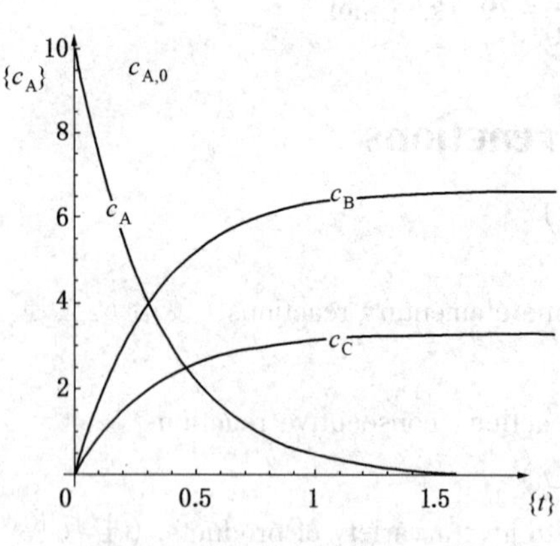

Fig. 10.9 $c_\mathrm{A}\sim t$ curve of the parallel first-order reaction

级数相同的平行反应，其特征就是产物的浓度之比等于速率常数之比，而与反应物的初始浓度及时间无关。但若平行反应级数不同，则没有以上特征。

几个平行反应的活化能往往不同，温度升高有利于活化能大的反应，反之温度低有利于活化能小的反应。不同的催化剂，有时也能加速不同的反应。所以可通过改变温度、改变催化剂等因素来选择性地加速所希望的反应，以提高产品产率。

10.5.2 Reversible first-order reaction(对行一级反应)

Reversible reaction—the forward and the backward/reverse reaction take place simultaneously. (定义：正向和逆向同时进行的反应称为对行反应或称对峙反应。)

对于单向反应，反应结束时，只有产物没有反应物；对于对行反应，当反应进行足够长时间后，反应达到平衡态，反应物与产物浓度都达到各自的平衡浓度，浓度不可能为零。

Reversible first-order reaction

$$\mathrm{A} \underset{k_{-1}}{\overset{k_1}{\rightleftharpoons}} \mathrm{B}$$

$t=0$	$c_{A,0}$	0
$t=t$	c_A	$c_{A,0}-c_A$
$t=\infty$	$c_{A,e}$	$c_{A,0}-c_{A,e}$

The rate of change of the concentration of A has two contributions: A is depleted by the forward reaction at a rate $v_f=k_1c_A$, but is replenished by the backward reaction at a rate $v_b=k_{-1}c_B$. The total rate of the change of concentration of A is therefore,

$$-\frac{dc_A}{dt}=k_1c_A-k_{-1}(c_{A,0}-c_A) \qquad (10.5.9)$$

In the limit as $t\to\infty$, the system reaches equilibrium, the rate of the forward and reverse reactions become equal. At equilibrium the concentration of each species is constant, and

$$-\frac{dc_{A,e}}{dt}=k_1c_{A,e}-k_{-1}(c_{A,0}-c_{A,e})=0 \qquad (10.5.10)$$

From(10.5.10), we get

$$\frac{(c_{A,0}-c_{A,e})}{c_{A,e}}=\frac{k_1}{k_{-1}}=k_e \qquad (10.5.11)$$

(10.5.9) substrate(10.5.10), we have:

$$-\frac{d(c_A-c_{A,e})}{dt}=k_1(c_A-c_{A,e})+k_{-1}(c_A-c_{A,e})=(k_1+k_{-1})(c_A-c_{A,e}) \qquad (10.5.12)$$

Separating the variables and integrating, we get

$$-\int_{c_{A,0}}^{c_A}\frac{d(c_A-c_{A,e})}{(c_A-c_{A,e})}=\int_0^t(k_1+k_{-1})dt$$

$$\ln\frac{(c_{A,0}-c_{A,e})}{(c_A-c_{A,e})}=(k_1+k_{-1})t \qquad (10.5.13)$$

A plot of $\ln(c_A-c_{A,e})$ vs. t gives a straight line of slop $-(k_1+k_{-1})$. Combine with $K_c=k_1/k_{-1}$ we can get k_1 and k_{-1}.

Concentration vs. time for reversible first-order reaction is shown in figure 10.10

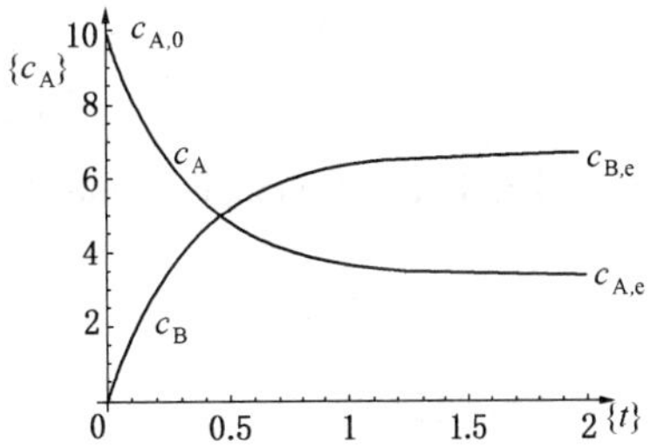

Fig. 10.10 $c_A\sim t$ curve of the reversible first-order reaction

Definition of half-life:

The time required for $\Delta c_{A,0}=c_{A,0}-c_{A,e}$ to drop half its value is called the reaction's half-life.

That is the time required for $c_A - c_{A,e} = \frac{1}{2}(c_{A,0} - c_{A,e})$

$$t_{1/2} = \frac{\ln 2}{k_1 + k_{-1}}$$

Example 10.7

For reversible reaction $A \underset{k_{-1}}{\overset{k_1}{\rightleftharpoons}} B$, known to be first-order reactions in both the forward and reverse directions, and the rate constants are equal, $k_1 = k_{-1} = 1.9 \times 10^{-6}\ s^{-1}$. If the original reactant is pure substance, Please calculate the time required when conversion is 10%.

Answer:

$$\frac{k_1}{k_{-1}} = \frac{c_{A,0} - c_{A,e}}{c_{A,e}} = 1$$

We can get:

$$c_{A,e} = 0.5c_{A,0} \tag{1}$$

$$\alpha = \frac{c_{A,0} - c_A}{c_{A,0}} = 10\%$$

$$c_A = 0.9c_{A,0} \tag{2}$$

For the first-order reversible reaction

$$\ln \frac{c_{A,0} - c_{A,e}}{c_A - c_{A,e}} = (k_1 + k_{-1})t \tag{3}$$

Substitution (1) and (2) into (3), we can get

$$t = 980\text{min}$$

10.5.3 Consecutive first-order reaction(连串反应)

Consecutive reaction—a product of one reaction becomes a reactant in a subsequent reaction.(当一个反应的部分或全部生成物是下一个反应的部分或全部反应物时的反应称为连串反应。)

This is true in multistep reaction mechanism. We shall consider only the simplest case, that of two consecutive irreversible first-order reactions

$$A \xrightarrow{k_1} B \xrightarrow{k_2} C$$

$t=0$	$c_{A,0}$	0	0
$t=t$	c_A	c_B	c_C

The rates of change in the A, B, and C concentrations are

$$-\frac{dc_A}{dt} = k_1 c_A \tag{10.5.14}$$

$$\frac{dc_B}{dt} = k_1 c_A - k_2 c_B \tag{10.5.15}$$

$$\frac{dc_C}{dt} = k_2 c_B \tag{10.5.16}$$

Since B is formed by the first reaction and destroyed by the second reaction, the expression for

dc_B/dt has two terms. Let only A be present in the system at $t = 0$:

$$c_{A,0} \neq 0,\ c_{B,0} = 0,\ c_{C,0} = 0$$

We have three coupled differential equations. Integration of (10.5.14) gives

$$\ln \frac{c_{A,0}}{c_A} = k_1 t \quad \text{or} \quad c_A = c_{A,0} e^{-k_1 t} \tag{10.5.17}$$

Substitution of (10.5.17) into (10.5.15) gives

$$\frac{dc_B}{dt} = k_1 c_{A,0} e^{-k_1 t} - k_2 c_B$$

$$\frac{dc_B}{dt} + k_2 c_B = k_1 c_{A,0} e^{-k_1 t} \tag{10.5.18}$$

Integration results

$$c_B = \frac{k_1 c_{A,0}}{k_2 - k_1}(e^{-k_1 t} - e^{-k_2 t}) \tag{10.5.19}$$

For

$$c_A + c_B + c_C = c_{A,0}$$

We get

$$c_C = c_{A,0}\left[1 - \frac{1}{k_2 - k_1}(k_2 e^{-k_1 t} - k_1 e^{-k_2 t})\right] \tag{10.5.20}$$

The maximum concentration in the intermediate specie occurs when $\frac{dc_B}{dt} = 0$.

$$c_B = \frac{k_1 c_{A,0}}{k_2 - k_1}(e^{-k_1 t} - e^{-k_2 t})$$

$$\frac{dc_B}{dt} = 0,\ \text{that is},\ e^{(k_1 - k_2)t} = \frac{k_1}{k_2}$$

$$t_{max} = \frac{\ln(k_1/k_2)}{k_1 - k_2},\quad c_{B,max} = c_{A,0}\left(\frac{k_1}{k_2}\right)^{\frac{k_2}{k_2 - k_1}} \tag{10.5.21}$$

10.6 Approximation methods of rate equation of complex reaction (复合反应速率的近似处理法)

Usually, the rate law is derived under simplifying assumptions. One often used assumption is that a rate-determining step exists.

10.6.1 Rate-determining step method (选取控制步骤法)

Rate-determining step is the slowest step which controls the rate of the reaction, and then the overall reaction rate is equal to that of the rate-determining step.

控制步骤的反应速率常数越小，其他各串联步骤的速率常数越大，则此规律就越准确。提高了控制步骤的速率，整个反应就能加速进行。

10.6.2 Equilibrium-state approximation method(平衡态近似法)

$$A+B \underset{k_{-1}}{\overset{k_1}{\rightleftharpoons}} C \qquad \text{(quick)}$$

$$C \xrightarrow{k_2} D \qquad \text{(slow)}$$

The last step is the rate-determining step. A consequence of this bottleneck situation is that the step preceding it is at equilibrium.

So we can get $k_1 c_A c_B = k_{-1} c_C$

$$c_C = \frac{k_1}{k_{-1}} c_A c_B = K_c c_A c_B$$

The overall reaction rate is equal to that of the rate-determining step

$$v = \frac{dc_D}{dt} = k_2 c_C = K_c k_2 c_A c_B = \frac{k_1 k_2}{k_{-1}} c_A c_B$$

Let
$$k = \frac{k_1 k_2}{k_{-1}}$$

We can get
$$v = \frac{dc_D}{dt} = k c_A c_B$$

10.6.3 Steady-state approximation method(稳态近似法)

For consecutive reaction

$$A \xrightarrow{k_1} B \xrightarrow{k_2} C$$

The steady-state approximation assumes that the rate of formation of a reaction intermediate essentially equals its rate of the destruction, so as to keep it at near-constant steady-state concentration.

That is
$$\frac{dc_B}{dt} = 0$$

These intermediates are very reactive and therefore do not accumulate to any significant extent during the reaction.

For example:
$$A \xrightarrow{k_1} B \xrightarrow{k_2} C \qquad (k_2 \gg k_1)$$

$$\frac{dc_B}{dt} = k_1 c_A - k_2 c_B = 0$$

Example 10.8

The mechanism of $COCl_2 = CO + Cl_2$ is as follows:

(1) $Cl_2 \underset{k_{-1}}{\overset{k_1}{\rightleftharpoons}} 2Cl\cdot$ (quick)

(2) $Cl\cdot + COCl_2 \xrightarrow{k_2} CO + Cl_3\cdot$ (slow)

(3) $Cl_3\cdot \underset{k_{-3}}{\overset{k_3}{\rightleftharpoons}} Cl_2 + Cl\cdot$ (quick)

Try to prove: $\frac{dCO}{dt} = k[COCl_2][Cl_2]^{1/2}$

***Answer*:**

$$\frac{d[CO]}{dt}=k_2[Cl\cdot][COCl_2] \qquad (a)$$

For reaction(1) at equilibrium

$$\frac{[Cl\cdot]^2}{[Cl_2]}=K=\frac{k_1}{k_{-1}}$$

$$[Cl\cdot]=\left(\frac{k_1}{k_{-1}}\right)^{1/2}[Cl_2]^{1/2}$$

Substitution of $[Cl\cdot]$ in(a), we get

$$\frac{d[CO]}{dt}=k_2[Cl\cdot][COCl_2]=k_2\left(\frac{k_1}{k_{-1}}\right)^{1/2}[Cl_2]^{1/2}[COCl_2]=k[COCl_2][Cl_2]^{1/2}$$

Where
$$k=k_2\left(\frac{k_1}{k_{-1}}\right)^{1/2}$$

Example 10.9

Example 10. 9

The mechanism of $C_2H_6+H_2=2CH_4$ is as follows:

(1) $C_2H_6 \rightleftharpoons 2CH_3\cdot$ $\qquad K$

(2) $CH_3\cdot +H_2 \rightarrow CH_4+H\cdot$ $\qquad k_2$

(3) $H\cdot +C_2H_6 \rightarrow CH_4+CH_3\cdot$ $\qquad k_3$

Reaction(1) is at equilibrium, H· is reactive substance, so steady state approximation method can be used. Try to prove

$$\frac{d[CH_4]}{dt}=2k_2K^{1/2}[C_2H_6]^{1/2}[H_2]$$

***Answer*:**

$$\frac{d[CH_4]}{dt}=k_2[CH_3\cdot][H_2]+k_3[H\cdot][C_2H_6] \qquad (a)$$

For reaction(1) is at equilibrium, we have

$$\frac{[CH_3\cdot]^2}{[C_2H_6]}=K \Rightarrow [CH_3\cdot]=K^{1/2}[C_2H_6]^{1/2} \qquad (b)$$

Assuming steady states for $[H\cdot]$, we have for the net reaction rate of $[H\cdot]$

$$\frac{d[H\cdot]}{dt}=k_2[CH_3\cdot][H_2]-k_3[H\cdot][C_2H_6]=0$$

$$k_2[CH_3\cdot][H_2]=k_3[H\cdot][C_2H_6] \qquad (c)$$

Substitution(b) and(c) into(a), we have

$$\frac{d[CH_4]}{dt}=2k_2[CH_3\cdot][H_2]=2k_2K^{1/2}[C_2H_6]^{1/2}[H_2]$$

10.6.4 Relation between the activation energy of the overall reaction and the elementary reaction(非基元反应的表观活化能与基元反应活化能之间的关系)

阿伦尼乌斯方程不但适用于基元反应，也适用于大多数非基元反应。

For overall reaction

$$A+B \underset{k_{-1}}{\overset{k_1}{\rightleftharpoons}} C \xrightarrow{k_2} D$$

Relation between the activation energy and rate constant of the overall reaction is

$$k=Ae^{-E_a/(RT)}$$

where E_a is called the apparent activation energy of the overall reaction. k is rate constant of the overall reaction

Use the equilibrium-state approximation method, we deduced that

$$k=\frac{k_1k_2}{k_{-1}} \quad (10.6.1)$$

The logarithmic form of the equation(10.6.1) as

$$\mathrm{dln}\ k=\mathrm{dln}\ k_1+\mathrm{dln}\ k_2-\mathrm{dln}\ k_{-1} \quad (10.6.2)$$

So we can get

$$\frac{\mathrm{dln}\ k}{\mathrm{d}T}=\frac{\mathrm{dln}\ k_1}{\mathrm{d}T}+\frac{\mathrm{dln}\ k_2}{\mathrm{d}T}-\frac{\mathrm{dln}\ k_{-1}}{\mathrm{d}T} \quad (10.6.3)$$

Substitution the Arrhenius equation in(10.6.3), we can get

$$\frac{E_a}{RT^2}=\frac{E_{a,2}}{RT^2}+\frac{E_{a,1}}{RT^2}-\frac{E_{a,-1}}{RT^2}$$

Relation between the activation energy of the overall reaction and the elementary reaction is

$$E_a=E_{a,2}+E_{a,1}-E_{a,-1} \quad (10.6.4)$$

10.7 Chain reaction
(链反应)

A chain reaction contains a series of steps in which a reactive intermediate is consumed, reactants are converted to products, and the intermediate is regenerated. Regeneration of the intermediate allows this cycle to be repeated over and over again. Thus a small amount of intermediate leads to production of a large amount of product.

Chain reaction is divided into straight chain reaction and branching chain reaction.

10.7.1 Common procedure of chain reaction(链反应的基本步骤)

The mechanism of $H_2+Cl_2 \longrightarrow 2HCl$ is believed to be

$$Cl_2+M \xrightarrow{k_1} 2Cl\cdot +M \quad (1)$$

$$Cl\cdot +H_2 \xrightarrow{k_2} HCl+H\cdot \quad (2)$$

$$H\cdot +Cl_2 \xrightarrow{k_3} HCl+Cl\cdot \quad (3)$$

$$\vdots$$

$$2Cl\cdot +M \longrightarrow Cl_2+M \quad (4)$$

Chain initiation(链的引发): In step (1), a Cl_2 molecule collides with any species M, thereby gaining the energy to dissociate into two Cl·, Cl· is the chain-carrying reactive species.

Chain propagation(链的增长): In steps (2) and (3) form a chain that consumes Cl·, converts H_2 and Cl_2 to HCl, and regenerates Cl·. For each Cl· produce by step (1), we get many repetitions of steps (2) and (3). The reactive intermediates Cl· and H· that occur in the chain propagating steps are called chain carriers. Adding step (2) and (3), we get $H\cdot + Cl\cdot + H_2 + Cl_2 \xrightarrow{k_2} 2HCl + Cl\cdot + H\cdot$, which agrees with the overall stoichiometry $H_2 + Cl_2 \longrightarrow 2HCl$.

Chain termination(链的终止): Two chain carriers collide with any species M, thereby loading the energy to form Cl_2.

In chain reactions, the carriers are commonly free radicals(including atoms). Ions can also serve as chain carriers.

10.7.2 Types of chain reaction(链反应的类型)

Straight chain reaction—each chain transfer step consumes one chain carrier and produces only one chain carrier.

$$\longrightarrow\longrightarrow\longrightarrow\longrightarrow$$

Branching chain reaction—each chain transfer step consumes one chain carrier and produces two or more chain carriers.

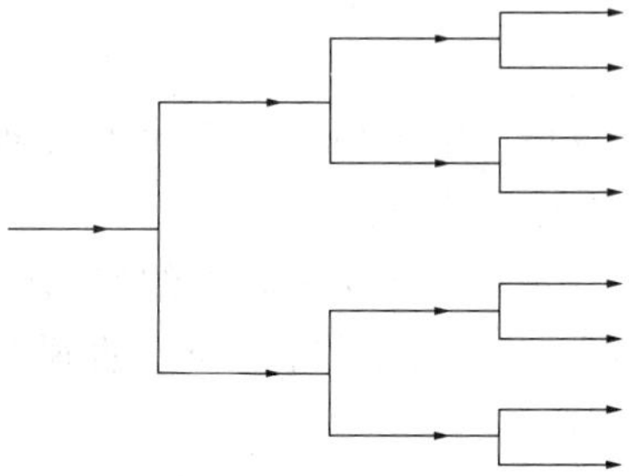

10.7.3 Rate equation of straight chain reaction(直链反应的速率方程)

The mechanism of $H_2 + Cl_2 \longrightarrow 2HCl$ is

$$Cl_2 + M \xrightarrow{k_1} 2Cl\cdot + M \quad (1)$$

$$Cl\cdot + H_2 \xrightarrow{k_2} HCl + H\cdot \quad (2)$$

$$H\cdot + Cl_2 \xrightarrow{k_3} HCl + Cl\cdot \quad (3)$$

$$\vdots$$

$$2Cl\cdot + M \longrightarrow Cl_2 + M \quad (4)$$

Depending on the reaction mechanism, we can get

$$\frac{dc(HCl)}{dt} = k_2 c(Cl\cdot)c(H_2) + k_3 c(H\cdot)c(Cl_2) \quad (10.7.1)$$

Applying the steady state approximation to the chain carriers, we can get

$$\frac{dc(H\cdot)}{dt} = k_2 c(Cl\cdot)c(H_2) - k_3 c(H\cdot)c(Cl_2) = 0 \quad (10.7.2)$$

$$\frac{dc(Cl\cdot)}{dt}=k_1c(Cl_2)c(M)-k_2c(Cl\cdot)c(H_2)+k_3c(H\cdot)c(Cl)-k_4[c(Cl_2)]^2c(M)=0 \tag{10.7.3}$$

From (10.7.2) we can get

$$k_2c(Cl\cdot)c(H_2)=k_3c(H\cdot)c(Cl_2) \tag{10.7.4}$$

Substitute (10.7.2) into (10.7.1), we can get

$$\frac{dc(HCl)}{dt}=2k_2c(Cl\cdot)c(H_2) \tag{10.7.5}$$

From (10.7.3) we can get

$$k_1c(Cl_2)=k_4[c(Cl\cdot)]^2$$

$$c(Cl\cdot)=(k_1/k_4)^{1/2}[c(Cl_2)]^{1/2} \tag{10.7.6}$$

Substitute (10.7.6) into (10.7.5), we can get

$$\frac{dc(HCl)}{dt}=2k_2(k_1/k_4)^{1/2}c(H_2)[c(Cl_2)]^{1/2}$$

$$=kc(H_2)[c(Cl_2)]^{1/2}$$

where $k=2k_2(k_1/k_4)^{1/2}$

10.7.4 Chain explosion and explosion limit of chain explosion reaction(链爆炸与链爆炸反应的界限)

爆炸是瞬间完成的高速化学反应。它的研究对于化工安全生产，对于经济建设与国防建设都具有重要意义。从原因上，爆炸反应分两类：

(1) Heat explosion(热爆炸)：放热反应在一局限的空间中进行，反应热散失不掉，使温度升高，则反应速率增大，放热更多，温升更快，如此恶性循环，使反应速率在瞬间大到无法控制，导致爆炸。

(2) Chain explosion(链爆炸)：由支链反应引起的，随着支链的发展，链传递物(活性质点)剧增，反应速率愈来愈大，最后导致爆炸。

链爆炸反应的温度、压力、组成有一定的爆炸区间，称为爆炸界限。

以 $H_2+(1/2)O_2 \longrightarrow H_2O$ 的反应为例，它是一个支链反应，机理如下：

① $H_2+O_2 \xrightarrow{k_1} 2HO\cdot$ chain initiation (链的引发)

② $HO\cdot+H_2 \xrightarrow{k_2} H_2O+H\cdot$ (快) chain propagation(链的增长)

③ $H\cdot+O_2 \xrightarrow{k_3} HO\cdot+\dot{O}\cdot$ (慢)
④ $\dot{O}\cdot+H_2 \xrightarrow{k_4} HO\cdot+H\cdot$ (快)
} chain branching(链的分支)

⑤ $H\cdot \xrightarrow{k_5}$ wall chain termination(链的终止)(低压下)

The propagation reaction is exothermic and fast. The third and fourth reactions are called branching reactions because two radicals are formed from one. If the rate of branching is greater than the rate of termination, the number of radicals increases exponentially with time, and an explosion results. Reaction ⑤ is endothermic and slow below 700K. The conditions of the first explosion limit are governed by the relative rates for branching 2 $k_3[H][O_2]$ and termination $k_5[H]$. As the concentration of oxygen is increased, the rate of branching becomes greater than the rate of termination and an explosion occurs.

To explain the second limit, above which there is no explosion, it is necessary to invoke a new termination step to prevent the exponential increase in the number of radicals. For a new termination step to become more important as the pressure is increased, the new step must be higher order than the branching reaction. Thus, to explain the second limit, the following reaction must be added to the previous reactions:

⑥ $H\cdot + O_2 + M \xrightarrow{k_6} HO_2\cdot + M$ chain termination(链的终止)(高压下)

In a stoichiometric mixture of oxygen and hydrogen, M may be hydrogen or oxygen, but these two gases have different efficiencies in this reaction. The HO_2 radical is relatively unreactive and does not produce another radical before it is quenched on the wall.

The third explosion limit results from the fact that the following reaction diminishes termination:

$$\text{Propagation } HO_2\cdot + H_2 + M \xrightarrow{k_6} H_2O + OH\cdot$$

If a propagation step in a chain reaction produces two or more radicals from one, there is a possibility of a rapid increase in rate and, for an exothermic reaction, an explosion. The reaction of hydrogen with oxygen can be explosive above about 773K. The ranges of temperature and pressure within which there are spontaneous thermal explosions are shown in figure 10.11. For example, at about 773K stoichiometric hydrogen-oxygen mixtures react very slowly at pressures below the 1st(lower) limit. As the pressure is increased, the reaction rate increases slowly, but at a pressure, depending on the volume of the vessel, there is a sudden explosion. On the other hand, if the gases are at a considerably higher pressure(2nd limit), the rate is again quite low.

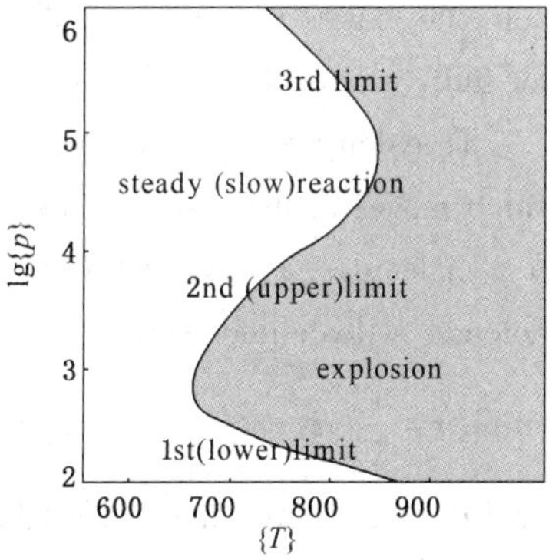

Fig. 10.11 Explosion limits of a stoichiometric mixture of H_2+O_2 ($n_{H_2} : n_{O_2} = 2:1$)

Finally, as the total pressure is increased above the explosion zone, the reaction rate increases until it becomes so fast that the reaction mixture may be said to explode. The fact that the exact limits depend on the vessel surface and the vessel diameter indicates that radical chains may be terminated by reaction at the wall.

10.8 Simple collision theory of gas reaction(SCT) (气体反应的简单碰撞理论)

The two theories most used to calculate Arrhenius parameters for elementary reactions are simple collision theory and transition-state theory. It is concerned with microscopic kinetics, that is, elementary reactions at the molecular level. The calculation of rate constants from properties of individual atoms and molecules is challenging because reactions occur as a result of collisions with a variety of energies, angles of approach, and states of reactants and products. Simple collision theory of bi-molecular reaction is based on consideration of collisions of rigid spherical molecules. To go further and take electronic structure into account, it is necessary to use the concept of the potential energy surface for a reaction.

(1) Fundamental assumption of collision theory(碰撞理论的基本假设)

① SCT assumes that molecules can be taken as rigid ball without inner structure.

② For a reaction to occur between molecules A and B, the two molecules must collide.

③ The collision can be either non-reactive (elastic) collision or reactive collision. Only the molecules posses energy exceeds threshold energy (E_c) can lead to reactive collision. The reaction rate should also be in proportion to the fraction of reactive collision(q).

④ The Maxwell-Boltzmann equilibrium distribution of molecular speeds is maintained during the reaction.

(2) Collision number(碰撞数)—Z_{AB}

For a mixture of A and B, let z_{AB} be the number of collisions per unit time that one particular A molecule makes with B molecules. Let Z_{AB} be the the total number of A−B collisions per unit time per unit volume. Let C_A and C_B be the molecule concentrations of A and B molecules.

To calculate Z_{AB}, we present that all molecules are at rest except one particular A molecule, which moves at the constant speed $\langle u_{AB}\rangle$, where $\langle u_{AB}\rangle$ is the average speed of A molecule relative to B molecules. Let d_A and d_B be the diameters of A and B and r_A and r_B their radii. The moving A molecule will collide with a B molecule whenever the distance between the centers of the pair is within $r=\frac{1}{2}(d_A+d_B)=(r_A+r_B)$ (figure 10. 12). Imagine a cylinder of radius r centered about the moving A molecule(figure 10. 13). In time dt, the moving molecule will travel a distance $\langle u_{AB}\rangle \mathrm{d}t$ and will sweep out a cylinder of volume $V=\pi\gamma^2\langle u_{AB}\rangle \mathrm{d}t$. The moving molecule will collide with all B molecules are uniformly distributed throughout the V, so we can get

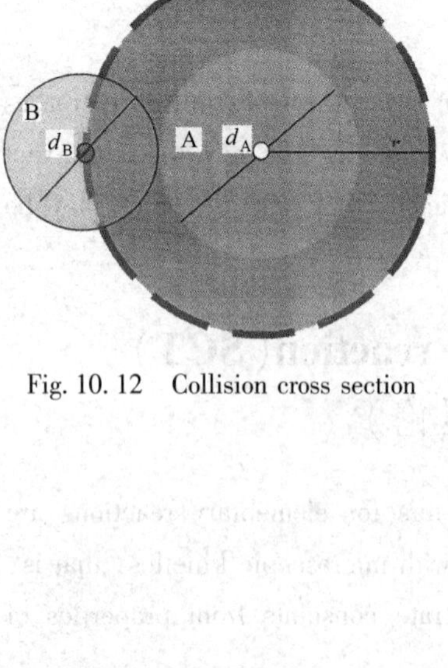

Fig. 10. 12 Collision cross section

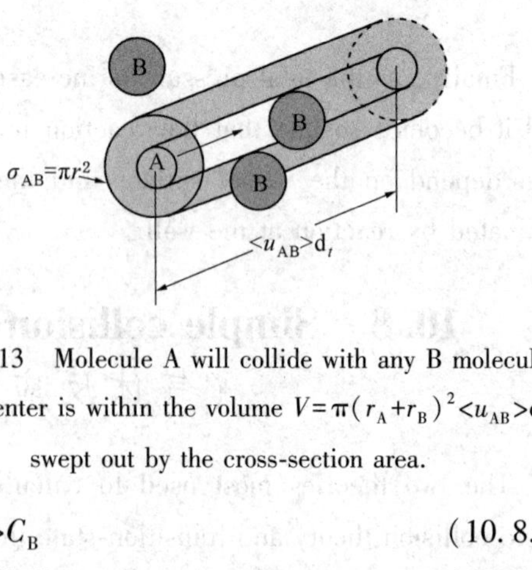

Fig. 10. 13 Molecule A will collide with any B molecule whose center is within the volume $V=\pi(r_A+r_B)^2\langle u_{AB}\rangle \mathrm{d}t$ swept out by the cross-section area.

$$z_{AB}=\pi r^2\langle u_{AB}\rangle C_B \qquad (10.8.1)$$

$$Z_{AB}=\pi r^2\langle u_{AB}\rangle C_A C_B \qquad (10.8.2)$$

The average speed of A molecules relative to B molecules $<u_{AB}>$

$$<u_{AB}>=\left(\frac{8k_B T}{\pi\mu}\right)^{1/2} \qquad (10.8.3)$$

Where, k_B is boltzman constant, $\mu=\frac{m_A m_B}{m_A+m_B}$

Substitution(10.8.3) in(10.8.2), we can get

$$Z_{AB}=\pi r^2\left(\frac{8k_B T}{\pi\mu}\right)^{1/2} C_A C_B \qquad (10.8.4)$$

(3) Activated collision fraction(活化碰撞分数) q

Definition of activated collision fraction

$$q=\frac{N^{\neq}}{N}=e^{-E_c/RT} \qquad (10.8.5)$$

Where, $N^{\neq}$ and N are the number of the active molecules and the total molecules, E_c is the threshold energy, $E_e=L\cdot\varepsilon_c$。

(4) Rate equation(反应速率方程)

The rate equation is

$$\frac{-dC_A}{dt}=Z_{AB}\times q=Z_{AB}e^{-E_c/RT} \qquad (10.8.6)$$

Substitution(10.8.4) in(10.8.6), we can get

$$\frac{-dC_A}{dt}=(r_A+r_B)^2\left(\frac{8\pi k_B T}{\mu}\right)^{\frac{1}{2}} e^{E_c/RT} C_A C_B \qquad (10.8.7)$$

10.9 Transition state theory (过渡状态理论)

Transition-state theory attempts to simplify the problem by making a"dynamic bottleneck assumption." Transition-state theory is not exact, but is based on a series of approximations. However, it has been useful since its inception in 1935.

Kinetic theory is helpful in that it tells us about molecular collisions, but it does not deal with the changes that take place on a molecular level when reactants are converted to products. When two molecules are very close to each other, they cannot be considered separately because their wave functions overlap. Thus, from the time the reactant molecules are close to each other until the products are well separated, the system is a kind of supermolecule. This is different from an ordinary molecule because it is in the process of change, but it is a molecule in the sense that its energy and electron distribution can be calculated for each nuclear configuration by use of quantum mechanics.

The simplest quantitative account of reaction rates is in terms of collision theory, which can be used only for the discussion of reactions between simple species in the gas phase.

Reactions in solution are classified into two types: diffusion-controlled and activation controlled.

The former can be expressed quantitatively in terms of the diffusion equation. In the theory, it is assumed that the reactant molecules form a complex that can be discussed in terms of the population of its energy levels. Transition state theory inspires a thermodynamic approach to reaction rates, in which the rate constant is expressed in terms of thermodynamic parameters. This approach is useful for parameterizing the rates of reactions in solution. The highest level of sophistication is in terms of potential energy surfaces and the motion of molecules through the experimental study. We also use transition state theory to examine the transfer of electrons in homogeneous systems and at electrodes.

If a reaction involves N nuclei, there are $3N$ nuclear coordinates, but the group of nuclei has three translational coordinates of the center of mass and two or three rotational coordinates (about the center of mass) that do not affect the potential energy. Thus, the potential energy is a function of $3N-5$ nuclear coordinates if the nuclei are constrained to a straight line and $3N-6$ nuclear coordinates in general. For the simplest type of reaction,

$$A + B - C \longrightarrow A - B + C$$

where A, B, and C are atoms, three coordinates are required. It is not possible to plot the potential energy as a function of three coordinates, but if the angle of approach of A to BC is fixed, the potential energy of the system can be plotted as a function of R_{AB} and R_{BC}, where R is intermolecular distance. Such a plot is shown in figure 10. 14. If R_{AB} is rather large, as on the left face of the diagram, the potential energy is essentially that of the BC molecule. Similarly, the right face gives the potential energy of AB. Thus, figure 10. 14 omits two monotonous valleys to the left and right that extend indefinite distances. Initially, the distance is very large. As A approaches BC, the lowest-energy path is given by the dashed line from reactants R to products P. This dashed line gives the minimum energy path, which is sometimes referred to as the reaction coordinate. We will soon see that the configuration of the system does not actually move along the reaction coordinate in the reaction, but the reaction coordinate does help us visualize the surface. The highest point along the reaction coordinate is a saddle point. At the saddle point the potential energy is a maximum along the reaction coordinate, but it is a minimum in the direction perpendicular to the reaction coordinate. The reaction system at this point is said to be in the in transition state. Figure 10. 14, D is a high plateau giving the potential energy of three atoms well separated from each other.

As a first simple example, consider what happens when A approaches a nonvibrating BC molecule along the internuclear axis. The point representing the configuration of the system moves along the minimum energy path, the dashed line in figure 10. 14. As R_{AB} decreases, kinetic energy is converted to potential energy as the point representing the system of three nuclei moves up the valley from the left. If there is initially enough kinetic energy for the system to go over the saddle point, AB and C are formed and gain energy as the system goes down the valley to the right. If the kinetic energy is too low, the system returns down the valley to the left, and we would say that the reactants bounced off each other.

The dashed lines show the quantum mechanical zero point energies for reactants, products, and the actived complex at the top of the potential energy barrier. Figure 10. 14 applies only when the nuclei are constrained to a line, and the potential energy surface will be different if there is a different angle of approach.

A + B—C	⟶	A···B···C	⟶	[A···B···C]$^{\neq}$
A 和 BC 迎面运动		A 与 BC 相碰撞 B—C 键拉长而减弱		A 与 B 更近，B—C 键更拉长，将断而未断（形成活化络合物）

A···B···C	⟶	A—B + C
A—B 成键 AB 与 C 将离开		AB 与 C 离开

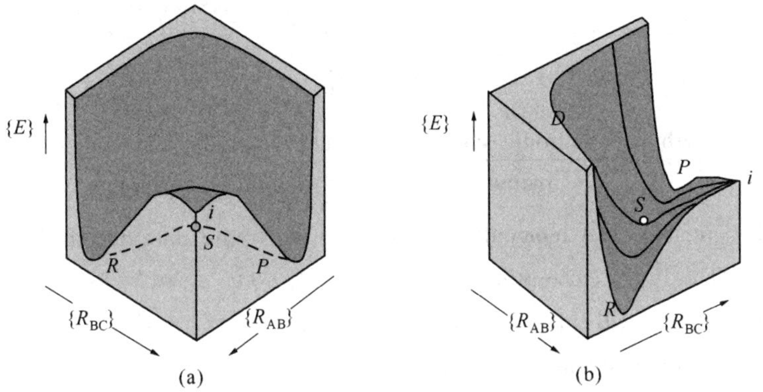

Fig. 10. 14 Potential energy surface

This surface is normally depicted as a contour diagram (figure 10. 15). The path of least potential energy is the one marked c, corresponding to R_{BC} lengthening as A approaches and begins to form a bond with B. The B–C bond relaxes at the demand of the incoming atom, and the potential energy climbs only as far as the saddle-shaped region of the surface, to the **saddle point** marked c. The encounter of least potential energy is one in which the atoms take route c up the floor of the valley, through the saddle point, and down the floor of the other valley as C recedes and the new A–B bond achieves its equilibrium length.

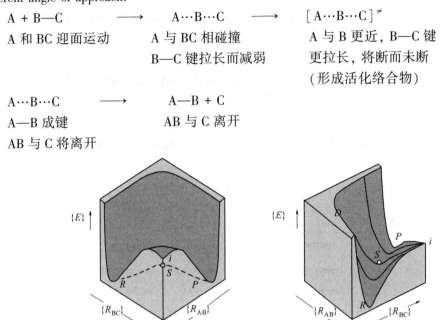

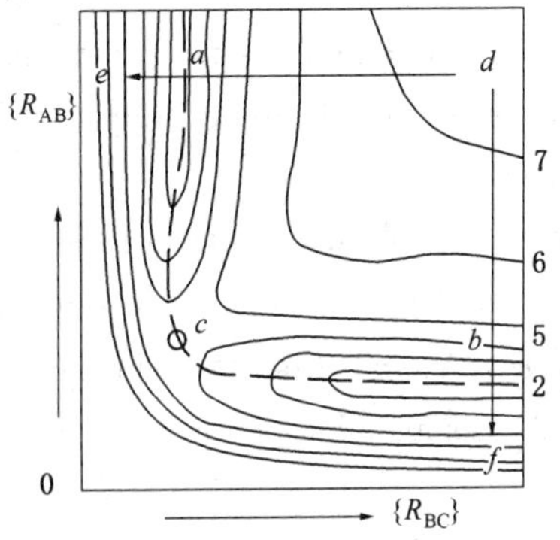

Fig. 10. 15 The energy profile for the transformation of the reactants to products

10. 10 Photochemistry

（光化学）

10. 10. 1 Photon and photochemistry（光与光化学反应）

（1）Wave-particle duality of light（光的波粒二象性）：

Light has wave and particle aspect.

The energy of a photon：$\varepsilon = hv$；

The energy of 1mol photons: $E_m = Lh\nu = Lhc/\lambda$

Where h is the Plank constant, c is the velocity of light in vacuum, λ is the wave-length of the light, and ν is the wave number.

(2) Photochemistry(光化学)

Photochemistry—the branch of chemistry which deals with the study of chemical reaction initiated by light.(光化学反应：由于吸收光量子而引起的化学反应称为光化学反应)。

Absorption of a photon of light of sufficient energy may raise a molecule to an excited electronic state. In this high-energy state it will be more likely to undergo a chemical reaction than in the ground electronic state.

$$NO_2 \xrightarrow{h\nu} NO_2^{\neq} \longrightarrow NO + \frac{1}{2}O_2$$

ground electronic state　　excited electronic state

primary step　　secondary process

Let A^* and A_0 denote an A molecule in an excited electronic state and in the ground electronic state, respectively. The initial absorption of radiation is $A_0 + h\nu \longrightarrow A^*$. In most cases, the ground electronic state is a singlet with all electron spins paired.

Following light absorption, many things can happen.

For example:

The A^* molecule can lose its electronic energy by spontaneously emitting a photo, thereby falling to a lower singlet state, which may be the ground electronic state: $A^* \longrightarrow A_0 + h\nu$. Emission of radiation by an electronic transition in which the total electronic spin doesn't change is called fluorescence.

The A^* molecule can transfer its electronic energy to another molecule during collision, thereby returning to the ground electronic state, a process called radiationless deactivation: $A^* + B \longrightarrow A_0 + B^*$, where A_0 and B on the right have extra translational, rotational, and vibrational energies.

(3) Comparation between the heat chemistry and photochemistry(光化学反应与热化学反应比较)

In thermal reactions, the activation energy is supplied by intermolecular collisions. The reaction rate changes sensitively with temperature.

In photochemical reactions, the activation energy is supplied by absorption of light. The reaction rate changes insensitively with temperature.

10.10.2 The basic laws of photochemistry and quantum yield(光化学基本定律及量子效率)

(1) The basic laws of photochemistry(光化学基本定律)

The first law of photochemistry: Only the light that is absorbed by a substance is effective in producing a photochemical change.(光化学第一定律:只有被系统吸收的光才可能产生光化学反应。)

The second law of photochemistry: The quantum of radiation absorbed by a molecule activates one molecule in the primary step of photochemical process.

(光化学第二定律:在初级过程中,一个光量子活化一个分子。)

(2) Quantum yield and energy efficiency(光化学的量子效率)

Quantum yield or quantum efficiency(ϕ) is the number of moles of reactant consumed divided by the number of moles of photons absorbed.

$\phi < 1$, the physical deactivation is dominant

$\phi = 1$, product is produced in primary photochemical process

$\phi > 1$, initiate chain reaction.

$$\phi \overset{\text{def}}{=\!=\!=} \frac{\text{moles of reactant consumed}}{\text{moles of photons absorbed}}$$

Quantum yields vary from 0 to 10^6. Quantum yields less than 1 are due to deactivation of A^* molecules by the various physical process discussed above and to recombination of fragments of dissociation. The quantum yield of the photochemical reaction $H_2 + Cl_2 \longrightarrow 2HCl$ with 400-nm radiation is typically 10^5. Absorption of light by Cl_2 puts it into an excited electronic state that "immediately" dissociates into Cl atoms; the Cl atoms then start a chain reaction, yielding many, many HCl molecules for each Cl atom formed.

Photosensitive reaction—reaction initiated by photosensitizer. When reactants themselves do not absorb light energy, photosensitizer can be used to initiate the reaction by conversion of the light energy to the reactants.(若反应物对光不敏感,但可以引入能吸收光的原子或分子,使它变为激发态,然后再将能量传给反应物,使反应物活化。能起这样作用的物质叫光敏物质或光敏剂。)

10.11 Catalysts and catalysis (催化剂和催化作用)

10.11.1 Definition of catalyst(催化剂的定义)

Catalysts—substances that can change the speed of chemical reactions without any chemical change of themselves.(催化剂——存在少量就能显著改变反应速率而本身最后并无损耗的物质。)

Catalysis—the phenomenon of acceleration or retardation of the speed of a chemical reaction by adding small amount of catalysts to the reactants.(催化作用——催化剂显著加速或减缓反应速率的作用。)

Autocatalysis—the phenomenon that the intermediate or product of a reaction acts as catalyst for the reaction.(自动催化作用——反应产物之一对反应本身起的催化作用。)

10.11.2 Types of catalysis(催化作用的分类)

(1) Homogeneous catalysis(均相催化)

Homogeneous catalysis—the catalyst is present in the same phase as the reactants and products.(反应物、产物及催化剂都处于同一相内,即为均相催化。)例如,酯的水解,加入酸或碱则反应加快,即为单相催化。

(2) Heterogeneous catalysis(多相催化)

Heterogeneous catalysis—the catalyst is present in the different phases as the reactants and products.(催化剂在反应系统中自成一相，其中，以气-固相催化应用最广。例如用固体的铁催化剂将氢与氮合成为氨。)

10.11.3 General characteristics of catalyst(催化剂的基本特征)

(1) Catalyst takes part in the reaction, alters the reaction path and cause significant change in apparent activation energy and reaction rate.

(2) Catalyst can shorten the time for reaching equilibrium but has no impact on the thermodynamic features of the reaction.

(3) Catalyst can't change the initial and the final state of the reaction. Therefore, catalyst has no impact on $\Delta_r H_m$ of the reaction.

(4) Selectivity of catalysts.

10.11.4 Mechanism of the catalyst reaction and the rate constant(催化反应的一般机理及速率常数)

Catalyst can speed up the reaction rate because the catalyst and the reactants can form unstable intermediate compounds, which change the reaction pathway and reduce the apparent activation energy, or increase the apparent pre-exponential factor. The reduction in activation energy, in particular, is significant, because the activation energy is in the exponential term of the Arrhenius equation.

For reaction A + B→AB, the mechanism of the reaction is

$$A + K \underset{k_{-1}}{\overset{k_1}{\longrightarrow}} AK(\text{I})$$

$$AK + B \xrightarrow{k_2} AB + K(\text{II})$$

Where K is the catalyst, A and B are the reactants, AB is the product.

Assuming the reaction(Ⅰ) reaches equilibrium, so

$$\frac{k_1}{k_{-1}} = K_c = \frac{c_{AK}}{c_A \cdot c_K}$$

$$c_{AK} = \frac{k_1}{k_{-1}} c_K c_A$$

The rate equation is

$$\frac{dc_{AB}}{dt} = k_2 c_{AK} c_B = k_2 \frac{k_1}{k_{-1}} c_K c_A c_B = k c_A c_B$$

$$k = k_2 \frac{k_1}{k_{-1}} c_K$$

k is rate coefficient of the overall reaction.

10.11.5 Activation energy of catalytic reaction(催化反应的活化能)

The activation energies of both the elementary reactions and the overall reaction can be expressed by Arrhenius equation.

$$k = k_{0,2} \frac{k_{0,1}}{k_{0,-1}} c_K e^{-\frac{E_1 - E_{-1} + E_2}{RT}} = k_0' c_k e^{-E/RT}$$

So we can get

$$E = E_1 - E_{-1} + E_2$$

$$k_0 = \frac{k_{0,1} k_{0,2}}{k_{0,-1}}$$

Where E is the apparent activation energy of the overall reaction, E_1, E_{-1}, E_2 are the activation energies of the elementary reactions. k'_0 is the pre-exponential of the overall reaction, $k_{0,1}$, $k_{0,-1}$, $k_{0,2}$ are the pre-exponentials of the elementary reactions.

The mechanism above has an overall activation energy that is less than that for the mechanism in the absence of catalyst K.

10.11.6 Composition of solid catalysts(固体催化剂的组成)

(1) Principal catalyst(主催化剂)—Substance having catalytic activity.

(2) Promoter(助催化剂)—Substance which has little or no catalytic activity, but can improve the activity of the main catalyst or prolong the life of the main catalyst.

(3) Carrier(载体)—Substance which can carry and disperse the principal catalyst. The carrier is often some natural or artificial porous material, such as natural zeolite, silica gel, artificial molecular sieve, etc.

10.11.7 Basic knowledge about catalysts(催化剂的基础知识)

(1) Activity of catalyst and active centers(催化剂的活性和活性中心)

Activity of a catalyst(催化剂的活性)—The ability to speed up the reaction rate.

Active center(活性中心) —The active part of the surface of a solid catalyst that is capable of catalyzing. It is a small part of the solid surface of the catalyst.

(2) Life, position, regeneration of catalyst(催化剂的寿命、中毒与再生)

The catalyst can be used for a certain period of time, from induction period to maturity period to decay period is the whole life of the catalyst.

Some impurities in the reaction system often makes the catalyst toxic, which can be divided into temporary poisoning and permanent poisoning. Temporary poisoning can be regenerated by some means to restore its activity, but permanent poisoning cannot be regenerated.

Supplementary Examples of Chapter 10

EXERCISES

(习题)

1. For the first-order reaction A→products, the initial rate is 1×10^{-3}mol · dm^{-3} · min^{-1}, the rate is 0.25×10^{-3}mol · dm^{-3} · min^{-1} after 1h. Calculate k, $t_{1/2}$ and initial concentration $c_{A,0}$.

Answer: $k = 0.0231\text{min}^{-1}$; $t_{1/2} = 30\text{min}$; $c_{A,0} = 0.0433\text{mol} \cdot \text{dm}^{-3}$

2. For first-order reaction, try to prove the time that the degree of dissociation reaches at 87.5% is three times that of 50%. For second-order reaction, how many times is it?

Answer: 7

3. For the second-order reaction A + B→C, the rate equation is $-\frac{dc_A}{dt}=k_A c_A c_B$, both the initial concentration of the two reactants are 1mol · dm^{-3}, the reaction is 25% complete after 10min. Calculate k.

Answer: $k=0.0333\text{dm}^3 \cdot \text{mol}^{-1} \cdot \text{min}^{-1}$

4. A gas phase reaction A(g) + B(g)→Y(g) generated in a closed container at a constant temperature when $T=300\text{K}$, the rate equation was determined to be:

$$-\frac{dp_A}{dt}=k_{A,p}p_A p_B$$

Assuming there were only reactants A(g) and B(g) at the beginning of the reaction (the ratio of initial volume was 1 : 1), the initial total pressure was 200kPa. when $t = 10\text{min}$, the total pressure was 150kPa, what is the rate coefficient of reaction at 300K? and what is the total pressure in the container when $t=20\text{min}$?

Answer: $k_{A,p}=0.001\text{kPa}^{-1} \cdot \text{min}^{-1}$; $p_t=133.3\text{kPa}$

5. At 500℃ and the initial pressure 101.325kPa, the half-life of the gas-phase thermo-decomposition of the hydrocarbon is 2s. If the initial pressure decreased to 10.133kPa, the half-life increased to 20s. Calculate the rate constant.

Answer: 4.93MPa^{-1} · s^{-1}

6. Isochoric reaction: A + 2B→Y, the rate equation was determined to be:

$$-\frac{dc_A}{dt}=k_A c_A c_B$$

the relationship between rate coefficient k_B and temperature T is:

$$\ln(k_B/\text{dm}^3 \cdot \text{mol}^{-1} \cdot \text{s}^{-1}) = 24.00-\frac{9622}{T/\text{K}}$$

(1) Calculate the activation energy of reaction;

(2) If $c_{A,0}=0.1\text{mol} \cdot \text{dm}^{-3}$, $c_{B,0}=0.2\text{mol} \cdot \text{dm}^{-3}$, how to control temperature to make the conversion arrives 90% in 10min.

Answer: $E_a=80.0\text{kJ} \cdot \text{mol}^{-1}$; $T=371.5\text{K}$

7. At 65℃, the rate coefficient of the gas-phase decomposition of N_2O_5 is 0.292min^{-1}, the activation energy is 103.3kJ · mol^{-1}. Calculate k and $t_{1/2}$ at 80℃.

Answer: $k=1.39\text{min}^{-1}$; $t_{1/2}=0.498\text{min}$

8. The decomposition of diphosgene $ClCOOCCl_3(g) \longrightarrow 2COCl_2(g)$ was first-order reaction. Certain diphosgene was put into a vessel quickly at 280℃, the system pressure was 2.710kPa after 751s; the pressure changed to 4.008kPa when the reaction proceeded completely after a long time. Repeated the experiment at 305℃, the system pressure was 2.838kPa after 320s; the pressure was changed into 3.554kPa when the reaction proceeded completely. Calculate the activation energy.

Answer: 169kJ · mol^{-1}

9. For the reaction: A + 2B→D, the rate equation is $-\frac{dc_A}{dt}=kc_A^{0.5}c_B^{1.5}$

(1) When $c_{A,0}=0.1\text{mol}\cdot\text{dm}^{-3}$, $c_{B,0}=0.2\text{mol}\cdot\text{dm}^{-3}$, at 300K the reaction proceeds 20s, $c_A=0.01\text{mol}\cdot\text{dm}^{-3}$. What is c_A after the reaction proceeds 20s continuously.

(2) At the same initial concentration, $c_A=0.003918\text{mol}\cdot\text{dm}^{-3}$ after 20s at 400K, please calculate the activation energy E_a.

Answer: (1) $0.00526\text{mol}\cdot\text{dm}^{-3}$; (2) $10^4\text{J}\cdot\text{mol}^{-1}$

10. If $A \underset{k_{-1}}{\overset{k_1}{\rightleftharpoons}} B$ is the first-order reversible reaction, the initial concentration of A is $c_{A,0}$; when the reaction time is t, the concentrations of A and B are $c_{A,0}-c_B$ and c_B.

(1) Try to show $\ln\dfrac{\dfrac{c_{A,0}}{k_1+k_{-1}}}{c_{A,0}-\dfrac{k_1+k_{-1}}{k_1}c_B}=(k_1+k_{-1})t$

(2) Given $k_1=0.2\text{s}^{-1}$, $k_{-1}=0.01\text{s}^{-1}$, $c_{A,0}=0.4\text{mol}\cdot\text{dm}^{-3}$, find the conversion of A when the reaction proceeds 100s.

Answer: (2) 95%

11. For the parallel reaction:

$$A\begin{cases}\xrightarrow{k_1} B\\ \xrightarrow{k_2} C\end{cases}$$

If the activation energy of overall reaction is E, try to prove $E=\dfrac{k_1E_1+k_2E_2}{k_1+k_2}$

12. Find the rate equation having follow mechanism of a gas phase reaction:

$$A \underset{k_{-1}}{\overset{k_1}{\rightleftharpoons}} B \qquad B + C \xrightarrow{k_2} D$$

B is reactive substance, so steady-state approximation method can be used. Please prove this reaction is first order at high pressure, while at low-pressure law is second order.

13. If the reaction mechanism of $3HNO_2 \longrightarrow H_2O+2NO+H^++NO_3^-$ is as follows,

$$2HNO_2 \overset{k_1}{\rightleftharpoons} NO+NO_2+H_2O \qquad \text{(at equilibrium)}$$

$$2NO_2 \overset{k_2}{\rightleftharpoons} N_2O_4 \qquad \text{(at equilibrium)}$$

$$N_2O_4+H_2O \xrightarrow{k_3} HNO_2+H^++NO_3^- \qquad \text{(slowest step)}$$

Try to find the rate equation that is expressed by $v[NO_3^-]$.

Answer: $\dfrac{d[NO_3^-]}{dt}=\dfrac{k_3K_1^2K_2[HNO_2]^4}{[NO]^2[H_2O]}$

APPENDIX Ⅰ　SI units

国际单位制是我国法定计量单位的基础，一切属于国际单位制的单位都是我国的法定计量单位。国际单位制的国际简称为SI。

Tab. 1　Basic units of SI

Name of quantity	Name of unit	Symbol of unit
Length(长度)	米	m
Mass(质量)	千克(公斤)	kg
Time(时间)	秒	s
Current(电流)	安[培]	A
Thermodynamic temperature(热力学温度)	开[尔文]	K
Amount of substance(物质的量)	摩[尔]	mol
Luminous intensity(发光强度)	坎[德拉]	cd

注：① 圆括号中名称，是它前面的名称的同义词，下同。

② 无方括号的量的名称与单位的名称均为全称。方括号中的字，在不致引起混乱、误解的情况下，可以省略。去掉方括号中的字即为其名称的简称，下同。

Tab. 2　Prefix of SI

Dimension	Name of prefix		Symbol
	English	Chinese	
10^{24}	yotta	尧[它]	Y
10^{21}	zetta	泽[它]	Z
10^{18}	exa	艾[可萨]	E
10^{15}	peta	拍[它]	P
10^{12}	tera	太[拉]	T
10^{9}	giga	吉[咖]	G
10^{6}	mega	兆	M
10^{3}	kilo	千	k
10^{2}	hecto	百	h
10^{1}	deca	十	da
10^{-1}	deci	分	d
10^{-2}	centi	厘	c
10^{-3}	milli	毫	m
10^{-6}	micro	微	μ
10^{-9}	nano	纳[诺]	n
10^{-12}	pico	皮[可]	p
10^{-15}	femto	飞[母托]	f
10^{-18}	atto	阿[托]	a
10^{-21}	zepto	仄[普托]	z
10^{-24}	yocto	幺[科托]	y

APPENDIX Ⅱ Greek characters

Name	正体		斜体	
	大写	小写	大写	小写
alpha	Α	α	A	α
beta	Β	β	B	β
gamma	Γ	γ	$\mathit{\Gamma}$	γ
delta	Δ	δ	$\mathit{\Delta}$	δ
epsilon	Ε	ε	E	ε
zeta	Ζ	ζ	Z	ζ
eta	Η	η	H	η
theta	Θ	θ	$\mathit{\Theta}$	θ
iota	Ι	Ι	I	I
kappa	Κ	κ	K	κ
lambda	Λ	λ	$\mathit{\Lambda}$	λ
mu	Μ	μ	M	μ
nu	Ν	ν	N	ν
xi	Ξ	ξ	$\mathit{\Xi}$	ξ
omicron	Ο	ο	O	o
pi	Π	π	$\mathit{\Pi}$	π
rho	Ρ	ρ	P	ρ
sigma	Σ	σ	$\mathit{\Sigma}$	σ
tau	Τ	τ	T	τ
upsilon	Υ	υ	$\mathit{\Upsilon}$	υ
phi	Φ	φ	$\mathit{\Phi}$	φ
chi	Χ	χ	X	χ
psi	Ψ	ψ	$\mathit{\Psi}$	ψ
omega	Ω	ω	$\mathit{\Omega}$	ω

APPENDIX Ⅲ Basic constants

Name of quantity	Symbol	Value and unit
Acceleration of gravity(重力加速度)	g	$9.80665 m \cdot s^{-2}$
Dielectric constant of vacuum(真空介电常数)	ε_0	$8.854188 \times 10^{-12} F \cdot m^{-1}$
Velocity of light in vacuum(真空中的光速)	c_0	$299792458 m \cdot s^{-1}$
Avogadro constant(阿伏加德罗常数)	L	$(6.0221367 \pm 0.0000036) \times 10^{23} mol^{-1}$
Molar gas constant(摩尔气体常数)	R	$(8.314510 \pm 0.000070) J \cdot K^{-1} \cdot mol^{-1}$
Boltzmann constant(玻耳兹曼常数)	k_B	$(1.380658 \pm 0.000012) \times 10^{-23} J \cdot K^{-1}$
Electric charge(元电荷)	e	$(1.60217733 \pm 0.00000049) \times 10^{-19} C$
Faraday constant(法拉第常数)	F	$(9.6485309 \pm 0.0000029) \times 10^{4} C \cdot mol^{-1}$
Planck constant(普朗克常量)	h	$(6.6260755 \pm 0.0000040) \times 10^{-34} J \cdot s$

APPENDIX Ⅳ Conversion factor

Name of non-SI unit	Symbol	Conversion factor
Pound-force per square inch(磅力每平方英寸)	$lbf \cdot in^{-2}$	$lbf \cdot in^{-2} = 6894.757Pa$
Standard atmospheric pressure(标准大气压)	atm	1atm = 101.325kPa
Kilogram force per square metre(千克力每平方米)	$kgf \cdot m^{-2}$	$1kgf \cdot m^{-2} = 9.80665Pa$
Torr(托)	Torr	1Torr = 133.3224Pa
Engineering atmospheric pressure(工程大气压)	at	1at = 98066.5Pa
Appointed millimetre mercury column(约定毫米汞柱)	mmHg	1mmHg = 133.3224Pa

APPENDIX V Atomic weights of the elements(1997)

Symbol	Name	Atomic weight	Symbol	Name	Atomic weight
Ac	锕	227.0278	Ga	镓	69.723(1)
Ag	银	107.8682(2)	Gd	钆	157.25(3)
Al	铝	26.981538(2)	Ge	锗	72.61(2)
Am	镅	243	H	氢	1.00794(7)
Ar	氩	39.948(1)	He	氦	4.002602(2)
As	砷	74.92160(2)	Hf	铪	178.49(2)
At	砹	210	Hg	汞	200.59(2)
Au	金	196.96655(2)	Ho	钬	164.93032(2)
B	硼	10.811(7)	I	碘	126.90447(3)
Ba	钡	137.327(7)	In	铟	114.818(3)
Be	铍	9.012182(3)	Ir	铱	192.217(3)
Bi	铋	208.98038(2)	K	钾	39.0983(1)
Bk	锫	247	Kr	氪	83.80(1)
Br	溴	79.904(1)	La	镧	138.9055(2)
C	碳	12.0107(8)	Li	锂	6.941(2)
Ca	钙	40.078(4)	Lr	铹	260
Cd	镉	112.411(8)	Lu	镥	174.967(1)
Ce	铈	140.116(1)	Md	钔	258
Cf	锎	251	Mg	镁	24.3050(6)
Cl	氯	35.4527(9)	Mn	锰	54.938049(9)
Cm	锔	247	Mo	钼	95.94(1)
Co	钴	58.93320(9)	N	氮	14.00674(7)
Cr	铬	51.9961(6)	Na	钠	22.989770(2)
Cs	铯	132.90543(2)	Nb	铌	92.90638(2)
Cu	铜	63.546(3)	Nd	钕	144.24(3)
Dy	镝	162.50(3)	Ne	氖	20.1797(6)
Er	铒	167.26(3)	Ni	镍	58.6934(2)
Es	锿	254	No	锘	259
Eu	铕	151.964(1)	Np	镎	237.048
F	氟	18.9984032(5)	O	氧	15.9994(3)
Fe	铁	55.845(2)	Os	锇	190.23(3)
Fm	镄	257	P	磷	30.973761(2)
Fr	钫	223	Pa	镤	231.03588(2)

Continued

Symbol	Name	Atomic weight	Symbol	Name	Atomic weight
Pb	铅	207.2(1)	Sr	锶	87.62(1)
Pd	钯	106.42(1)	Ta	钽	180.9479(1)
Pm	钷	145	Tb	铽	158.92534(2)
Po	钋	209	Tc	锝	98
Pr	镨	140.90765(2)	Te	碲	127.60(3)
Pt	铂	195.078(2)	Th	钍	232.0381(1)
Rb	铷	85.4678(3)	Ti	钛	47.867(1)
Re	铼	186.207(1)	Tl	铊	204.3833(2)
Rh	铑	102.90550(2)	Tm	铥	168.93421(2)
Rn	氡	222	U	铀	238.0289(1)
Ru	钌	101.07(2)	V	钒	50.9415(1)
S	硫	32.066(6)	W	钨	183.84(1)
Sb	锑	121.760(1)	Xe	氙	131.29(2)
Sc	钪	44.955910(8)	Y	钇	88.90585(2)
Se	硒	78.96(3)	Yb	镱	173.04(3)
Si	硅	28.0855(3)	Zn	锌	65.39(2)
Sm	钐	150.36(3)	Zr	锆	91.224(2)
Sn	锡	118.710(7)			

注：相对原子质量(Atomic weight)后面括号中的数字表示末位数的误差范围。

APPENDIX Ⅵ Critical parameters of substance

Substance	Critical temperature t_c/℃	Critical pressure p_c/MPa	Critical density ρ_c/kg · m^{-3}	Critical compression factor Z_c
He(氦)	-267.96	0.227	69.8	0.301
Ar(氩)	-122.4	4.87	533	0.291
H_2(氢)	-239.9	1.297	31.0	0.305
N_2(氮)	-147.0	3.39	313	0.290
O_2(氧)	-118.57	5.043	436	0.288
F_2(氟)	-128.84	5.215	574	0.288
Cl_2(氯)	144	7.7	573	0.275
Br_2(溴)	311	10.3	1260	0.270
H_2O(水)	373.91	22.05	320	0.23
NH_3(氨)	132.33	11.313	236	0.242
HCl(氯化氢)	51.5	8.31	450	0.25
H_2S(硫化氢)	100.0	8.94	346	0.284
CO(一氧化碳)	-140.23	3.499	301	0.295
CO_2(二氧化碳)	30.98	7.375	468	0.275
SO_2(二氧化硫)	157.5	7.884	525	0.268
CH_4(甲烷)	-82.62	4.596	163	0.286
C_2H_6(乙烷)	32.18	4.872	204	0.283
C_3H_8(丙烷)	96.59	4.254	214	0.285
C_2H_4(乙烯)	9.19	5.039	215	0.281
C_3H_6(丙烯)	91.8	4.62	233	0.275
C_2H_2(乙炔)	35.18	6.139	231	0.271
$CHCl_3$(氯仿)	262.9	5.329	491	0.201
CCl_4(四氯化碳)	283.15	4.558	557	0.272
CH_3OH(甲醇)	239.43	8.10	272	0.224
C_2H_6OH(乙醇)	240.77	6.148	276	0.240
C_6H_6(苯)	288.95	4.898	306	0.268
$C_6H_5CH_3$(甲苯)	318.57	4.109	290	0.266

APPENDIX Ⅶ van der Waals constants of gas

Gas	$10^3 a$/Pa · m^6 · mol^{-2}	$10^6 b$/m^3 · mol^{-2}
Ar(氩)	136.3	32.19
H_2(氢)	24.76	26.61
N_2(氮)	140.8	39.13
O_2(氧)	137.8	31.83
Cl_2(氯)	657.9	56.22
H_2O(水)	553.6	30.49
NH_3(氨)	422.5	37.07
HCl(氯化氢)	371.6	40.81
H_2S(硫化氢)	449.0	42.87
CO(一氧化碳)	150.5	39.85
CO_2(二氧化碳)	364.0	42.67
SO_2(二氧化硫)	680.3	56.36
CH_4(甲烷)	228.3	42.78
C_2H_6(乙烷)	556.2	63.8
C_3H_8(丙烷)	877.9	84.45
C_2H_4(乙烯)	453.0	57.14
C_3H_6(丙烯)	849.0	82.72
C_2H_2(乙炔)	444.8	51.36
$CHCl_3$(氯仿)	1537	102.2
CCl_4(四氯化碳)	2066	138.3
CH_3OH(甲醇)	946.9	67.02
C_2H_6OH(乙醇)	1218	84.07
$(C_2H_5)_2O$(乙醚)	1761	134.4
$(CH_3)_2CO$(丙酮)	1409	99.4
C_6H_6(苯)	1824	115.4

APPENDIX Ⅷ Relation between $C_{p,m}$ and T

$$C_{p,m}=a+bT+cT^2$$

Gas	$\frac{a}{J \cdot mol^{-1} \cdot K^{-1}}$	$\frac{10^3 b}{J \cdot mol^{-1} \cdot K^{-2}}$	$\frac{10^6 c}{J \cdot mol^{-1} \cdot K^{-3}}$	$\frac{\text{Temperature scope}}{K}$
H_2(氢)	26.88	4.347	-0.3265	273~3800
Cl_2(氯)	31.696	10.144	-4.038	300~1500
Br_2(溴)	35.241	4.075	-1.487	300~1500
O_2(氧)	28.17	6.297	-0.7479	273~3800
N_2(氮)	27.32	6.226	-0.9502	273~3800
HCl(氯化氢)	28.17	1.810	1.547	300~1500
H_2O(水)	29.16	14.49	-2.022	273~3800
CO(一氧化碳)	26.537	7.6831	-1.172	300~1500
CO_2(二氧化碳)	26.75	42.258	-14.25	300~1500
CH_4(甲烷)	14.15	75.496	-17.99	298~1500
C_2H_6(乙烷)	9.401	159.83	-46.229	298~1500
C_2H_4(乙烯)	11.84	119.67	-36.51	298~1500
C_3H_6(丙烯)	9.427	188.77	-57.488	298~1500
C_2H_2(乙炔)	30.67	52.810	-16.27	298~1500
C_3H_4(丙炔)	26.50	120.66	-39.57	298~1500
C_6H_6(苯)	-1.71	324.77	-110.58	298~1500
$C_6H_5CH_3$(甲苯)	2.41	391.17	-130.65	298~1500
CH_3OH(甲醇)	18.40	101.56	-28.68	273~1000
C_2H_6OH(乙醇)	29.25	166.28	-48.898	298~1500
$(C_2H_5)_2O$(乙醚)	-103.9	1417	-248	300~400
HCHO(甲醛)	18.82	58.379	-15.61	291~1500
CH_3CHO(乙醛)	31.05	121.46	-36.58	298~1500
$(CH_3)_2CO$(丙酮)	22.47	205.97	-63.521	298~1500
HCOOH(甲酸)	30.7	89.20	-34.54	300~700
$CHCl_3$(氯仿)	29.51	148.94	-90.734	273~773

APPENDIX Ⅸ Standard thermodynamic properties at 25℃ (standard pressure $p^{\ominus}$ = 100kPa)

Substance	$\Delta_f H_m^{\ominus}$ / kJ · mol^{-1}	$\Delta_f G_m^{\ominus}$ / kJ · mol^{-1}	$S_m^{\ominus}$ / J · mol^{-1} · K^{-1}	$C_{p,m}$ / J · mol^{-1} · K^{-1}
Ag(s)	0	0	42.55	25.351
AgCl(s)	-127.068	-109.789	96.2	50.79
Ag_2O(s)	-31.05	-11.20	121.3	65.86
Al(s)	0	0	28.33	24.35
Ag_2O_3(α, corundum)	-1675.7	-1582.3	50.92	79.04
Br_2(l)	0	0	152.231	75.689
Br_2(g)	30.907	3.110	245.463	36.02
HBr(g)	-36.40	-53.45	198.695	29.142
Ca(s)	0	0	41.42	25.31
CaC_2(s)	-59.8	-64.9	69.96	62.72
$CaCO_3$(s)	-1206.92	-1128.79	92.9	81.88
CaO(s)	-635.09	-604.03	39.75	42.80
$Ca(OH)_2$(s)	-986.09	-898.49	83.39	87.49
C(graphite)	0	0	5.740	8.527
C(diamond)	1.895	2.900	2.377	6.113
CO(g)	-110.525	-137.168	197.674	29.142
CO_2(g)	-393.509	-394.359	213.74	37.11
CS_2(l)	89.7	65.27	151.34	75.7
CS_2(g)	117.36	67.12	237.84	45.40
CCl_4(l)	-135.44	-65.21	216.4	131.75
CCl_4(g)	-102.9	-60.59	309.85	83.30
HCN(l)	108.87	124.97	112.84	70.63
HCN(g)	135.1	124.7	201.78	35.86
Cl_2(g)	0	0	223.066	33.907
Cl(g)	121.679	105.680	165.198	21.840
HCl(g)	-92.307	-95.299	186.908	29.12
Cu(s)	0	0	33.150	24.435
CuO(s)	-153.7	-129.7	42.63	42.30
Cu_2O(s)	-168.6	-146.0	93.14	63.64

Continued

Substance	$\Delta_f H_m^\ominus$ / $kJ \cdot mol^{-1}$	$\Delta_f G_m^\ominus$ / $kJ \cdot mol^{-1}$	$S_m^\ominus$ / $J \cdot mol^{-1} \cdot K^{-1}$	$C_{p,m}$ / $J \cdot mol^{-1} \cdot K^{-1}$
$F_2(g)$	0	0	202.78	31.30
$HF(g)$	−271.1	−273.2	173.779	29.133
$Fe(s)$	0	0	27.28	25.10
$FeCl_2(s)$	−341.79	−302.30	117.95	76.65
$FeCl_3(s)$	−399.49	−334.00	142.3	96.65
$Fe_2O_3(s)$	−824.2	−742.2	87.40	103.85
$Fe_3O_4(s)$	−1118.4	−1015.4	146.4	143.43
$FeSO_4(s)$	−928.4	−820.8	107.5	100.58
$H_2(g)$	0	0	130.684	28.824
$H(g)$	217.965	203.247	114.713	20.784
$H_2O(l)$	−285.830	−237.129	69.91	75.291
$H_2O(g)$	−241.818	−228.572	188.825	33.577
$I_2(s)$	0	0	116.135	54.438
$I_2(g)$	62.438	19.327	260.69	36.90
$I(g)$	106.838	70.250	180.791	20.786
$HI(g)$	26.48	1.70	206.594	29.158
$Mg(s)$	0	0	32.68	24.89
$MgCl_2(s)$	−641.32	−591.79	89.62	71.38
$MgO(s)$	−601.70	−569.43	26.94	37.15
$Mg(OH)_2(s)$	−924.54	−833.51	63.18	77.03
$Na(s)$	0	0	51.21	28.24
$Na_2CO_3(s)$	−1130.68	−1044.44	134.98	112.30
$NaHCO_3(s)$	−950.81	−851.0	101.7	87.61
$NaCl(s)$	−411.153	−384.138	72.13	50.50
$NaNO_3(s)$	−467.85	−367.00	116.52	92.88
$NaOH(s)$	−425.609	−379.494	64.455	59.54
$Na_2SO_4(s)$	−1387.08	−1270.16	149.58	128.20
$N_2(g)$	0	0	191.61	29.125
$NH_3(g)$	−46.11	−16.45	192.45	35.06
$NO(g)$	90.25	86.55	210.761	29.844
$NO_2(g)$	33.18	51.31	240.06	37.20
$N_2O(g)$	82.05	104.2	219.85	38.45
$N_2O_3(g)$	83.72	139.46	312.28	65.61

Continued

Substance	$\Delta_f H_m^\ominus$ / $kJ \cdot mol^{-1}$	$\Delta_f G_m^\ominus$ / $kJ \cdot mol^{-1}$	$S_m^\ominus$ / $J \cdot mol^{-1} \cdot K^{-1}$	$C_{p,m}$ / $J \cdot mol^{-1} \cdot K^{-1}$
$N_2O_4(g)$	9.16	97.89	304.29	77.28
$N_2O_5(g)$	11.3	115.1	355.7	84.5
$HNO_3(l)$	-174.10	-80.17	155.60	109.87
$HNO_3(g)$	-135.06	-74.72	266.38	53.35
$NH_4NO_3(s)$	-365.56	-183.87	151.08	139.3
$O_2(g)$	0	0	205.138	29.355
O(g)	249.17	231.731	161.055	21.912
$O_3(g)$	142.7	163.2	238.93	39.20
P(α-white)	0	0	41.09	23.840
P(red)	-17.6	-12.1	22.80	21.21
$P_4(g)$	58.91	24.44	279.98	67.15
$PCl_3(g)$	-287.0	-267.8	311.78	71.84
$PCl_5(g)$	-374.9	-305.0	364.58	112.80
$H_3PO_4(s)$	-1279.0	-1119.1	110.50	106.06
S(rhombic system)	0	0	31.80	22.64
S(g)	278.805	238.25	167.821	23.673
$S_8(g)$	102.30	49.63	430.98	156.44
$H_2S(g)$	-20.63	-33.56	205.79	34.23
$SO_2(g)$	-296.830	-300.194	248.22	39.87
$SO_3(g)$	-395.72	-371.06	256.76	50.67
$H_2SO_4(l)$	-813.989	-690.003	156.904	138.91
Si(s)	0	0	18.83	20.00
$SiCl_4(l)$	-687	-619.84	239.7	145.31
$SiCl_4(g)$	-657.01	-616.98	330.73	90.25
$SiH_4(g)$	34.3	56.9	204.62	42.84
SiO_2(α-quartz)	-910.94	-856.64	41.84	44.43
$SiO_2(s)$	-903.49	-850.70	46.9	44.4
Zn(s)	0	0	41.63	25.40
$ZnCO_3(s)$	-812.78	-731.52	82.4	79.71
$ZnCl_2(s)$	-415.05	-369.398	111.46	71.34
ZnO(s)	-348.28	-318.30	43.64	40.25
$CH_4(g)$	-74.81	-50.72	186.264	35.309
$C_2H_6(g)$	-84.68	-32.82	229.6	52.63

Continued

Substance	$\Delta_f H_m^\ominus$ / kJ · mol^{-1}	$\Delta_f G_m^\ominus$ / kJ · mol^{-1}	$S_m^\ominus$ / J · mol^{-1} · K^{-1}	$C_{p,m}$ / J · mol^{-1} · K^{-1}
$C_2H_4(g)$	52.26	68.15	219.56	43.56
$C_2H_2(g)$	226.73	209.20	200.94	43.93
$CH_3OH(l)$	-238.66	-166.27	126.8	81.6
$CH_3OH(g)$	-200.66	-161.96	239.81	43.89
$C_2H_5OH(l)$	-277.69	-174.78	160.7	111.46
$C_2H_5OH(g)$	-235.10	-168.49	282.7	65.44
$(CH_2OH)_2(l)$	-454.80	-323.08	166.9	149.8
$(CH_3)_2O(g)$	-184.05	-112.59	266.38	64.39
$HCHO(g)$	-108.57	-102.53	218.77	35.4
$CH_3CHO(g)$	-166.19	-128.86	250.3	57.3
$HCOOH(l)$	-424.72	-361.35	128.95	99.04
$CH_3COOH(l)$	-484.5	-389.9	159.8	124.3
$CH_3COOH(g)$	-432.25	-374.0	282.5	66.5
$(CH_2)_2O(l)$	-77.82	-11.76	153.85	87.95
$(CH_2)_2O(g)$	-52.63	-13.01	242.53	47.91
$CHCl_3(l)$	-134.47	-73.66	201.7	113.8
$CHCl_3(g)$	-103.14	-70.34	295.71	65.69
$C_2H_5Cl(l)$	-136.52	-59.31	190.79	104.35
$C_2H_5Cl(g)$	-112.17	-60.39	276.00	62.8
$C_2H_5Br(l)$	-92.01	-27.70	198.7	100.8
$C_2H_5Br(g)$	-64.52	-26.48	286.71	64.52
$CH_2CHCl(g)$	35.6	51.9	263.99	53.72
$CH_3COCl(l)$	-273.80	-207.99	200.8	117
$CH_3COCl(g)$	-243.51	-205.80	295.1	67.8
$CH_3NH_2(g)$	-22.97	32.16	243.41	53.1
$(NH_3)_2CO(s)$	-333.51	-197.33	104.60	93.14

APPENDIX X $\Delta_c H_m^\ominus$ of organic compounds at 25℃ (standard pressure $p^\ominus$ = 100kPa)

Substance	$\frac{-\Delta_c H_m^\ominus}{kJ \cdot mol^{-1}}$	Substance	$\frac{-\Delta_c H_m^\ominus}{kJ \cdot mol^{-1}}$
CH_4(g, 甲烷)	890.31	C_2H_5CHO(l, 丙醛)	1816.3
C_2H_6(g, 乙烷)	1559.8	$(CH_3)_2CO$(l, 丙酮)	1790.4
C_3H_8(g, 丙烷)	2219.9	$CH_3COC_2H_5$(l, 甲乙酮)	2444.2
C_5H_{12}(l, 正戊烷)	3509.5	HCOOH(l, 甲酸)	254.6
C_5H_{12}(g, 正戊烷)	3536.1	CH_3COOH(l, 乙酸)	874.54
C_6H_{14}(l, 正己烷)	4163.1	C_2H_5COOH(l, 丙酸)	1527.3
C_2H_4(g, 乙烯)	1411.0	C_3H_7COOH(l, 正丁酸)	2183.5
C_2H_2(g, 乙炔)	1299.6	$CH_2(COOH)_2$(s, 丙二酸)	861.15
C_3H_6(g, 环丙烷)	2091.5	$(CH_2COOH)_2$(s, 丁二酸)	1491.0
C_4H_8(l, 环丁烷)	2720.5	$(CH_3CO)_2O$(l, 乙酸酐)	1806.2
C_5H_{10}(l, 环戊烷)	3290.9	$HCOOCH_3$(l, 甲酸甲酯)	979.5
C_6H_{12}(l, 环己烷)	3919.9	C_6H_5OH(s, 苯酚)	3053.5
C_6H_6(l, 苯)	3267.5	C_6H_5CHO(l, 苯甲醛)	3527.9
C_8H_{10}(s, 萘)	5153.9	$C_6H_5COCH_3$(l, 苯乙酮)	4148.9
CH_3OH(l, 甲醇)	726.51	C_6H_5COOH(s, 苯甲酸)	3226.9
C_2H_6OH(l, 乙醇)	1366.8	$C_6H_4(COOH)_2$(s, 邻苯二甲酸)	3223.5
C_3H_7OH(l, 正丙醇)	2019.8	$C_6H_5COOCH_3$(l, 苯甲酸甲酯)	3957.6
C_4H_9OH(l, 正丁醇)	2675.8	$C_{12}H_{22}O_{11}$(s, 蔗糖)	5640.9
$CH_3OC_2H_5$(g, 甲乙醚)	2107.4	CH_3NH_2(l, 甲胺)	1060.6
$(C_2H_5)_2O$(l, 二乙醚)	2751.1	$C_2H_5NH_2$(l, 乙胺)	1713.3
HCHO(g, 甲醛)	570.78	$(NH_3)_2CO$(s, 尿素)	631.66
CH_3CHO(l, 乙醛)	1166.4	C_6H_5N(l, 吡啶)	2782.4

REFERENCE BOOKS

1 Atkins P, Paula J D, Keeler J. Physical Chemistry (12th Ed). London: Oxford University Press, 2022.

2 Levine I N. Physical Chemistry (6nd Ed). New York: McGraw-Hill, Inc., 2008.

3 Brown T L, Eugene LeMay H Jr, Bursten B E. Chemistry-the Central Science (14th Ed). New Jersey: Prentice-Hall, Inc., 2017.

4 Atkins P, Trapp C, Cady M P, Giunta C. Student Solutions Manual for Physical Chemistry (7th Ed). London: Oxford University Press, 2002.

5 Atkins P, Friedman R. Molecular Quantum Mechanics (5th Ed). London: Oxford University Press, 2010.

6 Laidler K J. Chemical Kinetics (3rd Ed). New York: Harper & Row Publishers, 2003.

7 Adamso A W, Gasl A P. Physical Chemistry of Surfaces (6rd Ed). New York: Wiley, 1997.

8 天津大学物理化学教研室. 物理化学(第六版). 北京：高等教育出版社，2017.

9 冯霞，陈丽，朱荣娇编. 物理化学解题指南(第三版). 北京：高等教育出版社，2018.

10 纪敏，郝策编. 多媒体 CAI 物理化学(第六版). 大连：大连理工大学出版社，2013.

11 傅玉普，纪敏编. 物理化学考研重点热点导引与综合能力训练(第五版). 大连：大连理工大学出版社，2012.

12 王新平编. 物理化学. 北京：高等教育出版社，2022.

13 傅献彩，侯文华编. 物理化学(第六版). 北京：高等教育出版社，2022.

14 侯文华，吴强，郭琳等编. 物理化学学习辅导. 北京：高等教育出版社，2022.

15 朱志昂，阮文娟，郭东升编. 物理化学(第七版). 北京：科学出版社，2023.

16 朱志昂，阮文娟编. 物理化学学习指导(第三版). 北京：科学出版社，2018.

17 傅鹰编. 化学热力学导论. 北京：科学出版社，1976.

18 胡英编. 物理化学(第六版). 北京：高等教育出版社，2014.

19 朱文涛编. 基础物理化学. 北京：清华大学出版社，2011.

20 高执棣编. 化学热力学基础. 北京：北京大学出版社，2006.

21 印永嘉，奚正楷，张树永等编. 物理化学简明教程(第四版). 北京：高等教育出版社，2007.

22 许越编. 化学反应动力学. 北京：化学工业出版社，2005.

23 滕新荣编. 表面物理化学. 北京：化学工业出版社，2009.